Proceedings of the Topical Meeting on

LASER MATERIALS and LASER SPECTROSCOPY

(A SATELLITE MEETING OF IQEC '88)

Proceedings of the Topical Meeting on

LASER MATERIALS and LASER SPECTROSCOPY

(A SATELLITE MEETING OF IQEC '88)

Shanghai, China
July 25–27, 1988

Editors

Wang Zhijiang
*Director, Shanghai Institute of Optics &
Fine Mechanics, Academia Sinica, China*

Zhang Zhiming
*Director, Laboratory of Laser Physics &
Optics, Fudan University, China*

World Scientific
Singapore • New Jersey • London • Hong Kong

Published by

World Scientific Publishing Co. Pte. Ltd.
P O Box 128, Farrer Road, Singapore 9128

USA office: World Scientific Publishing Co., Inc.
687 Hartwell Street, Teaneck, NJ 07666, USA

UK office: World Scientific Publishing Co. Pte. Ltd.
73 Lynton Mead, Totteridge, London N20 8DH, England

LASER MATERIALS AND LASER SPECTROSCOPY

ISBN 9971-50-739-0

Printed in Singapore by Kim Hup Lee Printing Co. Pte. Ltd.

PREFACE

The Proceedings contain the majority of the papers presented at the Topical Meeting on Laser Materials and Laser Spectroscopy which was held at the Tien-Ma Hotel in Shanghai, China on July 25-27, 1988. This topical meeting is a post-conference meeting for the 16th International Quantum Electronics Conference in Tokyo, Japan and it focused on the relatively narrow subjects of laser materials and laser spectroscopy. The Chinese optics and laser community has greatly progressed in these fields in the past years, and hope to have opportunities to share their results with the outside world as well as to learn from the international optics and laser community.

The Topical Meeting was sponsored by the Chinese Optical Society and co-operatively supported by the Chinese Physical Society, Joint Council of Quantum Electronics, Laser and Electro-Optics Society of IEEE, Optical Society of America and SPIE. The organizations in Shanghai district: Fudan University, Shanghai Institute of Laser Technology, Shanghai Institute of Optics and Fine Mechanics, Shanghai Laser Society, were in charge of the detailed preparation for the successful meeting.

The Organizing Committee is much obliged for the kind financial support from ICTP, National Science Foundation of China, National Scientific and Technological Committee of China, North China Research Institute of Electro-Optics, and Shanghai Jiaotong University.

The purpose of this topical meeting is to review the state-of-the-art research achievements in the fields of laser materials and laser spectroscopy and to create an academic and harmonic environment of understanding and interaction between the international scientists in China. More than twenty leading scientists presented their recent achievements in these two fields and over hundreds of papers were presented, including the work by young graduate students. The high quality of their presented researches is reflected in the Proceedings which fulfilled the success of this meeting.

All the papers selected for publication in this volume have been carefully revised and compiled by the Editorial Board of the << ACTA OPTICA SINICA>>. The Proceedings were published in Singapore and distributed by World Scientific Publishing Co. We would like to thank the staff involved in editing and publishing these proceedings and also the many persons both in the Program Committee and elsewhere for reviewing the papers. Our thanks are due to all the authors whose contributions produced the contents of both the meeting and this volume.

Z. J. Wang
Z. M. Zhang

Shanghai, China
Sept. 1988

Topical Meeting on
Laser Materials & Laser Spectroscopy
(LM&LS '88)

A Satellite Meeting of IQEC '88

Tian Ma Hotel, Shanghai, China

July 25 - 27, 1988

Sponsored by

Chinese Optical Society

Organized by

Fudan University,

Shanghai Institute of Laser Technology,

Shanghai Institute of Optics and Fine Mechanics,
    Academia Sinica,

Shanghai Laser Society.

Cooperatively Supported by

Chinese Physical Society,

Joint Council on Quantum Electronics,

Laser and Electro-optics Society of IEEE

Optical Society of America

SPIE

Organizing Committee

Honorary Chairman

Wang, Daheng	President of the Chinese Optical Society.

General Co-chairmen

Wang, Zhijiang	Shanghai Institute of Optics & Fine Mechanics, Academia Sinica, China
SHEN, Y.R.	University of California, Berkeley, USA

Program Committee

Co-chairmen

LIAO, P.F.	Bell Communications Research, USA

ZHANG, Zhiming	Fudan University, China

Members

ARECCHI, T	National Institute of Optics, Italy
BYER, R.	Stanford U., USA
CHEBOTAYEV, V.P.	Institute of Thermophysics, USSR
CHEN, Chuangtian	Fujian Institute of Research on the Structure of Matter, China
FLYTZANIS, C.	Ecole Polytechnique, France
GAN, Fuxi	Shanghai Institute of Optics & Fine Mechanics, Academia Sinica, China
GAN, Zhizhao	Beijing U., China
HÄNSCH, T.W.	Max-Planck-Institute for Quantumoptics, FRG
JIANG, Minghua	Shangdong U., China
LEE, Chi H.	U. Maryland, USA
LOY, M.M.T.	IBM Yorktown Heights, USA
MOORADIAN, A.	MIT Lincoln Lab, USA
MOULTON, P.F.	Schwartz Electro-Optics, USA
POWELL, R.	Oklahoma State U., USA
SCHÄFER, F.P.	Max-Planck-Institute for Biophysical Chemistry, FRG
SHIMIZU, T.	U. Tokyo, Japan
SVANBERG, S.	Lund Institute of Technology, Sweden
SVELTO, O.	Polytechnic of Milan, Italy
TANG, C.L.	Cornell U., USA
WANG, Zhaoyong	Fudan U., China
WITKOWSKI, S.	Max-Planck-Institute for Quantumoptics, FRG

WU, Chunkai	Anhui Institute of Optics & Fine Mechanics, Academia Sinica, China
YE, Peixian	Beijing Institute of Physics, China
YEN, W.	U. Georgia, USA
YU, Zhenxin	Zhongshan U., China
ZHANG, Cunhao	Dalian Institute of Chemical Physics, Academia Sinica, China
ZHANG, Guangyin	Nankai U., China

Local Members

LI, Changlin	Fudan U., China
LI, Yufen	Fudan U., China
LIN, Fuchen	Shanghai Institute of Optics & Fine Mechanics, Academia Sinica, China
MA, Xiaoshan	Shanghai Institute of Optics & Fine Mechanics, Academia Sinica, China
QIU, Mingxin	Shanghai Institute of Laser Technology, China
SHEN, Guanqun	Shanghai Institute of Laser Technology, China
WANG, Yuzhu	Shanghai Institute of Optics & Fine Mechanics, Academia Sinica, China
XIA, Huirong	East China Normal U., China
XIE, Shengwu	Shanghai Jiaotong U., China
ZHANG, Yinghua	Shanghai Institute of Ceramics, Academia Sinica, China
ZHU, Shiyao	Shanghai Jiaotong U., China

LOCAL COMMITTEE

Co-chairmen

NIE, Baocheng — Shanghai Institute of Laser Technology, China

WANG, Runwen — Shanghai Institute of Optics & Fine Mechanics, Academia Sinica, China

Secretary-General

WO, Xinneng — Shanghai Institute of Optics & Fine Mechanics, Academia Sinica, China

Deputy Secretary-General

LI, Fuming — Fudan U., China

Subordinate Groups for Secretariat

Secretarial Group Leader

TIAN, Shouyun — Shanghai Institute of Optics & Fine Mechanics, Academia Sinica, China

Program Group Leader

LI, Yifeng — Shanghai Institute of Optics & Fine Mechanics, Academia Sinica, China

FINANCIALLY SUPPORTED BY

International Centre for Theoretical Physics, United Nations Educational, Scientific and Cultural Organization

National Natural Science Foundation, PRC

4th Bureau for Technology, State Committee for Defence Science and Industry, PRC

North China Research Institute of Electro-Optics, PRC

Shanghai Jiaotong University, PRC

SPEECH AT THE RECEPTION PARTY OF THE
TOPICAL MEETING ON LASER MATERIALS & LASER SPECTROSCOPY

Wang Daheng

Distinguished Guests
Dear Colleagues
Ladies & Gentlemen

First of all it is a great pleasure for me at this reception party to speak in the name of the China Association for Science and Technology to congratulate the opening of this Topical Meeting on Laser Materials and Laser Spectroscopy, and also in the name of the Chinese Optical Society, the host organization, to express our hearty and warm welcome to all attending this meeting. This is a satellite meeting of IQEC '88; the latter had just ended a week ago in Tokyo.

Besides having the Chinese Optical Society as sponsor this meeting is cooperatively supported by the Joint Council on Quantum Electronics, Laser and Electrooptics Society of IEEE, the Optical Society of America and the Society of Photooptical Instrumentation Engineers as well as by the Chinese Physical Society.

Up to now, there are 77 overseas participants and 135 domestic participants. Our foreign participants come from 12 different countries, namely, Australia, Austria, Federal Republic of Germany, France, Hungary, Israel, Italy, Japan, Romania, and U. S. S. R.

I am glad to have the honour to introduce to you our distinguished guests, Prof. Xie Xide, Chairman of the Shanghai Branch of the China Association for Science and Technology, President of Fudan University and Wang Naili, Vicechairman of the Shanghai Branch of CAST. I believe it is also appropriate on this accasion to pay tribute to Prof. Y. R. Shen who took the initiative for convening the present meeting, and to Prof. P. F. Liao who took up the real responsibility as Cochairman of the Program Committee, not only in organizing the evaluation of submitted papers, but also for inviting and organizing the invited papers to be presented by eminent scientists who are on the frontiers of the present field.

We have also the honour to receive the Delegation of the Optical Society of America headed by Prof. William B. Bridges, President of the present session. We are looking forward to cooperation between the Optical Society of America and the Chinese Optical Society.

On convening this meeting, 325 papers have been submitted, among them 242 papers from China and 83 from abroad. As a result of double peer review by 45 eminent scientists through the Program Committee 179 papers were accepted. Consequently, we expect that the academic level in this meeting ought to meet the qualification of IQEC tradition.

Speaking candidly, we regard this meeting taking place in China a great event for our Chinese colleagues. I should like to express our gratitude to it for providing us the opportunity for forthcoming academic exchanges and further acquaintances with our foreign scientists and experts. This would certainly be beneficial for promoting our developments in these fields.

Before concluding my speech I should mention we are much obliged for the financial support given by:

1. International Center for Theoretical Physics of UNESCO.
2. National Science Foundation of P. R. C.
3. North China Research Institute of Electro-optics.
4. Shanghai Jiatong University.

We are also indebted to the local organizers, the Shanghai Institute of Optics and Fine Mechanics, Academia Sinica, Fudan University, Shanghai Institute for Laser Technology and Shanghai Laser Society. Without their diligent and complicated efforts, the convening of this meeting would be impossible.

Finally, I wish to thank all who have contributed towards the success of this meeting.

Thank you for your attention.

CONTENTS

Laser Materials

Laser Spectroscopy

*Invited Papers

Laser Site Spectroscopy of Transition Metal Ions in Glass

GAN Fuxi and LIU Huimin

Shanghai Institute of Optics and Fine Mechanics, Academia Sinica
P. O. Box 8211, Shanghai, P. R. China

ABSTRACT Laser-excited and time resolved fluorescence spectra of low valence transition-metal ions (Cu^+, Ti^{3+}, Cr^{3+}) in various kinds of inorganic glasses (phosphate, fluoro-phosphate and fluoride) have been studied at low and room temperature. Dynamic lattice model and site structure of activated ions are discussed.

Inorganic glasses doped with transition metal ions are becoming more and more important and hopeful as a kind of tunable laser material, to which much attention has been paid. We systematically studied optical and ESR spectra of transition metal ions in various kinds of glasses [1,2]. A certain knowledge was aquired by us about its energy level structure and interaction with host. In this paper the laser-excited and time resolved fluorescence spectra were reported in detail, and it is attempted to examine the microscopic mechanism in emission process by means of narrow line excitation and site selection in the time category [3-5].

1. Cr^{3+} Ions

In glasses there are differences from site to site not only in the energy gap, but also in the radiative and nonradiative transition probability. By using time-resolved fluorescence measurements the peak shift and variation can be displayed in the spectra. As shown in Tab. 1 and 2, the spectra by 5106Å excitation are located at longer wavelength and that by 5782Å excitation are at shorter wavelength. With the increase of delay time the spectra in Tab. 1 and 2 move toward shorter wavelength. The moving rate of the peak λ_p at room temperature is faster than that at low temperature. In addition, the variation in the bandwidth ($\Delta\lambda$) with delay time seems to be interesting. At low temperature all the spectra are narrowed. On the contrary, at room temperature the band width increases.

Tab. 1 Time-resolved fluorescence spectra of Cr^{3+} in
fluorophosphate glass at low temperature.

Delay time	(a) 510.6nm excited	(b) 578.2nm excited
1. 10 μs	λ_p=886nm $\Delta\lambda$=188nm	λ_p=850nm $\Delta\lambda$=199nm
2. 20 μs	λ_p=878nm $\Delta\lambda$=181nm	λ_p=840nm $\Delta\lambda$=187nm
3. 30 μs	λ_p=868nm $\Delta\lambda$=128nm	λ_p=836nm $\Delta\lambda$=173nm

Tab. 2 Time-resolved fluorescence spectra of Cr^{3+} in
fluorophosphate glass at room temperature.

Delay time	(a) 510.6nm excited	(b) 578.2nm excited
1. 3 μs	λ_p=886nm $\Delta\lambda$=145nm	λ_p=850nm $\Delta\lambda$=152nm
2. 7 μs	λ_p=868nm $\Delta\lambda$=159nm	λ_p=846nm $\Delta\lambda$=158nm
3. 17 μs	λ_p=850nm $\Delta\lambda$=161nm	λ_p=838nm $\Delta\lambda$=170nm
4. 27 μs	λ_p=840nm $\Delta\lambda$=176nm	λ_p=830nm $\Delta\lambda$=192nm

These facts, phenomenally discribed above in detail, show the
multiple distribution of transition probability due to the sitemul-
tiplity. Because the different sites have different relaxation rates,
with the increase of delay time after excitation the luminesecence
becomes more and more dorminant by the contribution of longer-lived
ions. If the ions with different emission frequencies, they would
manifest themselves as a spectral shift in the time-resolved spectrum.

In fluorophosphate glass the $^4T_{2g}$ level of Cr^{3+} spreads over 5000
cm^{-1}, that is to say, the difference in the separation between the
ground and excited states of the ions which are located in different
sites will be above 5000 cm^{-1}. For a transition between two electronic
states, in general, the larger the separation, the greater the redia-
tive transtion probability, and thus the smaller the nonradiative
transition probability. Moreover, for different sites in glasses the
ions having greater D_q value will have smaller nonradiative transition
probability and longer lifetime, and it is contrary to the ions having
smaller D_q value. In fluorophosphate glass the difference in nonradiative transition assisted by multiphonon is estimated to have several
order of phonon. Therefore, with the increase of delay time the luminesecence comes mainly from contribution of the ions having greater
D_q value.

2. Ti^{3+} Ions

A broad-band luminescence of Ti^{3+} ion in fluorophosphate glass can be observed by ultra-violet excitation near 3000Å. Using XeCl excimer laser and quadruple-frequency pulsed Nd:YAG laser with a duration of less than 20ns, time-resolved spectra were studied.

Tab. 3 represents the fluorescence of Ti^{3+} at room temperature and 77K, varying with the delay time. The spectral line with the peak position at 5200-5300Å and band width of 1300-1400Å, is asymmetrical at room temperature, and the line shape varies less with time. At 77K however, the line moves greatly from the peak position of 5100Å to 5500Å, and band width from 2250Å to 2000Å within 20μs. At 30μs after the excitation the spectrum appears to be diffused due to selfabsorption. In addition, no obvious increase of the fluorescent intensity with temperature was found in the experiment down to 77K.

As the decay time measurement, the fluorescence decays faster as the emission moves from long wavelengths to shorter ones. Unconvoluted lifetimes are 5μs at 4800Å, 8μs at 5330Å and 12μs at 6700Å, respectively. At room temperature, however, all the fluorescence lifetimes at different emission wavelengths are about 13μs, similar to the one at 77K with the emission at 6700Å.

Tab. 3 Time-resolved fouorescence spectra of Ti^{3+} in fluorophosphate glass.

(a) room temperature

(b) 77K

Delay (μs)		(nm)	(nm)	Delay (μs)		(nm)	(nm)
1.	6	520	130	1.	10	510	225
2.	20	525	135	2.	20	550	200
3.	30	530	140	3.	30	640	diffused

3. Cu^{+} Ions

Cu^{+}-doped fluorophosphate glass has a d^{10}-d^{9}s absorption band at UV 2500Å and an emission at 4000-4200Å. Time-resolved fluorescence was measured at room temperature and 77K by excitation with the pulse duration of 2us. It is shown from the Tab. 4 that the fluorescence peak position shifts toward longer wavelength with either increasing delay time or moving the excitation to longer wavelength. The shifting rate rises with delay time.

Tab.4 Time-resolved fluorescence spectra of Cu^+ in fluorophosphate glass at room temperature (R.T.) and 77K.

λexc. R.T./77K Delay	220 nm		250 nm		280 nm		300 nm	
	λ_p	$\Delta\lambda$	λ_p	$\Delta\lambda$	λ_p	$\Delta\lambda$	λ_p	$\Delta\lambda$
0-2 µs	380/390	110/95	380/390	90/85	385/400	105/92	/420	/100
60-100 µs	390/400	105/103	390/935	98/92	400/420	105/103	/435	/140
140 µs	400/410	115/110	395/	103/	420/	127/		
180-200µs	/420	/130	405/400	120/100	440/435	165/142	/510/590	/*
250-230 µs	–	–	/420	/125	/490/590	/*	/520/590	/*

* Represents spectral diffusion

Table 5 shows the values of fluorescence lifetime of Cu^+ in fluorophosphate glass at different temperatures as a function of excitation wavelength or emission wavelength. Note that the entire period for fluorescence decay lasts for several dozens of to hundreds of microseconds. Contrary to the fluorescence spectra obtained by non-delay sampling, the fluorescence decay curves obtained include the contribution from energy transfer between sites.

Table 5 Fluorescence lifetime (µs) of Cu^+ in fluorophosphate glass at different temperatures for different excitation and emission wavelengths

Temperature	Room Temperature					77K				
Emission wavelength(nm)	360	420	500	560	640	360	420	500	560	640
Excitation wavelength (nm) 220	(27)	(36)	(50)	(60)	(80)	(29)	(43)	(60)		
250	33	37	42	48	–	(37)	52	72	88	115
280	35	43	62	84	90	40	55	85	125	130
330	–	44	62	106	114	41	55	95	135	145

Particularly for those ions having longer emission wavelength, which acts as acceptor in the process of energy transfer, their lifetime values are enlarged by the contribution of non-radiative energy transfer.

Summary

(1) The long range disorder of glass structure results in the site-multiplicity of doped paramagnetic ions, or so called the static

lattice effect. Meanwhile, the thermal vibration of the lattice results in the temporal disorder, or so called the dynamic lattice disorder effect. Both of these effects are able seriously to cause the energy spreading of the ions, especially for 3d ions, located in multiple sites.

(2) Radiative and nonradiative relaxation rates are different for ions in different sites. Some ions having large D_q value possess longer lifetime, higher emission frequencies and lower nonradiative transition probability than those having small D_q value.

(3) Following a excitation the ions located in the excited state were unstable. Before the ions come back to the ground state the energy transfer occurred through the coupling between ion and phonons.

(4) In different kinds of glasses the natures of chemical bonds between paramagnetic ions and ligands differ from each other, as well as the interaction between ions and lattice. In general, the more the covalent extent of the chemical bond between active ions and ligands, the stronger the coupling between ion and phonons, and the intenser the energy and nonradiative transition.

References

[1] Gan, Fuxi and liu, Huimin, "Spectroscopy of Transition Metal Ions in Inorganic Glasses", J. Noncryst. Solids 80, 20-33 (1986).
[2] Gan, Fuxi and Liu, Huimin, "ESR Study on Glass", J. Noncryst. Solids 95&96, 61-70 (1987).
[3] Liu, Huimin and Gan, Fuxi, "Laser Excited and Time-resolved Spectra of Cr^{3+} in Inorganic Glasses", Kexue Tongbao 31 447-454 (1987).
[4] Gan, fuxi and Liu, Huimin, "Time-resolved Fluorescence Spectra of Titanium-containing Fluorophosphate Glass", J. Noncryst. Solids 80, 422-428 (1986).
[5] Liu, Huimin and Gan, Fuxi, "Time-resolved Fluorescence and Energy Transfer of Cu^{+}-doped Fluorophoaphate Glass", J. Lumin. 40&41, 129-130 (1988).

SPECTROSCOPY OF CHROMIUM DOPED TUNABLE LASER MATERIALS

Richard C. Powell

Department of Physics, Oklahoma State University
Stillwater, Oklahoma 74078-0444
U.S.A.

ABSTRACT

Laser-induced grating spectroscopy was used to characterize the properties of exciton dynamics, radiationless relaxation, and excited state absorption of chromium ions in alexandrite, emerald, ruby, and several types of garner crystals.

1. INTRODUCTION

Four-wave mixing techniques were used to establish and probe excited state population gratings of chromium ions in various host crystals used for solid state laser applications. [1] These included ruby (Al_2O_3), alexandrite ($BeAl_2O_3$), emerald ($Be_3Al_2(SiO_3)_6$), and several different types of garnets, GGG ($Gd_3Ga_5O_{12}$), GSGG ($Gd_3Sc_2Ga_3O_{12}$), ($La_3Lu_2Ga_3O_{12}$), and crystals of GGG containing various optically inactive ions to disorder the lattice. The most significant difference among these materials is the strength of the crystal field at the sight of the chromium ion. This causes a shift in the positions of the quartet electronic energy levels with respect to the doubles which significantly alters the spectral dynamics of the materials. The most important excited levels for laser operation are the $^4T_2$ pump band and the ^2E metastable state. In high crystal field materials the ^2E level is the source of the sharp R-line laser emission while in low crystal fiels materials it acts as a storage level for laser emission from the $^4T_2$ level.

Figure 1 shows the fluorescence spectra of six of the samples at room temperature. The strength of the crystal field is lowest for the spectra in the upper part of the figure. As crystal field decreases, the $^4T_2$ emission becomes dominant over the ^2E emission.

Along with altering the optical spectra, the variation of the energy levels with crystal field also change the properties of energy migration radiationless relaxation, and excited state absorption in these samples. These properties are important in determining the laser chara-

cteristics of these materials. The results of laser-induced grating
(LIG) measurements have been used to characterize these properties[2-5)]
and this work is summarized here.

2. ENERGY MIGRATION AMONG Cr^{3+} IONS

The variation of the LIG signal decay pattern with crossing angle of
the laser excitation beams was monitored as a function of temperature
for each of the samples. The signal patterns were fit with a theoretical model that describes four-wave mixing in the presence of spatial
energy migration.[6)] The fitting parameters are the ion-ion interaction
rate and the exciton scattering rate. These microscopic parameters can
be combined to determine the exciton diffusion constant D. Only long
range diffusion over distances of the order of 100 nm or greater can
be detected by this technique.

No long range energy diffusion was observed for Cr^{3+} ions in ruby and
in inversion sites in alexandrite. For mirror site ions in alexandrite[2)] energy diffusion was found to increase as $T^{-\frac{1}{2}}$ at temperatures below
about 150 K while in emerald[3)] the diffusion coefficient was found to
increase approximately exponentially with temperature. Figure 2 shows
the temperature dependences of the exciton diffusion coefficient for
three of the garnet samples.[4,5)] In the GGG samples D decreases as temperature is raised while in GSGG D increases with temperature. The different variations of D with temperature can be explained by the changes
in the ion-ion interaction strengths.[2-5)]

3. RADIATIONLESS RELAXATION RATES

Measurements were also made of the efficiency of the four-wave mixing
signal as a function of the crossing angle of the laser write beams at
low temperatures.[2-5)] A numerical fit to the data was obtained with a
model describing the four interacting electric fields in the crystal,
treating the coupling coefficients as adjustable parameters. From the
coupling coefficients obtained by this procedure, the laser-induced
modulation of the real and imaginary parts of the complex refractive
index can be obtained. This provides information about the excited
state absorption cross section and the dephasing time of the atomic

system. The values obtained for the excited state absorption cross sections by this method are consistent with those found from direct excited state absorption measurements.

When the quartet level is resonantly pumped, the dephasing time is dominated by the radiationless relaxation to the ^2E level. The value obtained for the dephasing times are found to become shorter as the crystal field strength is decreased as shown in Fig.3. This indicates that the dominate dephasing process occurs through direct relaxation in the doublet potential well as opposed to initial relaxation in the quartet potential well and subsequent crossover to the doublet. This type of relaxation is predicted theoretically only if anharmonic effects are included.

4 SUMMARY

Four-wave mixing is a powerful spectroscopic technique which can be used to elucidate the properties of energy migration, radiationless relaxation, and excited state sbsorption as well as the nonlinear optical properties of materials. The information obtained from the results of these measurements is important in modeling the spectral dynamics of these materials during laser operation.

ACKNOWLEDGMENTS.

This research was sponsored by the U.S. Army Research Office. The experimental results described here were obtained by the authors of references 2-5.

REFERENCES

1. Eichler, H.J., Gunter, P., and Pohl, D.W., "Laser-Induced Dynamic Gratings" (Springer-Verlag, Berlin, 1986).

2. Gilliland, G.D., Suchocki, A., Ver Steeg, K.W., Powell, R.C., Heller, D.F., Phys. Rev. B. To be published; Suchocki, A, Gilliland, G.D., and Powell, R.C., Phys. Rev. B 35, 5830 (1987); Ghazzawi, A.M., Tyminski, J.K., Powell, R.C., and Walling, J.C., Phys. Rev. B 30, 7182 (1984).

3. Quarles, G.J., Suchocki, A., Powell, R.C., and Lai, S., Phys. Rev. B., to be published.

4. Suchicki, A. and Powell, R.C., Chem. Phys., to be published.

5. Durville, F., Powell, R.C., and Boulon, G., J. de Phys. supp. 12, 48, C7-517 (1987).

6. Y.M. Wong and V.M. Kenkre, Phys. Rev. B 22, 3072 (1980); V.M. Kenkre and D. Schmid, Phys. Rev. B 31, 2430 (1985).

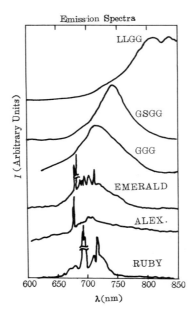

Fig. 1. Fluorescence spectra of Cr^{3+} in six different host crystals at 300 K.

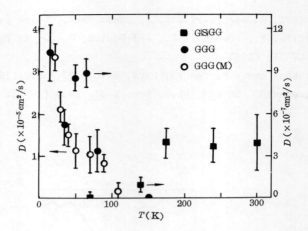

Fig. 2. Temperature dependences of the exciton diffusion coefficients in three different garnet crystals.

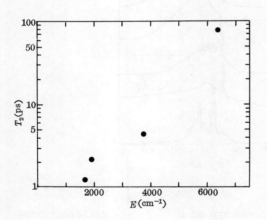

Fig. 3. Dephasing time of the laser-induced grating singnal at 12 K after pumping into the $^4T_2$ level as a function of the energy difference between the peak of the absorption transition and the lowest component of the ^2E level. In order of decreasing energy, the points represent alexandrite inversion site ions, ruby, alexandrite mirror site ions, and emerald.

SPECTROSCOPIC PROPERTIES OF Nd^{3+} IONS IN $LaMgAl_{11}O_{19}$ CRYSTAL

Zhang Xiurong and Ma Xiaoshan

Shanghai Institute of Optics and Fine Mechanics, Academia Sinica

Shanghai. P.R.C.

ABSTRACT

Spectroscopic properties of $LaMgAL_{11}O_{19}:Nd^{3+}$ were investigated experimentally at 77K,300 K and calculated using Judd-Ofelt theory. Lasing output has been obtained in the crystal at $1.054\,\mu m$, 300K.

Aluminates of rare-earth element are of of particular interest for new technology. To search for active media using aluminates with reduced quenching of neodymium luminescence, single crystal of $La_{1-x}Nd_xMgAl_{11}O_{19}$ doped Nd^{3+} concentration 5 at% was grown from Czochralski method. Spectroscopic properties of Nd^{3+} ions in the crystal were investigated and calculated using Judd-Ofelt theory. Lasing experiments were carried out at 1.054 μm.

Fluorescence Spectra

A.Fluorescence Spectra

The fluorescence and polarization spectra were measured at 77K ,300K by "Visible, Near Infrared Fluorescence Spectrometer"(Fig.1). Transition levels are from $^4F_{3/2}$-$^4I_{9/2,11/2,13/2}$, the peak wavelengths are 0.87 0.91, 1.054, 1.082, 1.30, 1.34, 1.38 μm respectively. The observed two strong luminescence lines (peak at 1.054 and 1.082 μm) correspond to the transition $^4F_{3/2}$-$^4I_{11/2}$. So, a single crystal can carry out lasing at both wavelengths (1.054 and 1.082 μm).

From Fig.2, we found the shift of fluorescence spectral lines depending on temperature is not obvious.

The observed fluorescence spectra at 77K and 300K are very similar as the $LaMgAl_{11}O_{19}$ single crystal had a hexagonal structure of magnetoplumbite $(PbFe_{12}O_{19})^{[1]}$. Each unit cell contains ten O^{2+} ions concentrated layers. The Nd^{3+} ions were limited by the concentrated layers. The effect of temperature on lattice vibration is not obvious and the interaction electron and photon is very weak. The splitting levels of Nd^{3+} in

$LaMgAl_{11}O_{19}$ crystal at 77K and 300K has little difference. The shift of fluorescence spectral lines is independent on temperature.

B. Polarization Spectra

Anisotropy of spectroscopic properties in the crystal is considerably weaker than those of other single axis crystals such as YAP, the cross section of the lasing transition ($^4F_{3/2}-^4I_{11/2}$) for E∥b orientation is the largest (Fig.3).

C. Calculation Results

Calculated results based on Judd—Ofelt theory show in table 1.The results are in agreement with the published ones [1].

D. Lasing Experiments

Lasing experiments were carried out using active elecment measured ϕ 4 x 29.5 mm, oriented along "a" axis pumped with a pulsed xenon lamp. The active element was put in a single elliptical cavity. The resonator consisted of two plane mirrors having transmission T=0 and T= 10% respectively at 1.054μm, and 57 mJ pulse output has been obtained.

Conclusion

Hexagonal $LaMgAl_{11}O_{19}:Nd^{3+}$ crystal has good physico—chemical and spectroscopic properties. The crystal has higher doped concentration of Nd^{3+} ions, but lower concentration quenching proability of Nd^{3+} luminescence lifetime is longer than that of YAG. The crystal can carry out lasing at two wavelengths (1.054 and 1.082μm). The amplification coefficient at 1.054μm is the largest. It can be used in high power lasers. 1.082 μm wavelength is a resonant transition point of triplet term spectra of He, it can be used as exciting source of He^3, He^4 and studying tools of resonant levels and controlling nuclear fission. It is a promising material for high strong energy lasers.

References

[1]   Kh.S.Bagsarov et al.;Soviet J.Quantum Electron.,13(8) (1983),

[2]   B.R.Judd;Phys.Rev.,127(1962)750.

[3]   G.S.Ofelt;J.Chem.Phys.,37(1962)37.

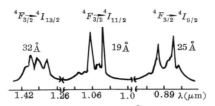

Fig 1 Fluorescence spectra of Nd^{3+} in LaMgAl$_{11}$O$_{19}$ crystal at 300K

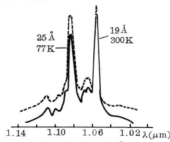

Fig 2 Fluorescence spectra of Nd^{3+} in LaMgAl$_{11}$O$_{19}$ crystal at 77 and 300K ($^4F_{3/2}$ $^4I_{11/2}$)

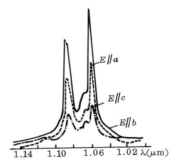

Fig 3 Polarization fluorescence spectra of Nd^{3+} in LaMgAl$_{11}$O$_{19}$ crystal at 300K ($^4F_{3/2}$ $^4I_{11/2}$)

table 1 Spectral properties of Nd^{3+} in LaMgAl$_{11}$O$_{19}$ and YAG crystals

crystals	$^4F_{3/2}$→	peak (µm)	Δλ$_{flu}$ (Å)	τ$_{flu}$ (µs)	τ$_{rad}$ (µs)	η	N$_{Nd}$ at%	σ$_{19}$ x10	β(ij)				Ω(246)		
									9/2	11/2	13/2	15/2	Ω(2)	Ω(4)	Ω(6)
LaMgAl$_{11}$O$_{19}$:Nd^{3+}	$^4I_{11/2}$	1.054	12	315	338	0.93-1	5	3.2	0.35	0.54	0.15	0.009	0.21	0.48	1.5
YAG:Nd^{3+}	$^4I_{11/2}$	1.064	6.5	240	420	0.56	1	4.5	0.32	0.58	0.11	0.003	0.2	2.7	5.0

SPECTRAL STUDY AND 2.938 μm LASER EMISSION OF Er^{3+} IN THE $Y_3Al_5O_{12}$ CRYSTAL

Yuan-qi Lin, Bao-cheng Yang, Yong-fang Li,

Zhong-jun Shen, & Feng-yu Li*

Physics Department, East China Normal University,
200062, Shanghai, China
*Shanghai Institute of Optics and Mechanics,
Academia Sinica

For a free TR^{3+} ion, electric-dipole transitions between state of the same configuration ($4f^n$) are strictly parity forbidden. In crystals they are allowed due to mixing of wave functions of the $4f^n$ configuration, caused by crystal field odd parity terms. Properties such as the structures of the energy levels , the transition strength and dynamic mechanism of transfer can be revealed through the study of the absorption and emission spectra of the rare-earth ions doped crystals. The fluorescence spectrum of the crystal Er:YAG pumped by flashlamp is obtained (Fig.1), labeled and compared with those of Er:LN, $KBiEr(MoO_4)_4$ and $Er:P_5O_{14}$ and the others already studied by the authors. It can be seen that among the crystals the degeneracy of the level $^4I_{13/2}$as well as $^4I_{15/2}$ in Er:YAG is removed in the highest degree, which is also confirmed by the absorption spectrum. This enables the crystal easier to emit stimulated emission, because the Stark splitted sublevels separate from one another far enough that mixing of the wave functions of the sublevels is unlikely to occur. Therefore, population inversion between some particular pairs of the levels is possible. As for 2.938 μm stimulated emission, the low level $^4I_{13/2}$ has seven sublevels which are populated according to Boltzmann law, so the sublevel $^4I_{13/2}(7)$ is least populated; The upper level $^4I_{11/2}$ has six sublevels with $^4I_{11/2}(1)$ most populated for the same reason. Population inversion is relatively more quickly and easily reached between this pair of sublevels, which accounts for the single line laser oscillation.

To achieve stable laser emission, it is generally required that the lifetime of the upper level be much longer than that of the low level,

which, however, is not the case for the levels of Er^{3+} ions involved
in the emission of 2.938 μm laser, where bottle neck effect may block
normal population circulation. Fortunately, there are two transitions
associated with $^4I_{13/2}$, one being $^4I_{13/2}$--$^4I_{15/2}$, and the other $^4I_{9/2}$
--$^4I_{13/2}$, with their transition energy nearly same. Because of the
interaction between active ions, reasonance energy transfer can take
place. As a result of the cross relaxation a active ion at $^4I_{13/2}$ loses
energy dropping to $^4I_{15/2}$, wheares another acquires energy and tran-
sits to $^4I_{9/2}$ and the returns to $^4I_{11/2}$ by nonradiative transition.
This helps to eliminate the bottle neck effect and favors the population inversion. For such a process to prevail the transition probability of the reasonance energy transfer needs to be larger than the non-
energy-transfer transition probability. To that end higher Er^{3+} density is necessary. So 2.938 μm laser is easier to obtain in high density
erbium doped crystals. This is also verified by the fact that the effect
of density quenching of fluorescence is strong for the transition
$^4I_{13/2}$--$^4I_{15/2}$ in our experiment.

Fig.2 shows the spectrum of the 2.938 μm laser output of Er:YAG obtained. YAG crystal with Er^{3+} concentration of 50 % is adopted. The rod
is ϕ 6 mm x 90 mm which is pumped by flashlamp. The reflectivity of
the output mirror is 93 %. The laser has a line width of about 1 cm^{-1}.
The wavelength is 2.938 μm and the threshold pumping energy is about
160 J.

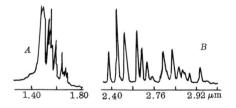

Fig.1 A The absorption spectrum for Er
ions in Er:YAG crystal B Fluoresconce
spectrum of Er YAG recorded on the
$^4I_{11/2}$-$^4I_{13/2}$

Fig.2 Laser radiation
spectra of Er:YAG
crystals.

RAMAN-INFRARED SPECTRA AND RADIATIONLESS RELAXATION OF LASER CRYSTAL $NdAl_3(BO_3)_4$

Hong Shuili

Physics Department, Fuzhou University,
Fuzhou, P.R. China

Luo Zundu

Center of Theoretical Physics, CCAST (World Laboratory),
Beijing and Fujian Institute of Research on the
Structure of Matter, Chinese Academy of Science,
Fuzhou, P.R. China

Neodymium Aluminum Borate $NdAl_3(BO_3)_4$ crystals (for short NAB) have two kinds of structure corresponding to R32 (D_3^7) and C2/c (C_{2h}^6) space group respectively. There are two $NdAl_3(BO_3)_4$ molecules in each unit cell. It can be shown that the $(BO_3)^{3-}$ groups are sited at C_3 symmetry positions, which is subgroup of both the factor group D_3 of space group D_3^7 and the molecular point group D_{3h}. There are 117 optical phonon modes with $\vec{K} \cong 0$ (except the three acoustical branches) can be symmetrical classified by using site group analysing method.

The Raman spectra of NAB crystals were measured in different conditions with different geometric configurations. On the other hand, the infrared absorption spectrum of NAB crystal powder was recorded. In order to investigate the effect of optical phonon modes on the radiationless relaxation of the luminescent ions Nd^{3+}, we first study the function of ν_3 mode which has a infrared absorption activity in the six internal vibration modes $\Gamma_g = A_1' + 2E' + A_2''$ of $(BO_3)^{3-}$ groups. It is shown that symmetry character E' (x,y) - E (x,y) - E(x,y) corresponds to the $D_{3h}-C_3-D_3$ point group transfer. The basic X-ray data[1] indicate that the unit cell of NAB contains two non-equivalent borate molecules, therefore the boron ions occupy the C_3 and C_2 symmetry positions respectively. The degenerate E' mode split into four peaks with wave number $1253 cm^{-1}$, $1316 cm^{-1}$, $1358 cm^{-1}$ and $1410 cm^{-1}$, all have rather strong absorption. These peaks can also be seen in Raman spectra, but they are very weak in the latter case.

Weak crystal field is at the Nd^{3+} sites. According to previous result[1], electron matrix elements of the radiationless transition produced by ligand internal vibration are proportional to the products of the crystal field parameters and the variation of these parameters during the ligand internal vibration. It can be seen that the radiationless transition rates results from the internal vibration of $(BO_3)^{3-}$ is weaken, although their strong internal vibrations have high frequency. On the other hand, the Nd^{3+} ions occupy the center of distorted oxygen trigonal prisms and have six nearest oxygen atoms at a distance of 2.371 Å. The investigation shows that the NdO_6 stretching mode A_1, which has the most important effect on the radiationless transition of Nd^{3+} ions, has a lower frequency.

reference
[1] H.Y-p. Hong and K. Dwight, Mat. Res. Bull., 9, 1665(1974).
[2] Luo Zun Du, Chem. Phys. Lett., 94, 498(1983).

A STUDY ON HB AND FLN IN BaFCl$_{0.5}$Br$_{0.5}$:Sm^{2+}AT 77K

Changjiang WEI, Shihua HUANG and Jiaqi YU

Changchun Institute of Physics, Acadimia Sinica, Changchun,

China

ABSTRACT

Spectral hole burning(HB) and fluorescence-line-narrowing(FLN) were studied in BaFCl$_{0.5}$Br$_{0.5}$:Sm^{2+} at 77K. The results indicated that the random arrangement of Cl and Br around Sm^{2+} ions do not affect the local symmetry of Sm^{2+} ions, but disturb the center of gravity of its energy levels. Tens of holes can be burned in this material at 77K, and this might be important for the application of HB for frequency domain optical storage.

1. Introduction

HB and FLN have been used as a tool for high resolution spectroscopy [1,2] and the research has been stimulated by the novel scientific phenomena involved as well as by the possible application of HB for frequency domain optical storage[3]. One of the main obstacles to the pratical application of optical storage is that HB can be done only at very low temperature (usually, at LHeT). We report the results of HB and FLN in BaFCl$_{0.5}$Br$_{0.5}$:Sm^{2+} at 77K(LNT). By adding Br into BaFCl, the linewidth of Sm^{2+} ions transition was much broadened, therefore, HB can be studied at quite high temperature.

2. Results And Discussion

A dye laser pumped by a pulsed Nd:YAG laser is used for the study of HB and FLN in BaFCl$_{0.5}$Br$_{0.5}$:Sm^{2+} powder sample which are immersed in liquid nitrogen. We find that the linewidth of $^5D_J^- {}^7F_J$ of Sm^{2+} in BaFCl$_{0.5}$Br$_{0.5}$ is much broader than that in BaFCl, which is attributed to the disordering effect. The hole is probed by excitation spectrum, as shown in Fig.1. The homogeneous width is about 0.47 cm^{-1}. Our results show that tens of holes can be burned at 77K. The hole can remain at least several hours with some broadening and shallowing.

The excitation spectra obtained by monitoring different position in the $^5D_0-{}^7F_0$ transition in BaFCl$_{0.5}$Br$_{0.5}$:Sm^{2+} are similar with that in BaFCl:Sm^{2+}[4] and the FLN results are shown in Fig.2. The results indicate that the random arrangment of Cl and Br around Sm^{2+} ions do not change the local symmetry, but affect the strength of crystal field, as a result, the inhomogeneous linewidth is much broadened. The un-expected two lines structure in FLN are interpreted by accidental coincidence[5].

References

[1] R.M.Macfarlane and R.M.Shelvy , J.Luminesc. 36, 179 (1987).

[1] W.M.Yen and R.T.Brundage, J. Luminesc. 36, 209 (1987).

[3] G.Castro et al, U.S.Patent 4, 101, 976 (1978).

[4] Changjiang Wei, Shihua Huang and Jiaqi Yu, Proc. 3rd Asia Pacific Phys, Conf. (Hongkong, 20-24 June, 1988).

[5] Changjiang Wei, Shihua Huang and Jiaqi Yu, to be published

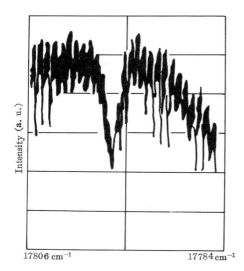 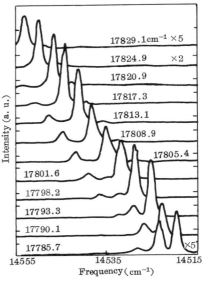

Fig 1 Spectral hole burning observed in excitation spectrum at 77K in $BaFCl_{0.5}Br_{0.5}:Sm^{2+}$

Fig 2 FLN spectra in $BaFCl_{0.5}Br_{0.5}$:Sm^{2+} at 77K

PAIR-PUMPED UPCONVERSION SOLID STATE LASERS

Stephen C. Rand, Depts. of Physics and Electrical Engineering, University of Michigan, Ann Arbor, MI 48109.

Recent work has revealed an interesting class of nonlinear optical phenomena in which weak, electronic coupling between clusters of a few atoms has an important influence on coherent optical interactions. In this paper we consider primarily the observation of pair-pumped upconversion laser action in rare-earth-doped solids. However we also describe briefly several other new effects which are similarly due to multi-atom processes. These include optical suppression of the Van der Waal's force, the occurrence of seven-line resonance fluorescence spectra in stationary pair systems, and the influence of dark pair processes in four-wave mixing spectroscopy.

Cooperative effects in the radiation from small groups of atoms develop as the result of short-range multipole-multipole or exchange interactions. They are quite different from effects like superradiance in which interatomic coupling is furnished by the radiation field itself or multi-photon processes in which sequential linear processes or nonlinear response of single atoms is involved. The mechanism for the results reported here depends on simple electric dipole-dipole coupling between atoms combined with the driving force of intense coherent light.

In the past, pair interactions had significant impact only on the operation of lasers in which energy downconversion played a role. For example, the N-line ruby laser operated on a difference pair transition near 703 nm instead of the usual 694 nm transition. The more recent Tm,Ho:YSGG and Tm:YSAG lasers have quantum efficiencies close to two, rather than unity, because single incident photons are able to promote two ions to the upper laser level via nearly resonant energy transfer processes involving dopant pairs. However, we are principally concerned here with pair interactions in solids which furnish energy upconversion through energy "pooling". Of course multiphoton processes relying on linear or nonlinear susceptibilities of individual atoms are commonly used for frequency upconversion of radiation and even laser pumping. However, extremely efficient upconversion can also occur by cooperative multi-atom processes involving simultaneous transitions on two or more atoms.

Cooperative pair pumping in 5% Er:YLF has recently been shown to produce laser action from an upconverted energy state following pumping at 1.54 microns with an Er:glass laser on the ground state (J=15/2) to first resonance (J=13/2) transition. In Er crystals with low rare earth concentrations, the lifetime of the second excited state (J=11/2) normally exceeds that of J=13/2 and laser action on the 11/2 to 13/2 transition is quenched. However, at high Er concentrations pair interactions contribute to depletion of the lower level and laser action becomes possible. In the Er system, two 13/2 ions decay cooperatively to yield one high energy ion in the 9/2 state, which relaxes rapidly to the upper laser level (J=11/2). The cooperative process has a very high rate because of strong inter-atom coupling and the long 13/2 lifetime in YLF. Its rate greatly exceeds the spontaneous decay from the same state as well as the filling rate from higher states at modest input powers. The upconversion process is shown for the first time to be so effective in Er:YLF that this mechanism alone can invert 11/2 with respect to lower states. Cw operation is also predicted, despite the fact that atoms are excited directly to the lower laser level by pump radiation. The interatomic coupling merely has to be strong enough so that depletion exceeds the filling rate in the lower level. Analysis of saturation and coherence effects in laser systems of this kind will be presented.

Other effects of pair interactions on coherent optical phenomena which have been observed or calculated recently are also reported. The pair resonance fluorescence spectrum for example is predicted to consist of seven lines, rather than the usual three, because transitions occur between triplets of dressed states rather than doublets. Also, coherent dips are predicted in the stimulated pair emission spectrum when the Rabi frequency reaches a value determined by the strength of the interatomic coupling. These coherent features are due to dynamic decoupling of the atomic pair interactions by the light field, a phenomenon called the electric magic angle effect. At the magic angle condition, the Van der Waal's force between the coupled atoms is suppressed.

In addition, experimental results from nearly degenerate four-wave mixing spectroscopy (NDFWM) have revealed that the coherent signal intensity is sensitive to non-radiative (dark) processes involving cross-relaxation of ionic pairs in crystals. In the promising laser host material Beta"-Alumina, weak pair processes are observed at high concentrations of trivalent Nd dopant ions with NDFWM, but are obscured in conventional fluorescence studies by overwhelming decay contributions from other impurities. Lineshape analysis of the (1 kHz wide) four-wave mixing spectrum reveals the onset of cross-relaxation at 6×10^{20} Nd/cm^3, although excited state decay times measured in fluorescence are constant throughout this doping range. Saturation intensities determined from the NDFWM results also decrease by an order of magnitude in this density range, establishing that the maximum useful density of Nd ions in this material for laser applications is limited by a dark pair process.

Hence pair interactions can have major effects on practical laser operation and in future will permit construction of efficient lasers with operating wavelengths much shorter than the excitation wavelengths. It is now known that pair processes can be enhanced in special crystal systems where the majority of absorbers consist of pairs or trios of atoms. In such solids it is to be expected that higher upconversion efficiencies and further novel aspects of such interactions will be encountered in studies of stimulated emission and de-localized coherences induced on spatially separated atoms.

This work was supported by AFOSR.

CW Upconversion Laser Action in Neodymium and Erbium Doped Solids

R. M. Macfarlane, A. J. Silversmith[a], F. Tong[a], and W. Lenth

IBM Research Division, Almaden Research Center, San Jose, California 95120

Abstract: Upconversion pumping of miniature solid state lasers has produced single mode cw output of several mW at 550nm ($Er:YAlO_3$), 380nm ($Nd:LaF_3$) and 730nm ($Nd:YLiF_4$). In $Nd:YLiF_4$ the output at 413nm arises from a spin forbidden transition and $11\mu W$ is obtained. In $Er:YLiF_4$ a self-pulsed output with an average power of 5mW was achieved at 550nm.

1. INTRODUCTION

Upconversion processes leading to spontaneous emission at frequencies higher than that of the excitation have been studied rather extensively in rare-earth doped materials. The two most important physical mechanisms leading to upconversion are two-step, sequential photon absorption [1] and energy transfer in which at least two ions are excited.[2-5] Stimulated emission by upconversion excitation was first reported by Johnson and Guggenheim [6] in BaY_2F_8 doubly doped with Yb/Er and Yb/Ho. They used flashlamp excitation and observed pulsed output at 77K.

A revival of interest in upconversion processes that lead to laser action has occurred recently.[7-11] This has been stimulated by the availability of reliable, tunable and often compact near-infrared laser sources and by the increasing interest in the applications of short wavelength solid state lasers. A number of schemes for the generation of green and blue radiation by frequency doubling [12] and mixing[13] in diode-laser-pumped Nd:YAG lasers has been demonstrated. We have investigated another approach, that of nonlinear or upconversion pumping of solid state lasers. Our studies have concentrated on two ions: Er^{3+} and Nd^{3+} in a variety of oxide and fluoride hosts.

2. EXPERIMENTAL SETUP

Erbium and neodymium doped (1%) laser crystals typically 1-3mm long, were placed in a variable temperature cryostat, where temperatures from 10K-300K could be maintained.[7,10,11] In all cases, cooling of the crystals to temperatures $\leq 90K$ was necessary to achieve laser operation with several hundred milliwatts of cw pump power. Mirror coatings were applied directly to the spherically polished faces of the crystals, and chosen to give an output coupling of 1%. Pump light from cw dye lasers was focussed to a $\sim 15\mu m$ beam waist in the crystals. In the most general setup, two different pump wavelengths were used to achieve two-step doubly resonant pumping. The two beams were combined using a polarization beam splitter.

---

[a]IBM Visiting Scientist

3. ERBIUM UPCONVERSION LASERS

The energy level diagram for the Er^{3+} ion in $YAlO_3$ is shown in Fig. 1. CW laser action was observed on the $^4S_{3/2}(1) \rightarrow ^4I_{15/2}(4)$ transition at 549.8 nm following two-step absorption from $^4I_{15/2}$ to $^4I_{9/2}$ (792.1 nm) and $^4I_{11/2}$ to $^4F_{5/2}$ (839.8 nm) at temperatures up to 77K.[7] The energy levels are labeled by $^{2S+1}L_J(n)$, where the number in parenthesis denotes the crystal component for a given J. The intermediate $^4I_{11/2}$ state has a lifetime of 1.2 ms, and that of the $^4S_{3/2}$ upper laser level is $160\mu s$. Some contribution to the inversion density comes from phonon assisted energy transfer processes in which two ions in the $^4I_{11/2}$ state transfer their energy to $^4S_{3/2}$. However, at our pump levels of $\lesssim 40kW/cm^2$ the lasing threshold was not reached with a single pump source at 792.1nm. The output power as a function of the input is shown in Fig. 2. Maximum power output occurred at a temperature of $\sim$30K, at which the temperature-dependent cavity mode frequency is matched to the peak of the narrow ($1.1cm^{-1}$ width) Er^{3+} emission line.[7] At temperatures above 77K, the reabsorption losses associated with the population of the lower laser level at $218cm^{-1}$ increase rapidly and prevent lasing. In $YLiF_4:Er^{3+}$, upconversion laser action was obtained using only one cw pump laser at 802nm. Efficient energy transfer processes involving two ions in the long-lived $^4I_{11/2}$ state ($\tau = 2.9ms$) lead to population of $^4S_{3/2}$. This excitation mechanism was demonstrated clearly by the persistence of the green laser output after the pump laser was turned off since energy-transfer pumping continues until the population in $^4I_{11/2}$ decays. The 550-nm laser output consisted of a series of semi-random pulses of 150-80 ns duration and an average repetition rate of 40-250 kHz over the range of average laser output powers shown in Fig. 3.

4. NEODYMIUM UPCONVERSION LASERS

Upconversion pumping has allowed the observation of lasing on three new transitions of Nd^{3+} in the violet, blue and far-red spectral regions. In $LaF_3:Nd^{3+}$, we observed cw laser action on $^4D_{3/2}(1) \rightarrow ^4I_{11/2}(2)$ at 380.06nm and $^4D_{3/2}(1) \rightarrow ^4I_{11/2}(3)$ at 380.52nm.[10] The lifetime of the upper state is $20\mu s$. The most efficient mechanism for excitation of $^4D_{3/2}$ was sequential two-step absorption using IR ($^4I_{9/2} \rightarrow ^4F_{5/2}$) and yellow ($^4F_{3/2} \rightarrow ^4D_{3/2}$) pump photons (Fig. 4a). In addition, lasing was obtained with a single pump source around 578nm using doubly resonant, sequential absorption of two yellow photons as schematically illustrated in Fig. 4b. Single mode, cw output of 12mW at 380.06nm was obtained with pump powers of several hundred mW and 1% output coupling at 20 K (Fig.5). Reduced powers were obtainable at higher temperatures, that at 77K being 4mW. The main cause for this temperature-dependent output is the reduction in the peak emission cross section arising from line broadening.[10]

The $YLiF_4:Nd^{3+}$ system exhibits some unusual behavior and is quite different from $LaF_3:Nd^{3+}$. The $^4D_{3/2}$ lifetime is $1\mu sec$ and nonradiative decay to $^2P_{3/2}$ at $26259 cm^{-1}$ quenches much of the violet emission. Lasing is now observed from $^2P_{3/2}$ (τ = 39μs) although the output power in the blue, corresponding to the spin forbidden $^2P_{3/2}(1) \rightarrow\ ^4I_{11/2}(3)$ transition at 413nm is very low: 11μW for 300mW of yellow pump light at 12K using 1% output coupling (see Fig. 6). The most efficient pump wavelength is 603.64nm which is only resonant with the second, excited-state absorption step [$^4F_{3/2}(1) \rightarrow\ ^4D_{3/2}(1)$] but is not resonant with an electronic transition from the ground state. The first step is only weakly absorbed, and the overall absorption of pump light increases with time over several milliseconds as the $^4F_{3/2}$ population builds up. The mechanism responsible for this unusual, delayed excitation process requires further investigation. Another unusual property of this system is that lasing occurs also on a $^2P_{3/2}(1) \rightarrow\ ^2H_{9/2}/\ ^4F_{5/2}$ transition at 729.52nm where the reflectivity of the coatings is very low. Figure 5 shows the double-sided laser output obtained from the front (R = 17%) and rear (R = 6%) crystal facets.

5. CONCLUSION

We have shown that upconversion pumping of miniature solid state lasers can provide cw single mode output of several mW in the green to violet spectral regions with good efficiency. Table I summarizes the properties of the upconversion laser systems which we have studied so far. Currently these lasers operate at temperatures below 90K partially due to the reduction of the peak emission cross sections with increasing temperature. Further optimization of the device parameters can be carried out.

REFERENCES

1. N. Bloembergen, Phys. Rev. Lett. 2 , 84 (1959)
2. F. Auzel, C. Acad. Sci. (Paris) 262 , 1016 (1966)
3. V. V. Ovsyankin and P. P. Feofilov, Zh. E. T. F. Pis'ma 4 , 471 (1966) [JETP Lett. 4 , 317 (1966)].
4. F. Auzel, Proc. IEEE 61 , 758 (1966).
5. V. V. Ovsyankin and P. O. Feofilov, Izv. Akad. Nauk SSSR Ser. Fig. 37 , 262 (1973) [Bull. Acad. Sci USSR Ser. Plup. 37 , No.2, 37 (1973).]
6. L. F. Johnson and H. J. Guggenheim, Appl. Phys. Lett. 19 , 44 (1971).
7. A. J. Silversmith, W. Lenth, and R. M. Macfarlane, J. Opt. Soc. Am. A3 , p128 (1986); A. J. Silversmith, W. Lenth, and R. M. Macfarlane, Appl. Phys. Lett. 51 , 1977 (1987).
8. S. A. Pollack, D. B. Chang, and M. Birnbaum, Proc. Topical Meeting on Tunable Solid State Lasers, post-deadline paper PD4-1, Williamsburg, VA, Oct 26-28 (1987).
9. G. Kintz, L. Esterowitz, and R. Allen, Proc. Topical Meeting on Tunable Solid State Lasers, paper WE5-1, Williamsburg, VA, Oct 26-28 (1987).
10. R. M. Macfarlane, F. Tong, A. J. Silversmith and W. Lenth, Appl. Phys. Lett. 16 , 1300 (1988).
11. W. Lenth, J.-C. Baumert, G. C. Bjorklund, R. M. Macfarlane, W. P. Risk, F. M. Schellenberg, and A. J. Silversmith, SPIE 898 ,61 (1988).
12. T. Baer and M. S. Keirstead, CLEO Digest post-deadline paper THZZ1, Optical Society of America, 1985; T. Y. Fan, G. J. Dixon, and R. L. Byer, Opt. Lett., 11 , 204 (1986); W. J. Kozlovsky, C. D. Nabors, and R. L. Byer, IEEE J. Quantum Electronics 24 , 913 (1988).
13. J.-C. Baumert, F. M. Schellenberg, W. Lenth, W. P. Risk, and G. C. Bjorklund, Appl. Phys. Lett. 51 , 2192 (1987).

TABLE I: CW-Pumped Upconversion Lasers

Host	Er^{3+}		Nd^{3+}	
	YAlO$_3$	YLiF$_4$	LaF$_3$	YLiF$_4$
Transition	$^4S_{3/2}(1) \rightarrow {}^4I_{15/2}(4)$	$^4S_{3/2}(1) \rightarrow {}^4I_{15/2}(5)$	$^4D_{3/2}(1) \rightarrow {}^4I_{11/2}(2)$	$^2P_{3/2}(1) \rightarrow {}^4I_{11/2}(3)$ $(^2P_{3/2}(1) \rightarrow {}^2H_{9/2}/{}^4F_{5/2})$
Wavelength	549.8nm	550.0nm	380.06nm	412.96nm (729.52nm)
Upper state lifetime (μs)	160	400	20	39
Pump (nm)	792.1 + 839.8	802	~596 + 788,579	603.64
Output (T = 1%) for ~200mW input	1mW	5mW (self-pulsed)	12mW	11μW (blue) (30mW,red,see text)

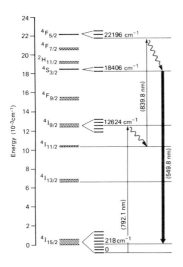

Figure 1: Energy level diagram of YAlO$_3$:Er^{3+} showing the pumping and lasing transitions.

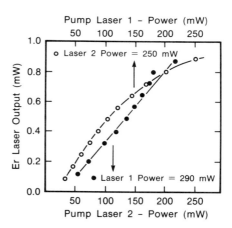

Figure 2: Output power of the 550-nm YAlO$_3$:Er^{3+} laser as a function of the power of pump laser 1 with the power of pump laser 2 held constant and vice versa.

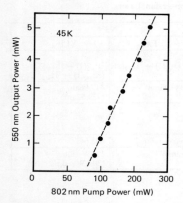

Figure 3: Average output power of the YLiF$_4$:Er^{3+} laser at 45K as a function of the cw dye laser power.

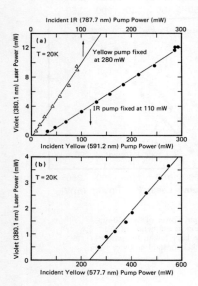

Figure 5: Violet output of the LaF$_3$:Nd^{3+} upconversion laser as a function of pump power at 20K: (a) two-step excitation with the IR power fixed and yellow power varied and vice versa; (b) single pump laser at 577.7nm.

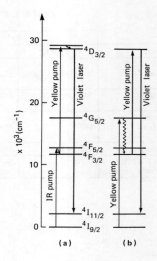

Figure 4: Energy level diagram of LaF$_3$:Nd^{3+} showing upconversion pumping and lasing transitions. (Not all levels of Nd^{3+} are shown.)

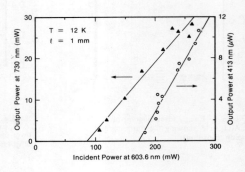

Figure 6: Blue (413nm) and far-red (730nm) output power of the YLiF$_4$:Nd^{3+} laser as a function of the yellow pump power at 12K.

ULTRA-HIGH SENSITIVE UPCONVERSION FLUORESCENCE*
OF YbF$_3$ DOPED WITH TRACE Tm^{3+} and Er^{3+}

Zhang Heyi,   He Xuehua,   Zhao Liang

Dept. of Physics, Peking Univ.

William M. Yen

Dept. of Physics, Univ. of Geogia

The upconversion fluorescence is a interest subject not only for applications but also for basic physics research. Many publications concerned with this subject especially the upconversion fluorescence of rare earth ions doped in crystals. But very few paper concerned with the sensitivity of upconversion fluorescence. The very high sensitive upconversion fluorescence of YbF$_3$ doped with trace Tm^{3+} and Er^{3+} was reported here for first time to our knowledge.

The YbF$_3$ crystal was supported by Ames Lab. It contains trace rare earth ions Tm^{3+} 0.2ppm, Er^{3+} 0.2ppm. For excite upconversion fluorescence, the infrared laser's wavelength λ=9900A which coincides with the absorption line of Yb^{3+} in YbF$_3$ crystal. When the infrared laser beam was focused into YbF$_3$ sample, the weak blue and green light emission was observed. By using a 1/4 meter grating monochromator, we got the upconversion fluorescence spectra. There are three lines λ=4750A, 5450A, 6500A. The λ=4750A and λ=6500A come from same upper excited energy state $^1G_4$ of Tm^{3+}, because these two fluorescence lines have same lifetime and same decay curve. The λ=5450A spectra line is the upconversion fluorescence of Er^{3+} ions. The energy transfer processes are shown in Fig.1. From Fig.1 one can find that the λ=4750A is the three step process. From the rate equation we got

$$I \propto P^3 (N_g^d)^3 N_0$$

in which P is the incident laser power, No is the population of Tm^{3+} ions in the ground state, N_g^d is the population of ground state of donor Yb^{3+} ions. While the λ=5450A

30

fluorescence is a kind of two step upconversion processes.
The intensity will be

$$I \propto P^2 (N_g^d)^2 No$$

in which No is the population of ground state of Er^{3+} ions.
Fig.2 showed the logI $\sim$ logP diagram. The slope of λ=4750A
r=2.7 close to 3, the slope of λ=5450A r=1.9 almost equal
to 2, both are agree with theory analysis of rate equation.

From the rate equation one can easily found that the
sensitivity I/No $\propto P^3 (N_g^d)^3$. For Tm^{3+} ion upconversion fluorescence is very high, because it is proportional to $P^3 \cdot$
$(N_g^d)^3$. The concentration of Yb ions in YbF_3 is 100%, so
for these kind of crystal one can easily get very high sensitive upconvertion fluorescence. These upconversion fluorescence may have some applications for purity analysis.

ACKNOWLEDGEMENT

Some parts of this work are finished in University of
Wisconsin-Madison, Professor W.M. Yen Lab.

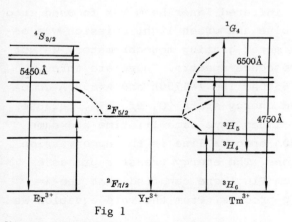

Fig 1

Upconversion fluorescence processes of
Yb^{3+} Tm^{3+} Yb^{3+} Er^{3+} system in YbF_3
crystal

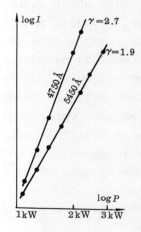

Fig 2

logI logP diagram of
upconversion fluorecence

* This work is partially supported by NSF of China

THE GROWTH AND PROPERTIES OF
NYAB AND EYAB MULTIFUNCTIONAL CRYSTAL

Lu Baosheng,   Wang Jun,   Pan Hengfu,
Liu Mingguo,   Jiang Minhua

Institute of Crystal Materials,
Shandong University, Jinan, P.R. China

I. Introduction

Yittrium Aluminum Borate (YAB) is a kind of optical crystal with high nonlinear coefficient. Neodymium Aluminum Borate (NAB) and Erbium Aluminum Borate (EAB) are self-activated laser materials with low concentration quenching at high active ion concentrations.. Therefore, the crystals that consist of YAB and NAB (or EAB) will certainly be a new kind of multifunctional crystal. The fundamental feature of NYAB crystal realizing laser self-frequency-doubling from 1.06μ to 0.53μ has been measured for the first time by us in 1986 [1]. Now, we are engaging in 1.53μ and 1.67μ laser operation in EYAB crystal, because it has low losing energy in optical fibers for these wavelengths. Meantime, we are also carrying out studies on the possibilities of realizing laser self-frequency-doubling at some wavelengths for EYAB crystal.

II. Crystal growth of NYAB, EYAB

Due to the fact that NYAB, EYAB are incongruent compounds, so the selection of flux system for growing good quality crystal is crux. Some flux systems have been reported, e.g. $PbO+PbF_2$, $Li_2B_4O_7$, $MoO_3+K_2SO_4$... But, using these flux systems, we cannot grow large and transparent crystals. A new flux system for growing NYAB, EYAB crystals has been found by us.

2-1 According to the solubility curve of NYAB, EYAB, the solute concentration is 20wt.%. Melting temperature of NYAB, EYAB is $1170-1200^{\circ}C$.

2-2 Saturated temperature is about $1050^{\circ}C$ (by seeded measuring)..

2-3 The temperature lowering rate is about $2-3^{\circ}C$/day. Final
   growth temperature is about $900^{\circ}C$.

III.   Spectral properties of NYAB, EYAB crystals

   According to the measuring results of absorption and
fluorescence spectra, we use Judd-Ofelt theory to calculate
the oscillator strengths and intensity parameters of NYAB,
EYAB crystals. Then, from these results, the Einstain spontaneous radiative coefficients, radiative lifetimes, branching ratios and integrated emission cross sections of different transitions can all be obtained. Based on these coefficients, the transition probabilities of realizing laser
operation in NYAB, EYAB are discussed.

3-1 Some spectral coefficients of $^4F_{3/2} \rightarrow {}^4I_{11/2}$ transition
   on NYAB crystal.
   * Measured bandwidth is $30cm^{-1}$.
   * Calculated emission cross section is $2.01 \times 10^{-18} cm^2$.
   * Calculated lifetime of metastable state $^4F_{3/2}$ is
     $409.6\mu s$.
   * Measured lifetime of metastable state $^4F_{3/2}$ is $60\mu s$,
     it is two times larger than that of NAB crystal.
   * Branching ratio is 14.6%.

3-2 Some spectral coefficients of $^4I_{13/2} \rightarrow {}^4I_{15/2}$ (1.53μ),
   $^4S_{3/2} \rightarrow {}^4I_{9/2}$ (1.67μ) and $^4I_{11/2} \rightarrow {}^4I_{13/2}$ (2.75μ) transi-
   tion on EYAB crystal.
   For $^4I_{13/2} \rightarrow {}^4I_{15/2}$ transition
   * Calculated integrated emission cross section is
     $2.575 \times 10^{-18} cm$.
   * Calculated lifetime of metastable state $^4I_{13/2}$ is
     3.81ms.
   For transition $^4S_{3/2} \rightarrow {}^4I_{9/2}$
   * Calculated integrated emission cross section is
     $1.36 \times 10^{-18} cm$.
   For transition $^4I_{11/2} \rightarrow {}^4I_{13/2}$
   * Calculated integrated emission cross section is
     $1.51 \times 10^{-18} cm$.

* Calculated lifetime of upper state $^4I_{11/2}$ is 3.10ms.
We compare the above results with those for other laser materials (given in Table 1,2).

Table 1

	Nd:YAG	Nd:YAP	NYAB
emission cross section (cm^2)	$3x10^{-19}$	$4-8x10^{-19}$	$1.80x10^{-19}$
concentration (cm^{-3})	$1.2x10^{20}$	$2x10^{20}$	$5-10x10^{20}$
measured ($^4F_{3/2}$) lifetime μs	200	150	60
branching ratio	0.55	0.58	0.50

* all these results for transition $^4F_{3/2} \rightarrow {}^4I_{11/2}$

Table 2

host	transition	wave length (nm)	oscillator strength $Px10^6$	radiative transition rate (S^{-1})	integrated emission cross section $10^{-18}cm$	note
Er:YAG EYAB	$^4S_{3/2} \rightarrow {}^4I_{9/2}$	1670	1.21 1.57	107 117	1.03 1.36	pulse at RT
Er:YAG EYAB	$^4S_{3/2} \rightarrow {}^4I_{13/2}$	850	1.61 3.64	557 1056	2.61 3.15	output at 77K
Er:YAG EYAB	$^4I_{11/2} \rightarrow {}^4I_{13/2}$	2750	1.28 1.76	41 48.8	1.24 1.54	pulse at RT
Er:YAG EYAB	$^4I_{13/2} \rightarrow {}^4I_{15/2}$	1530	2.04 3.10	211 263	2.00 2.58	pulse at RT

3-3 The conclusion drawn from the spectral parameters:
A) NYAB crystal is both an excellent laser material and a promising laser self-frequency-doubling material. These properties have been realized by us in 1986.
B) The transitions of realizing laser operation are possible. It is possible to realize laser self-frequency-doubling effect at some wavelengths on EYAB

crystal.

IV. The measurement of optical properties of NYAB crystal

4-1 Refractive index.

The refractive index of 0 and e light have been measured by using V-type prism at seven wavelengths. Given in Table 3.

Table 3

wavelength (nm)	706.5	656.3	589.3	546.1	486.1	435.8	404.7
n_o	1.7723	1.7728	1.7769	1.7800	1.7854	1.7927	1.8067
n_e	1.6991	1.6987	1.7018	1.7146	1.7089	1.7143	1.7192

then the main refractive indicices of basic wave and harmonic wave obtained by extrapolation and interpolation respectively: n_o^{ω} =1.7553, $n_o^{2\omega}$=1.7808, n_e^{ω}=1.6869, $n_e^{2\omega}$=1.7051.

4-2 The calculation of phase matching angles.

According to the following equations:

$$n_o^{\omega} = \frac{n_o^{I\omega} n_e^{2\omega}}{[(n_o^{2\omega})^2 Sin^2\theta_m^I + (me^{2\omega})^2 Cos^2\theta_m^I]^{\frac{1}{2}}} \tag{1}$$

$$n_o^{\omega} + \frac{n_o^{\omega} n_e^{\omega}}{[(n_o^{\omega})^2 Sin^2\theta_m^{II} + (n_e^{\omega})^2 Cos^2\theta_m^{II}]} = \frac{2n_o^{2\omega} n_e^{2\omega}}{[(n_o^{2\omega})^2 Sin^2\theta_m^{II} + (n_e^{2\omega}) Cos^2\theta_m^{II}]^{\frac{1}{2}}} \tag{2}$$

θ_m^I (Type I phase matching angle) = $34^{\circ}32'$

θ_m^{II} (Type II phase matching angle) = $50^{\circ}34'$

Measured values: $\theta_m^I = 32^{\circ}54'$, $\theta_m^{II} = 51^{\circ}21'$

From this, we can see that the theoretical calculated values and the experimental values are in good agreement.

4-3 The measurement of nonlinear coefficient.

Method: SHG phase matching method.

Standard samples: KDP crystal.

Results: $d_{11}^{NYAB} = 3.9 d_{36}^{KDP}$. Take $d_{36}^{KDP} = 1.04 \times 10^{-9}$ e.s.u., then $d_{11}^{NYAB} = 4.0 \times 10^{-9}$ e.s.u.

V. Operation of self-frequency-doubling

5-1 The direction of processing NYAB crystal.

According to crystal physics equation, we can obtain $d^I_{eff} = 2.04\ d^{II}_{eff}$. So the direction of processing crystal is according to Type-I phase matching angle.

5-2 The size of NYAB crystal: 5x3x3mm.

5-3 The selection of pumping light.

In consideration of the feature of Nd^{3+} ions' absorbing spectrum, we select tunable dye laser whose wavelength is 587.9nm as pumping light.

5-4 The parallel-parallel cavity consists of mirrors M_1 and M_2. The length of cavity is less than 3cm.

5-5 Results of measurements are as follows:

* The self-frequency-doubling threshold energy is less than 1 mj.
* The green light output energy at 0.53μ is 5mj when output energy is 40mj.
* The conversion efficiency to pumping light is about 12%.

REFERENCE

[1] Lu Baosheng et al., Chinese Phys. Lett., Vol.3, No.9, (1986).

STUDY ON FLUORESCENCE AND LASER LIGHT OF Er^{3+} IN GLASS

Qi Changhong, Zhang Xiurong, Jiang Yasi and Jiang Yanyan

Shanghai Institute of Optics and Fine Mechanics

Academia Sinica, Shanghai, P.R.C.

ABSTRACT

Luminescence characteristics of Er^{3+}, Cr^{3+}, and Yb^{3+} ions doped phosphate glass and the sensitization of Er^{3+} luminescence by Cr^{3+} and Yb^{3+} ions were investigated, and laser output of 2.3 J at 1.54 μm has been obtained.

Near infrared laser crystals doped with Er^{3+} ions have got wide practical applications[1]. Our purpose is to investigate the luminescence characteristics of Er^{3+} ions in doped Li Al phosphate glass and explore the possibility of free-operation laser at 1.54 μm.

The contents of rare earth or chromium oxides in the samples were (1) 0.07 wt% Cr$_2$O$_3$, (2) 0.07wt% Cr$_2$O$_3$ + 17 wt% Yb$_2$O$_3$, (3) 0.07 wt% Cr$_2$O$_3$ + Yb$_2$O$_3$ + 0.25 wt% Er$_2$O$_3$ respectively.

It is notable that the energy characteristics of erbium glass laser are considerably inferior to those of neodymium glass lasers[2] because the three-level nature of lasing due to $^4I_{13/2}-^4I_{15/2}$ resonant transition of Er^{3+} ions and the consequent high threshold. Meanwhile the efficiency of the direct pumping of Er^{3+} ions is quite low because of its weak and rare absorption bands, thus the excitation of Er^{3+} glass is mainly carried out through the sensitizer ions of Yb^{3+} ions which have one wide and intense absorption band in the range 880–1200 nm. However, the flashlamp radiation energy is utilized inefficiently in ytterbium glass laser .Ytterbium ion has no 4f absorption level in visible wave length range, whereas chromium ions(Cr^{3+}) in phosphate glasses have three high-intensity wide excitation bands in the UV and visible (see Fig 1). The emission spectrum of Cr^{3+} presented has shown that the chromium emission (a broad

-band fluorescence in the near infrared) overlaps the ytterbium absorption, resulting in efficient nonradiative energy transfer from Cr^{3+} to Yb^{3+} ions. Therefore Cr^{3+} can be used as a sensitizing ion for Yb^{3+} luminescence.

The Cr^{3+} fluorescence lifetime at 300k in an only chromium-doped sample is 15µs. Attempts were made to measure the lifetime and emission spectrum of Cr^{3+} in the sample codoped with ($Cr^{3+}+Yb^{3+}$), but without success. When Yb^{3+} is present, the combination of weaker Cr^{3+} emission and shorter lifetime make accurate measurments difficult. These results are the evidence of the nonradiative $Cr^{3+} \rightarrow Yb^{3+}$ energy transfer. The risetime of Yb^{3+} fluorescence following pulsed selective excitation of Cr^{3+} was less than a few µs and limited by the flashlamp pulse duration (~ 5µs). It can be estimated that $Cr^{3+} \rightarrow Yb^{3+}$ energy transfer processes would be completed within $\sim 10^{-6}$sec.

The measured lifetime of Yb^{3+} luminescence in the sample with three activators ($Cr^{3+}+Yb^{3+}+Er^{3+}$) is less than that in the sample with ($Cr^{3+}+Yb^{3+}$), $\tau_{Yb}^{Cr,Yb,Er}$ =0.12ms, $\tau_{Yb}^{Cr,Yb}$ =0.6ms. THe fluorescence lifetime of Er^{3+} doped ($Cr^{3+}+Yb^{3+}+Er^{3+}$) phosphate glass was measured τ_{Er}=4ms. As can be seen from Fig.2,the luminescence spectra show that the emission intensity (980nm) of Yb^{3+} in ($Cr^{3+}+Yb^{3+}+Er^{3+}$) doped glass has decreased by 8 times that of Yb^{3+} in doubly-doped ($Cr^{3+}+Yb^{3+}$) glass, and the emission intensity (1.54µm) of Er^{3+} in ($Cr^{3+}+yb^{3+}+Er^{3+}$) doped glass has increased by 24 times that of Er^{3+} in singly doped glass (o.25wt% Er_2O_3) (see Fig. 3).

The experimental results show that the sensitization of Er^{3+} luminescence by Yb^{3+} ions in the phosphate glass is efficient and there is a high probability of nonradiative energy transfer from Yb^{3+} to Er^{3+} ions. The laser experiment was carried out using an active element 6 mm in diameter and 80 mm long placed in a silver-coated single Xe flashlamp circular illumination enclosure. The resonator was formed by two plane dielectric mirrors (R_1=99.9%, R_2=55%) and the cavity length was about 30 cm.

The threshold was achieved at an electric pump energy of about 350J,and an output of 2.3 J was obtained with 1400 J pumping with a pulse duration of 3.3 ms.

REFERENCES

[1] Kaminskii,A.A., Laser 79 Opto-Electronic. 109(1979).

[2] Edwards, J.G. and Sandoe,J.N., J.Phys. D7, 1078(1974).

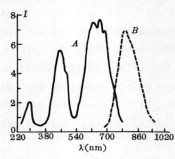

Fig 1 Excitation spectrum (A) (moni tored at 840nm) and luminescence spectrum (B) (excited at 450nm) of Cr^{3+} ions in singly Cr^{3+} doped Li Al phosphate glass

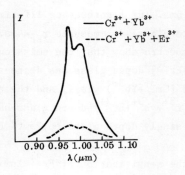

Fig 2 Emission spectra of Yb^{3+} ions in Li Al phosphate glasses at 300K

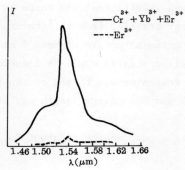

Fig 3 Emission spectra of Er^{3+} ions in Li Al phosphate glasses (300K)

GROWTH AND PROPERTIES OF SINGLE CRYSTAL
FIBERS FOR LASER MATERIALS

Ding Zuchang, Fu Senlin, Chen Jiqin, Dong Mianyu, Lu Zihong

Lab. for Joint Research in Single Crystal Optical Fiber and Laser,
Zhejiang University, Hangzhou, Zhejiang, P.R.C.

People pay more and more attention to various optical devices of single crystal fibers which can satisfy the development of fiber communication and sensor technology[1].

By means of double beam CO_2 laser-heated pedestal growth (LHPG), two typical single crystal fibers for the laser materials, i.e. Nd^{+3} doped $Y_3Al_5O_{12}$ and Cr^{+3} doped Al_2O_3 , have been grown. These crystal fibers are 10--20 cm long and 180 μm in the smallest diameter. Fiber diameter fluctuations are less than $\pm2\%$.

These fibers are indeed proved to be single crystal fibers by X-ray Laue method. The orientation of single crystal fiber is determined by that of the seed. The experimental results show: the best growing directions of crystal fibers for Nd^{+3}:YAG and Cr^{+3}:Al_2O_3 are respectively <111> and <001> orientations, their cross-section micrographs are all rounded hexagons.

In order to achieve uniform diameter fibers, the following aspects should be noted: (1) the diameter of source rod is very uniform when automatic diameter control is not available; (2) the diameter ratio between source rod and crystal fiber is in apropriate range, 1.4 to 2.7 for the above materials; (3) the growing velocity of crystal fiber is not too large, the usual velocity is less than 3.0 mm/min, as the diameter of crystal fiber becomes small, its growing velocity can be properly increased.

The fluorescent characteristics of Cr^{+3}:Al_2O_3 crystal fiber are measured. There are 6943Å and 6929Å characteristic peaks of Cr^{+3} ions in the fluorescent spectrum whether the Cr^{+3}:Al_2O_3 crystal fiber is pulled once or more than once (Fig. 1).

Since the balancing distribution coefficient of Nd^{+3} ions between YAG solid liquid is 0.2[2], it must be considered how many Nd^{+3} ions there are in Nd^{+3}:YAG fiber pulled by means of LHPG. With the help of X-ray fluorescent spectrograph, we analyzed the relative composition of Nd^{+3} ions in 2.6 mm diameter source rod and 0.9 mm diameter crystal fiber pulled only once. The fluorescent spectra of Nd^{+3} ions at different locations of Nd^{+3}:YAG crystal fiber are measured by means of Shimadzu RF-540 fluorescent spectrometer. (Fig.2). The calculated results for above measured data show: under the growing condition in this paper, the distribution coefficient of Nd^{+3} ions between YAG solid and liquid is about 0.7.

References

[1] R.S.Feigelson, W.L.Kway, R.K.Route; Optical Engineering, 24(6),
    P.1102,(1985).
[2] B.Cockayne; Phil.Mag., 12(119),P.943,(1965).

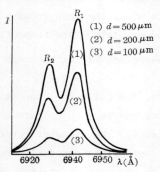

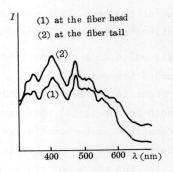

FIg.1 The fluorescent spectra of Cr^{+3}: Al_2O_3 crystal fiber.

Fig.2 The fluorescent spectra of Nd^{+3}: YAG crystal fiber.

A STUDY ON THE QUALITY OF SAPPHIRE, RUBY AND Ti^{3+} DOPED SAPPHIRE GROWN BY TEMPERATURE GRADIENT TECHNIQUE (TGT) AND CZOCHRALSKI TECHNIQUE (CZ)

Zhang Qiang and Deng Peizhen

Shanghai Institute of Optics and Fine Mechanics, Academia Sinica
P. O. Box 8211, Shanghai, P.R.China

ABSTRACT Optical method and Lang X-ray topography have been employed in the study on quality of TGT sapphire and CZ sapphire, ruby and Ti^{3+} doped sapphire crystals. The relation between crystal quality and growth method has been discussed briefly.

Sapphire is an important optical and laser window material, and Ti^{3+} doped sapphire is a tunable laser crystal material with fine properties at room temperature, which is being rapidly developed in recent years. Utilizing etching technique, Lang X-ray topography, optical and electron microscopy, the study on quality of sapphire, ruby grown by different growth methods has been made by many investigators over an extended period of time. However, much of this work has concerned the structure and behaviours of dislocation. The present paper is to study the macroscopic defects, dislocation density and arrangement as well as the quality in relation to growth methods.

In Czochralski sapphire, ruby and Ti^{3+} doped sapphire similar defects, dislocation density and arrangement have been observed. The primary defects are iridium particales, small- angle boundaries, gas bubbles and associated strains. The dislocation densities on (0001) basal plane and (1$\bar{1}$20), (1$\bar{1}$00) prismatic planes are about $10^3/cm^2$ to $10^4/cm^2$, in the region near seed about $10^5/cm^2$ or more. X-ray topographs revealed an arrangement of basal dislocation generally described as Frank-type network with node formatiom where dense cluster of dislocation appeared as 'tangle' by slipping, climbing and reaction of dislocation.

The TGT sapphire is of high quality and nearly absence of the macroscopic defects as in CZ sapphires. The dislocation density on (0001) basal plane is comparatively low, less than $10^2/cm^2$, and on

42

(11$\bar{2}$0), (1$\bar{1}$00) prismatic planes about $10^2/cm^2$ to $10^3/cm^2$. X-ray topographs displayed the arrangement of basal dislocation on (0001) slice shown in Figure. 1, in which three groups of basal tranglarcross-grid straight dislocation lines extended along <1$\bar{1}$00> directions respectively The existence of quite perfective region almost free from dislocation defects and pendelosung fringes at the wedge-shaped margin clearly manifested the perfection of TGT sapphire is relatively excellent.

For Czochralski sapphire, ruby and Ti^{3+} doped sapphire, there are no marked differences either in the dislocation density and arrangement or in the macroscopic defects. But comparing with the TGT sapphire the differences become more obviours. The TGT sapphire is of high quality and superior to the CZ sapphire, ruby and Ti^{3+} doped sapphire. Apparently ,the defects, dislocation density and arrangement and the crystal quality are necessarily related to the growth technogical characteristics. Furthermore, we believe that the temperature gradient technique is not only an important growth technique for obtaining high quality of sapphire, but also a practical growth technique for obtaining high quality and large diameter of ruby and Ti^{3+} doped sapphire single crystals.

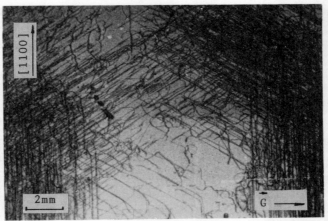

Figure. 1 Topograph of (0001) basal dislocation arrangement in the TGT sapphire; A$_g$K , G = [11$\bar{2}$0].

THE MEASUREMENT OF OUTPUT PROPERTY OF $Ti^{3+} Al_2O_3$ LASER CRYSTAL

Xie Sheng-wu, Liu Tao, Wang Jia-ji, Zhou Ting-ting, You Lu

Applied Physics Department Shanghai Jiao Tong University

ABSTRACT

The paper reports the measurements of output wavelength, time width, pulse energy, conversion efficiency and oscillation threshold of $Ti^{3+}:Al_2O_3$ laser crystal.

$Ti^{3+}:Al_2O_3$ crystal is one of the turnable laser crystal. Since 1982, people has paid more attention to this crystal for its wide wavelength width and small size of crystal rod. We grew $Ti^{3+}:Al_2O_3$ crystal successfully in 1985 and got laser output of this crystal in July of 1987. The pump laser wavelength is 532 nm. The fundamental wave is 1064 nm output of Quantel YAG laser model YG-58. The SHG crystal is KTP, its size is 3x3x5mm. The pump energy per pulse is about 2 to 20mj, which can be controlled by a stop of camera. The diameter of the stop can be changed, we may use it to measure the oscillation threshold of $Ti^{3+}:Al_2O_3$ crystal. The optical parts and instruments to measure the output property are shown in Fig.1. F1 is a infrared filter, T(1064nm)=0.3% T(532nm)=89%. L is a focal lens, f=300mm. F2 is a visible filter, T(532nm)=0.01%, T(750nm)98.7%. G is a grating monochromator model WDG-30. The detector element is a optoelectric diode model MRT-500. We use TEK-466 oscilloscope(100MHz) to measure the time width of pulse. The measurement results are follows.

1. The wavelength of output laser of $Ti^{3+}:Al_2O_3$ crystal is 7306-8268Å. The width is 962Å.

2. The time width of output laser is 15ns when the width of pump laser is 8ns. They are shown in Fig.2 and Fig.3.

3. The conversion efficiency of energy is from 1.8% to 6.4%.

4. We measure the energy using LPE-1A energy detector. The maximum out put energy is 0.61mj, the maximum efficiency is 6.4%. We have already got laser output of 6 rods of $Ti^{3+}:Al_2O_3$ which contain different density of Ti^{3+} ion. We found output energy is independence of polarization direction of pump laser.

5. The length of resonant cavity can be changed from 55 mm to 210 mm. The plane—parallel mirrors are used to avoid damage of mirrors for focal pump beam and to adjust cavity more convenient. M1 is input mirror, T(532nm)=87%, R(750nm)=95.9%. M2 is output mirror, R(532nm)=93%, T(750nm =27%.

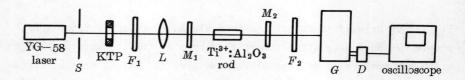

Fig 1 Optical Parts and Instruments for Measurements

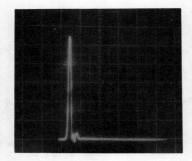

Fig 2 The Time Width of
Output Laser

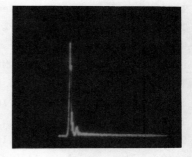

Fig 3 The Time Width of
Pump Laser

AN $X\alpha$ STUDY OF THE LASER CRYSTAL $MgF_2:V^{2+}$

Qiu Yuanwu

Department of Physics, Tongji University, Shanghai, P.R.C.

Zhu Jikang

Shanghai Software Center, Shanghai, P.R.C.

ABSTRACT

In this paper we report the results of multiple scattering $X\alpha$ calculations for the laser crystal $MgF_2:V^{2+}$. The calculated spin polarized splittings, value of energy 10 Dq, energies of the lowest doublet states and energy of the charge transfer transition are given. The relations between the laser properties and the electronic structures are discussed.

Divalent transition metal-doped solid-state lasers are now emerging as important sources of tunable radiation in the near infrared wavelength region. Because of its high efficiency as an oscillator, the $MgF_2:V^{2+}$ laser has been considered for use as a fusion driver[1,2].

In this paper we report the results of multiple scattering $X\alpha$ calculations for the laser crystal $MgF_2:V^{2+}$. The calculated spin polarized splittings, value of 10 Dq energies of the lowest doublet states and energy of the charge transfer transition were given. The relations between the laser properties and the electronic structure were discussed. The atomic coordinates of the cluster $(VF_6)^{4-}$ are taken from Ref.[3]. The site symmetry of Fe^{2+} is D_{2h}.

The calculated one-electron eigenvalues of the cluster $(VH_6)^{4-}$ are listed in Table 1. The splittings of spin-up orbitals and spin-down orbitals characteristize the effect of spin polarization. Our calculated spin polarization splittings are 2.06 and 1.29 for $t_{2g}^{*\uparrow} - t_{2g}^{*\downarrow}$ and $e_g^{*\uparrow} - e_g^{*\downarrow}$ respectively. The empirical expression for estimating the spin polarization splitting is $\delta_\varepsilon = n(3.5B + 1.4C)$[4], where n is the number of unpaired electrons, and B and C are the Racah parameters of the central ion. By using B=766 cm^{-1} and C=2855 cm^{-1}[5] the empirical value of the spin polarization splitting is 2.48. It is seen that the calculated values are in good agreement with the empirical value. The reason why the value of δ_ε for an e_g orbital is smaller than that for a t_{2g}

orbital is discussed.

The electron configuration of V^{2+} is d^3 for free ions and is (t_{2g}^3) under O_h site symmetry. $^4A_{2g}(t_{2g}^3)$ is the ground state, and $^4A_{2g}(t_{2g}^2 e_g)$ is the lowest quartet excited state. Experimentally the value of 10 Dq can be considered as the energy interval of $^4T_{2g}$ and $^4A_{2g}$, and is 10080 cm^{-1}[6]. In order to obtain the calculated value of 10 Dq, we calculated the three transition energies $a_{g1}^* \uparrow - b_{1g}^* \downarrow$, $b_{2g}^* \uparrow - b_{1g}^* \downarrow$, $b_{3g}^* \uparrow - b_{1g}^* \downarrow$ and took their average value 12226 cm^{-1} as the value of 10 Dq. It is seen that our calculated value agrees with measured value. The energies of the lowest doublet states ^2E_g, $^2T_{1g}$ and $^2T_{2g}$ of the configuration (t_{2g}^3) were also obtained.

The charge tramsfer transitions from the F_{2u} - antibonding orbitals of the ligands to the 3d crystal orbitals of V^{2+} lie in the ultraviolet wavelength region. we use the transition state theory to calculate the transition energy $(t_{1u}^* \uparrow - t_{2g}^* \downarrow)$. which corresponds to the UV absorption limit. The result is 6.8 eV.

Table 1 One-electron eigenvalues of 3d orbitals (eV)

O_h	D_{2h}	spin $\uparrow$	spin $\downarrow$
e_g^*	a_g^*	1.605	0.339
	b_{1g}^*	1.727	0.430
t_{2g}^*	b_{3g}^*	3.155	0.472
	b_{2g}^*	3.250	0.657
	a_g^*	3.318	0.907

References
[1] Moulton,P.F. and Mooradian,A.,J.O.S.A.,70,635(1980).
[2] Dickson D., Nature,288204(1980)
[3] Baur W.H.,Guggenheim,S., and Lin,J.-C.,Acta Cryst.,B38,351(1982)
[4] Adachi H. etal.,J.Phys.Soc.Japan,47,1528(1979)
[5] Griffith J.S., The Theory of Transition-Metal Ions, Cambridge University Press (1961).
[6] Johnson,L.F. and Guggenheim,H.J.,J.Appl.Phys.,38,4837(1967).

Q-SWITCHED NAB LASER

Yan Ping Deng Renliang

Optical Engineering Department, Beijing Institute of Technology,
P.O.Box. 327, Beijing, PRC

ABSTRACT

A giant pulse solid state mini-laser is realized by using a NAB crystal rod, the fluorescence lifetime of which is about 19 microsecond. In our experiment, with a ϕ 5 x 14.2 (mm x mm) rod of NAB crystal 5 mJ output energy and 3 ns pulse duration are obtained.

The Neodymium Aluminium Borate Crystal $(NdAl_3(Bo_3)_4)$ (NAB) is a self-activated solid crystal with high Nd^{3+} concentration (5.4×10^{21} cm^{-3}) , high emission cross section, low concentration quenching and short fluorescence lifetime ($19 \mu s$). Usually the short fluorescence lifetime of NAB makes it difficult to operate in Q-switched mode. In this paper, a Q-switched miniature NAB laser is presented, which was made by using a special pumping source. The results obtained have shown the perspective future of NAB laser.

In order to get a high initial inversion population (i.e. high Q-switched pulse output) a cold cathode thyratron switched electrical circuit is designed to pump the xenon flash lamp. Using this system, 0-12 J output of pumping energy and about 10 μs discharge duration of xenon lamp are made.

In the experiments, three NAB rods with different dimension are used, one with the dimension of ϕ 2.5 x 9.3 (mm x mm) was made by Fujian Institute of Research on the Structure of Matter, and the others were made by Institute of Crystal Materials Shandong University. Both end surfaces of the rods are coated with anti-reflection layer to reduce the parasitic oscillation. The initial transmissivity of saturated BDN dye is T (see table 1). A plane-parallel laser resonator with 42 mm separation is used. The two cavity mirrors were coated at $1.06 \mu m$ wavelength with reflectivity of 40 % and 100 % respectively. The pumping chamber is made of silver foil in a closed-coupled fashion. The laser output is measured by a energy meter and a oscilloscope. Table 1 shows the experiment results. The largest output energy of 5 mJ and

48

the shortest half width of the Q-switched laser pulse of 2.8 ns are
obtained. The Q-switched pulse waveform is shown in Fig.1. In order to
make comparision, an experiment No.4 on Q-switched mode is made.The
pumping duration is about 50 μs. The experiment result has shown that
it is efficient to get high peak power by means of short pulse pumping.

Table 1: experiment results of Q-switched NABlaser

No.	Dimension of laser rods (mm x mm)	Pump time	Dimension of xenon lamps (mm x mm)	T	Output energy	Q-pulse duration
1	Φ 5 x 14.2	10μs	Φ 1.8 x 13.72	50 %	5 mJ	3 ns
2	Φ 2.5 x 9.3	10μs	Φ 1.8 x 10.84	29 %	2 mJ	2.8 ns
3	Φ 3 x 11.2	10μs	Φ 1.8 x 11.82	31 %	3.2 mJ	3 ns
4	Φ 3 x 11.2	50μs	Φ 1.8 x 11.82	31 %	1.34mJ	4.2 ns

In conclusion, our experimental results have shown that NAB laser is a
promising type of laser. We believe that NAB laser will have pratical
applications in the future.

Acknowledgment
The authors wish to thank Prof. Wang Jiyang in Shandong University,
Prof. Wei Guanghui, Ding Zhigao and Ding Renqiang in our research group
for their discussions and assistance.

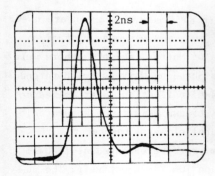

Fig 1 Q switched pulse
waveform

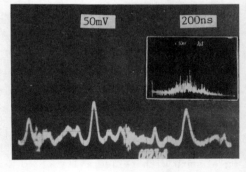

200 nsec
Fig 2 Laser oscillation in
free running mode

MINIATURE YAG LASERS

Zha Guigen, Ru Huayi, Jiang Senhua,

Jiang Shijie, Sheng Guojian, Wang Wunwen

Shanghai Institute of Optics and Fine Mechanics,

Academia Sinica

1. The study and construction of miniature xenon flashlamps

The xenon flash lamps must be not only miniaturized but also kept at effective operation energy and efficient radiation intensity. Xenon flashlamps[1] with efficient radiation intensity, low UV-radiation and lifetime sealed at high tempreture have been successfully developed in Shanghai Institute of Optics and Fine Mechanics, Academia Sinica. It solved one of key technical problems for solid state laser miniaturization. The main technical specifications[2] of these flash lamps have come up to   those of similar lamps made in other countries.

2. The design and construction of miniature repeative pulsed YAG lasers

In the past several years, extensive developments have been made on the miniaturization of solid state lasers in the world.[3][4][5].This year, we made experiments on different material and different shaped reflectors.   The experiments indicated that although the miniature ceramic and glass reflectors have the same codensing efficiency as the metallic reflectors

with some advantages of light weight, convenient construction, and easy to be produced at large scale etc., yet they are not suitable to operate at pulse repetition frequency f>1 pulse/sec. due to their lower efficiency of heat removal.

Our experimental comparison between metallic circular-crossing reflector and elliptical reflector shows that the circular-crossing reflector has higher condensing efficiency, more efficient heat removal and less heat-distortion of laser rod because it provides bigger spacing distance between the laser rod and the flashlamp at the same output energy.

An experimental comparison between laser output of different reflectors as an function of time is shown in Fig.2.

Based on the above experiments a new miniature YAG laser with metallic circular-crossing reflector without water cooling has been developed by us. Fig.3. shows the photograph of this laser. It can operate continualy at 2 pulses/sec. with dimensions of 80x30x26mm and weight of about 200g. The single-pulse output energy is 30mj, total efficiency about 1%, and the laser output energy with Q-switching film is 12mj, pulse width of 10 ns, peak power of 1 mW. Recently a newly developed pen sized laser with smaller dimensions of $\emptyset$ 22x70 mm and weight of about 70 g will be introduced.

The successful development of miniature lasers provides an ideal transmitting source for laser range-finders, new light source for portable light alignment device in muti-staged laser systems. It will be also useful in frequency-doubling lasers for educational demonstration.

References

[1] Zha Guigen et al., Chinese Physics Lasers 5. P.30-32, Vol.14 (1987)

[2] Г.А.Волкова; Ж П С,ВЫП.1,стр 30-33,ТОМ.43,1985.

[3] И.Т.Синцова и др; Ж П С,ВЫП.2,стр 294-296,ТОМ.41,1984.

[4] Lasers & Applications, 1986,6

[5] В.А.Беренберг и др;Квантовия Электроника,ВЫП.2,стр 375-377,ТОМ.12,1985.

[6] International Defence Review No 11 P 1868 Vol 18(1985)

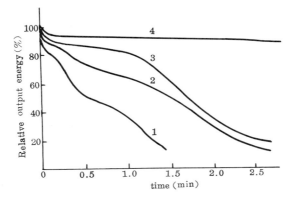

Fig. 1 Comparison of laser output energy
with following reflectors:
1. Ceramic elliptical reflector;
2. Glass elliptical reflector;
3. Metallic elliptical reflector;
4. Metallic circular crossing
   reflector used in out sample
   laser.

Fig. 2 Miniature frequency–doubling YAG laser.

STUDY OF HIGH EFFICIENCY LiF:F_2^- COLOR CENTER CRYSTALS

Li Shenghua, Wang Shulun

Shanghai Jiaotong University, P. R.C.

Fan Fuchang, Jin Derong

Shanghai Institute of Optical Instruments P. R.C.

Zhu Weizu

Shanghai University of Engineering and Technology, P.R.C.

Zhang Meizhen, Zhang Genchang

Shanghai Institute of Optics & Fine Mechanics

Academia Sinica,P. R.C.

The LiF:F_2^- center has the most attractive properties among the room-temperature tunable color center laser crystals. It has found widespread applications in near infrared tunning, 1.06 μm Q-switching and mode-locking purposes. The output energy of the wide range impulse laser is 1.85 J. THe parameters in obtaining LiF:F_2^- laser crystals are reported. High-quality LiF:F_2^- crystals can be acquired by using large-dimension LiF crystals and γ-irradiation.

1. The Properties of LiF:F_2^- Crystals

The LiF:F_2^- color center cryatal does not deliquence and has unlimited shelf life at room temperature. The corresponding laser operating time can be over 8 years. F_2^- centers have a broad absorption band from 0.83 μm to 1.13 μm, with peak at 0.96 μm. Having rather strong absorption at 1.06 um, (absorption cross section 2 x 10^{-17} cm^2), and the properties of saturation of the absorption, F_2^- centers can be used as Q switch and realize mode locking in neodymium solid-state lasers. The emission band of LiF:F_2^- centers at room temperature is from 1.03 μm to 1.28 μm with peak at 1.16 μm. Laser output has been achieved in the wavelength range 1.1–1.26 μm.

2. High-quality LiF Crystals and the Selection of Radiation Doses

The F_2^- center consists of three electrons shared by two adjacent anion vacancies in [110] direction. F_2^- centers tend to capture electrons when impurities such as "electron trap" exsist in LiF crystals.

Therefore, the purity has been strictly restricted in obtaining the laser active crystals with high LiF:F_2^- center concentration. By irradiation, the F_2^- centers were created in highly transparent LiF crystals with no scattering particles. In the irradiation, the formation of high aggregete will limit the generation of F_2^- centers. It can be known that the density of the F_2^- center can be increased and large output energy acquired by appropriate radiation dose, increasing the length of the crystals or using large-dimension crystals(55 x 65 x 75 mm^3). The density of colloid increased and anti-damage threshold of the color center decreased sharply with the increase of the radiation dose. The appropriate radiation dose is 2-3 x 10^8 Rad and its anti-damage threshold about 4 GW/cm^2.

3. Properties of Laser Operation

The absorption and emission spectra of the F_2^- four-level system are broad bands. The relaxation time of the absorption transition is 10^{-12} s and the luminescence decay time is about 10^{-8} s. It then takes 10^{-1} s to relax to the ground state. Using LiF:F_2^- color center crystal as a Q switch, pumping with ϕ 8 mm neodymium laser (output energy 8.5 J), a 1.16 μm broadly tunable laser output was achived. The energy has reached 1.85 J, light-light conversion efficiency 21.8 %.

STUDY ON THE FORMATION CONDITIONS AND OPTICAL PROPERTIES OF (F_2^+)$_H$ COLOR CENTER IN NaCl:OH$^-$ CRYSTALS

Wang Jiaxin, Li Xiaogang,
Chen Ruijin, Wan Liangfeng

Physics Department, Tianjin University,
Tianjin, China

Recently advance at a miraculous pace is making on study of a perturbed F color center in NaCl:OH$^-$ crystals. With auxiliary light source, in the NaCl system CW laser output powers at watt degree, and a tuning rang from 1.42 to 1.82 μm have been reported by Pinto et al. and Wandt et al.[1,2] Using a frequency doubling light of 1.06 μm as the auxiliary light, German invented an $(F_2^+)_H$ color-center laser in NaCl:OH$^-$ pumping only by a 1.06 μm YAG laser.[3] In this paper we report the experimental results of the formation conditions and spectral properties of $(F_2^+)_H$ center in NaCl:OH$^-$ crystals in our laboratory.

The crystal is NaCl+5x10^{-4} mol% NaOH and contains 120 ppm of OH$^-$. The actual concentration of the OH$^-$ impurity in the crystal was determined by pH titration curves and compared with the value obtained from the vibrational ir absorption line at 3654 cm^{-1}.[4]

Before coloration the crystals with dimensions of 3x12x-15 mm^3 were cleaved and polished. The samples were additive colored in a Na-vapor heat pipe apparatus disigned by ourselves at various Na pressures at 730 c for 30 min., rapidly quenched to room temperature following coloration (auxiliary annealing process does not necessary). If the cooling is sufficient rapid, the absorption band from the F_2 and F_3 centers is negligible, the background absorption is lower and the F center concentration is about 3x10^{17}cm^{-3}.

After coloration the crystals were repolished and were exposure of the crystal to a F band light for 30 min. at

room temperature. The light for optically induced aggregation was provided by a 200 W mercury arc lamp filtered
through a filter (transmitting region: 0.32-0.50 µm), and
focused with a compound Pyrex glass lenses. After additive
coloration (dashed curve) the crystal absorption correspond
mainly to the presence of the F centers (strong absorption
band at 450nm) and F_3 centers (600nm). After F aggregation
(dotted-dashed curve) the most pronounced change in the absorption is presence a peak at 1.05µm which correspond the
$(F_2^+)_{H1}$ center absorption, and a high reduction,slight red
shift of the absorption in the F band. Besides this, the
F_3 band rises and the F_2 band could faintly be seen. By an
added F band exposure at 77K (solid curve), the $(F_2^+)_{H1}$ absorption are shifted (from 1.05 to 1.09µm). We used $(F_2^+)_{H2}$
to characterize this configuration and called this process
an optically induced transformation in configuration.

When pumped by a 1.08µm YAP laser, the crystal displays
fluorescence peaking at 1.49µm which corresponds to an em-
mission of $(F_2^+)_{H1}$. Excitation of the 1.09µm band leads to
an emission at 1.59µm which corresponds to $(F_2^+)_{H2}$ emission.
Because of having a lower thermal stability, the $(F_2^+)_{H2}$
configuration can be reversed to $(F_2^+)_{H1}$ by warming the crystals (in the dark) to 300K.

In addition, without optically induced transformation at
77K, after F aggregation at room temperature the fluorescence peak at 77K is already at 1.61µm in two crystal
samples. Warming this crystal (in the dark) ten hours later
the fluorescence peak is still at 1.61µm at 77K. It is provided a trace that through certain process one might get
the $(F_2^+)_{H2}$-like type configuration which is stable on room
temperature.

REFERENCES
[1] J.F. Pinto et al., Opt. Lett., 11, 519 (1986).
[2] D. Wandt, W. Gelleman, and F. Luty, J. Appl. Phys., 61, 864 (1987).
[3] K.R. German, Opt. Lett., 12, 474 (1987).
[4] B. Wedding et al., Phys. Rev., 177-3, 1274 (1969)
*The project supported by National Natural Science Foundation of China.

NOVEL SPECTROSCOPIC PROPERTIES OF
LiF: F_3^+ — F_2 MIXED COLOR CENTERS LASER CRYSTALS

Hongen Gu, Lan Qi and Liangfeng Wan

Department of Physics, Tianjin University

Tianjin, P.R.China

Tunable lasers using F_2 centers and F_3^+ — F_2 mixed centers in LiF have been reported[1,2]. No emission peak has been observed between the peaks of F_3^+ centers (530nm) and F_2 centers (670nm) until now. In our experiment, we find some very strong emission peaks in LiF by improving the coloration method of electron beam radiation at liquid-nitrogen temperature (LNT). These peaks are at 592nm, 598nm, 602nm, 622nm and 628nm, respectively. It is possible that the crystals contain some new centers which produce these peaks. These new centers are very essential for obtaining yellow-green lasers in the same crystal.

A typical emission spectrum of a LiF excited by 450nm (see Fig.1) shows that there are a green emission band at 532nm, an orange-red one at 602nm and a red one at 628nm; the first peak corresponds to F_3^+ centers, the second and the third to the new centers. The emission peak of F_2 centers is very weak under these conditions. Spectroscopic analysis indicates that the absorption bands corresponding to 602nm and 628nm peaks are at M band in LiF. After many observations, we found that the quantum efficiency, optical gain and opto-and thermo-stability of these new centers were much the same as those of F_3^+ centers.

In this work LiF crystals polished were exposed to an 1.5Mev electron beam at LNT. Then the crystals were heated properly in the dark. After aggregating, several centers were formed in the same Laser crystal. The tunable ranges of these centers using a 3-mirror folded cavity[2] can be in the region of 510-580nm, 520-650nm (see Fig.2), 500-640nm or 520-720nm[2], which varies with coloration condition. Thresholds of 530, 580, 620, and 690nm lasers are 5.2, 8.3, 6.5 and 3.8 μJ (output mirror T=1.5%), respectively. The divergences of the laser beams are in the region of 0.75-1.4 mrad. The slope efficiencies of these lasers are in the region of 1.5-2.5%.

References

1. Mollenauer, L.F., Opt. Lett. 1, 164 (1977).
2. Gu, H.E., Wang, J.X. and Wan, L.F., Digest of '87 ICL, 13 (1987).

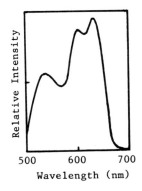

Fig.1 Emission spectrum excited by 450nm.

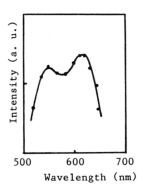

Fig.2 Relative output intensity as a function of laser wavelength.

TERRACED SUBSTRATE VISIBLE GaAlAs SEMICONDUCTOR LASERS WITH A LARGE OPTICAL CAVITY

Chen Guoying, Liu Wenjie

Semiconductor Device Laboratory,
Hebei Institute of Technology, Tianjin, P.R. China

Zhang Xingde

The Institute of Modern Optics, Changchun College
of Optics and Fine Mechanics, Changchun, P.R. China

Terraced substrate (TS) large optical cavity (LOC) visible CaAlAs semiconductor lasers, with either single dopants or no dopants, and in particular double dopants, have been fabricated. Wavelengths range from 7172Å to 7700Å. CW threshold current is 300 mA at room temperature. Stable transverse and single longitudinal mode operation are observed under 1.5 times threshold.

Visible semiconductor lasers have been required for various application, such as video-disk players and laser printers. In recent years, many articals about TS double heterostructure (DH) visible semiconductor lasers are reported. In order to obtain higher output power and less beam divergence, TS-LOC visible GaAlAs lasers have been fabricated by liquid epitaxial technique.

Structure and Device Fabrication

Fig.1 and 2, respectively show a schematic structure and a microscope photograph of the cleaved facet of the TS-LOC visible laser. Five layers of a TS-LOC visible laser were grown successively on the terraced substrate. The first n-$Ga_{0.4}Al_{0.6}As$ (Te-doped) cladding layer is about 2 µm and 1.3 µm respectively, at the terraced part and on both the outside flat areas. The second n-$Ga_{0.75}Al_{0.25}As$ (Te doped) guiding layer is about 0.5 µm and 0.3 µm. The third $Ga_{0.8}Al_{0.2}As$ (nondoped or Si-doped or Si-Zn-doped) active layer is about 0.3 µm and 0.2 µm. The fourth p-$Ga_{0.4}Al_{0.6}As$ (Zn-doped) cladding layer is 1.5 µm and 1.0 µm. The fifth p^+ - GaAs (Zn-doped) Ohmic contact layer is formed

till the surface is flat. The thickness of the optical cavity is the sum of the active layer's thickness and guiding
layer's thickness.

Prior to the liquid epitaxail growth, terraces were
formed by chemical etching. Terrace edges were parallel
to the <011> direction on the (100) ± 0.1 surface of an
n-GaAs substrate, A terrace was about 1.5 μm to 2.0 μm
high. The starting growth temperature and the cooling rate
were 860°C and 0.5°C/min., respectively.

Experimental Results

The wavelengths range from 7172A to 7700A. CW threshold
current is 300 mA at room temperature. The output
power obtained is 15 mW with non-kink without any facet
coating. The beam divergences of the FWHP in the direction parallel and perpendicular to the junction plane are
15° and 24° respectively. Some of the lasers can operate
in fundamental transverse mode and single longitudinal
mode under 1.5 times threshold.

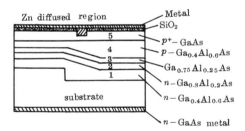

Fig.1 Schematic Structure.

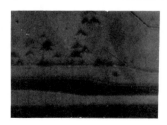

Fig.2 Microscope Photograph.

THE TEMPERATURE DEPENDENCE OF GAIN SPECTRA, THRESHOLD CURRENT AND AUGER RECOMBINATION IN InGaAsP-InP DOUBLE HETEROJUNCTION LASER DIODE

Yue Jingxing, Li Yuzhang, Xu Junying, Zheng Baozhen, Wang Xiaojie, Zhuang Weihua, Shen Dezhong*

Institute of Semiconductor, Academia Sinica, P.O. Box 912, Beijing, China

The temperature stability of a low threshold current and high output power InGaAsP/InP laser is an interesting problem from the point of view of both optical communication and physical process. To explain the strong dependence of threshold current on operating temperature, several mechanisms have been suggested, such as carrier leakage over the heterobarriar into the confining layer, intervalence band absorption and Auger recombination. We observed a 950nm electroluminescence emission band in both 1300nm and 1550nm InGaAsP/InP double heterojunction laser diode and proved this new emission band is due to the overflow of injected carriers from InGaAsP into InP by Auger recombination in InGaAsP[1].

In this work gain spectra measurement has been performed in order to study threshold current temperature behaviour of 1300nm InGaAsP/InP DH semiconductor laser diode.

We can show that $T > T_b$ ($T_b = 255k$ for our InGaAsP/InP laser diode) the Auger recombination is the most important factor in nonradiative recombination current by following different independent ways.

1) The current density dependence of $g(max)$ on $J(nom)$ [here $J(nom)$ is the nominal current density] coincides with the experiment points at room temperature, if we take quantum efficiency is determined mainly by Auger recombination i.e. $Eff-Eff(Aug)=1/[1+t(rad)/t(Aug)]$.

2) The internal quantum efficiency $Eff(T)$ coincides with the $Eff(Aug)$ from 270k to 300k (Fig.1).

3) We compare the emission intensity ratio $I(1300)/I(950)$

here I(1300) is radiative emission intensity at 1300nm and
I(950) is nonradiative emission intensity at 950nm, with
the calculated radiative to Auger nonradiative recombination probability t(Aug)/t(rad) given by K.N. Dutta[2]. Two
curves are coincident from 270k to 300k (Fig.2).

From above evidence we can conclude that in InGaAsP/InP
laser diode the Auger recombination is the main cause of
low character temperature in 270k to 300k range.

The gain spectra measurement is a very simple, practical
and useful methods. With measuring gain spectra we can get
a lot of information and parameters in laser diode i.e.
gain coefficient and its temperature and current dependence
, from which we can evaluate the quatity of semiconductor
laser diode.

REFERENCE

[1] W.H. Zhuang et al., IEEE J. Quantum Electronics, QE-21 712(1985).
[2] N.K. Dutta et al., ibid, QE-18, 871(1982).

* Permanent Address : Research Institute of Sythetic Crystal,
                      Beijing, China

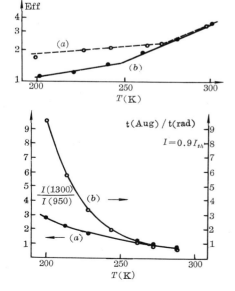

Fig 1

Temperature dependence of inter
nal quantum efficiency

Fig 2

Comparison of exp[I(1300)/I(950)]
($\bullet$) and cal t(Aug)/t(rad)(T) ($\circ$)
showa Auger recombination is the
main process when T>260K

TIME-RESOLVED PHOTOLUMINESCENE AND ENERGY TRANSFER OF
BOUND EXCITIONS IN GaP:N CRYSTALS

Lin Xiuhua, Jiang Bingxi, Ye Xinmin

Department of Physics, Xiamen University

Wang Naiguang, Chen Junde, Ye Lili, Lu Shiping

Anhui Institute of Optics and Fine Mechanics, Academia Sinica, China

Energy transfer of bound excitons in semiconductors is an interesting topic, which has attracted much attention of research workers. Previously, the energy transfer of bound excitons in GaP:N has been investigated by means of excitation spectra or photoluminescence under selective excitation[1]. The time-resolved luminescence spectrum of GaP:N has not been reported yet. In this paper we report investigation on time-resolved luminescnece spectra of GaP:N in the temperature range between 9.3 – 72.3 $^{\circ}$K. The measurements have been performed using the following experimental set-up: a Nd:YAG laser operating at 355 nm (the third harmonic generation), a double monochromator equipped with an EMI photomultiplier, a Boxcar averager and an X-Y recorder. The energy and duration of the laser were typically 3.4 mJ and 10 nsec, respectively. The samples used here were prepared by liquid-phase epitaxy technique and placed in a cryogenic refrigerator. The nitrogen was determined by optical absorption method and was estimated to be 10^{17} cm^{-3}. At low temperature under pulsed laser excitation the luminescence from GaP:N crystals was very strong: In the wavelength between 534 nm and 570 nm appeared a series of sharp lines due to radiative recombination of excitons bound to isolated N-traps and NN_i-traps. The spectra consisted of zero-phonon lines NN_i as well as their optical-phonon replicas NN_i^* as shown in Fig. 1.

The time evolution of luminescence due to excitons bound to isolated N-trap and NN_i-traps is shown in Fig.2. From Fig.2 we can see that the luminescence due to excitons bound to NN_i-traps ($i \geq 3$) decays rapidly.

For the NN_5-line the decay is most prominent. The decay is weakened in the sequence of NN_5, NN_4, NN_3. For A-line and A-LO line the decay consists of a fast exponential component (10 -40 nsec) and followed a slow component. Above mentioned results are due to exciton tunnelling from the sites of higher energy to the sites of lower energy. So the depopulation of excitons bound to the sites of higher energy may occur in two ways: radiative recombination and tunnelling of excitons[2]. The appearance of maxima in the intensities of NN_1 and NN_2 lines provide another evidence of exciton tunnelling. It has been pointed out that under above-band-gap excitation of GaP:N the tunnelling of excitons from N-traps to NN_i (i=1,2)-traps exist at temperatures lower than $20^{\circ}K$. Our results are consistent with that reported by Wiesner et al[3]. Who considered that excitons transfer must occur over at least 10 nm because of the large spatial extent of the exciton wave function.

Reference

[1] N.R. Nurtdinov, M.Munir, Soviet Physics Semicnd., 15, 1210(1981)
[2] J.H. Collect, J.A. Kash, Solid-state Phys.,16, 1288(1983)
[3] P.J. Wiesner et al., Phys. Rev. Lett., 36, 1366(1975)

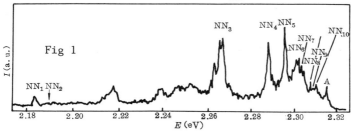

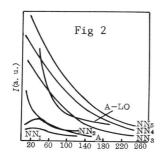

Fig 1 Luminescence spectra of GaP N at $93^{\circ}K$

Fig 2 Luminescence decay of GaP N at $93^{\circ}K$

OPTICAL LIMITING WITH SEMICONDUCTORS

E.W. Van Stryland, Y.Y. Wu, D.J. Hagan, M.J. Soileau and Kamjou Monsour

Center for Research in Electro-Optics and Lasers
University of Central Florida
Orlando, Florida 32816

The ideal optical limiter has a high linear transmission for low input (eg. energy E or power P), a variable limiting input E or P, and a large dynamic range defined as the ratio of the E or P at which the device damages (irreversibly) to the limiting input. While limiting can be obtained using a variety of nonlinear materials, it is very difficult to get the limiting threshold as low as is often required and at the same time have a large dynamic range. Because high transmission for low inputs is desired, we must have low linear absorption. These criteria lead to the use of two-photon absorption (2PA) and nonlinear refraction.[1] We present the detailed operational characteristics and a theoretical description of optical limiting devices based on 2PA and the subsequent photogenerated free-carrier defocusing in semiconductors. Such devices can be made to have low limiting thresholds, large dynamic ranges, and broad spectral responses. For example, a monolithic ZnSe device limits at inputs as low as 10nJ (300W), and has a dynamic range greater than 10^4 for 0.53μm, 30ps (FWHM) pulses. Also, the input/output characteristics of this device should not change significantly for input wavelengths from 0.5 to 0.85μm.[2] A monolithic optical limiter geometry is shown in Fig.1.

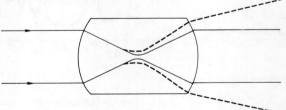

1. A schematic drawing of the monolithic limiter geometry. The solid line shows linear beam propagation for low inputs, and the dashed lines show the beam for high inputs.

By focusing into the bulk of the semiconductor the irradiance on the damage prone surface is reduced,and for *high* inputs the 2PA and defocusing reduce the irradiance in the bulk preventing damage. This happens while maintaining a low limiting threshold. Unfortunately, a quantitative description is difficult.[3] What happens can be qualitatively described as follows. At very low inputs, the thick limiter acts linearly. For higher inputs the limiter behaves linearly except in a region near the focal position as determined by linear optics. Only in this region does the irradiance become high enough to have significant 2PA along with subsequent carrier defocusing. At *higher* inputs the irradiance becomes large enough to have significant 2PA well in front of the linear focal position. This has two consequences. First, the beam at focus will be depleted making it more difficult to damage. Secondly, the negative phase change induced on the wave front by the photogenerated carriers negates the beam convergence prior to focus. The beam is defocused and damaging irradiances are never reached within the material. The dynamic range is now only limited by front surface damage. In principle this threshold can be made arbitrarily high by making the optics larger. Using a geometry similar to that shown in Fig.1 but using separate lenses, we have obtained limiting energies down to $\simeq$14nJ, corresponding to a peak power of $\simeq$400 W for the psec input pulses. With nanosecond pulses the limiting power is actually reduced to $\simeq$80 W. This is true because for a fixed irradiance longer pulses create more carriers which more effectively defocus the beam; thus, the fluence at some distance toward the far field will be limited. Again, the device was self protecting in the bulk. In order to increase the dynamic range we combined the optical elements into the monolithic limiter shown in Fig. 1 . This design takes the damage prone surface as far from focus as possible while maintaining high irradiance within the bulk. This monolithic limiter is self-protected against high irradiance picosecond pulses. Unfortunately, the device suffered bulk damage when nsec pulses were used. When we focus extremely tightly as in the monolithic device, the focal volume becomes so small that the temperature change due to nonlinear absorption of the more energetic nanosecond pulses may give rise to a thermal nonlinearity which in ZnSe results in self-focusing. For very tight focusing this self-focusing may overcome the carrier defocusing and cause beam collapse and damage. The problem can be overcome, however, by using materials with a negative thermal nonlinearity or by not focusing as tightly.

1. E.W. Van Stryland, H. Vanherzelle, M.A. Woodall, M.J. Soileau, A.L. Smirl, S. Guha and T.F. Boggess, Opt. Eng., *24*, 613 (1985).
2. D.J. Hagan, E.W. Van Stryland, M.J. Soileau, and Y.Y. Wu, Opt. Lett. 13, 315 (1988).
3. E.W. Van Stryland, Y.Y. Wu, D.J. Hagan, M.J. Soileau and Kamjou Mansour, to be published JOSA B, Jan. 1988.

A CRITICAL REVIEW OF HIGH-EFFICIENCY CRYSTALS
FOR TUNABLE LASERS

J. T. LIN

Center for Research in Electro-Optics & Lasers (CREOL)
University of Central Florida, Orlando, FL 32816 USA

ABSTRACT

Recent progress of high-efficiency nonlinear crystals and applications for the generation of tunable coherent sources ranging from deep-UV to mid-IR (0.19 to 5 microns) are presented. The critical issues of materials selections such as damage threshold, figure of merit, transparency range and phase-matchable range are discussed. Various frequency conversion techniques for both up- and down-conversion of lasers with fixed and tunable emissions are presented. Novel schemes for rapid beam steering and diode-pumped miniature lasers are analyzed. New directions and devices for potential applications using wave-guided, thin-film and laser-polled crystals are explored.

INTRODUCTION

Recent advances in the growth technology of high-efficiency nonlinear crystals had led to tunable coherence sources ranging from deep-UV to mid-IR (0.19 to 5 microns). These very wide spectral tuning ranges can only be partially covered by the direct output from tunable lasers. For examples, Ti:sapphire laser (tuning range of 0.65-1.2 um), Co:MgF$_2$ laser(1.5-2.3 um), Alexandrite laser (0.7-0.8 um) and dye laser (0.4-0.9 um). Using frequency conversion techniques such as up-conversion by second harmonic generations (SHG), third harmonic generation (THG), sum frequency generation (SFG), and down-conversion by optical parametric oscillation (OPO), nonlinear crystals provide an efficient tool for tunable sources from lasers either with fixed emission wavelengths (such ad Nd, Ho and Er-doped YAG) or with tunable emissions.

In this paper, we shall review the recent progress in high-efficiency crystals and their applications for the generation of tunable lasers. The key issues of materials selections and the conversion efficiency parameters are analyzed. Experimental results using various crystals and lasers are compared. Novel schemes using new crystals for new laser sources including diode-pumped lasers are discussed.

TUNABLE LASERS

As shown in Tables 1 and 2, nonlinear cyrstals with transparency ranges of 0.2-5.0 um and 0.5-20 um are compared for the figure of merits (FOM), phase-matchable ranges and the damage thresholds.[1-3]   We note that crystals shown in Table 2 have much higher FOM than those crystals in Table 1. However, applications are limited by their rather low damage thresholds.

The newly developed crystal of Beta Barium Borate (BBO) provides a variety of applications in the generation of UV and visible sources in both solid-state and dye lasers.   Deep-UV source at 189 nm has been recently reported by the SFG of dye lasers.[4]

DIODE-PUMPED LASERS

High-efficiency frequency conversions have also been recently achieved via noncritical phase-matching operation, where crystals of MgO-doped and wave-guide lithium niobate, potassium niobate and KTP have been used in low-power lasers.[5-7]

Organic crystals such as m-NA, POM, COANP and DAN (see Table 1) with extremely large FOM are the excellent candidates for frequency conversion of low-power lasers such as laser-diode. Thin films and guided-wave devices may also be fabricated by organic crystals.

RESULTS AND DISCUSSION

Figure 1 shows the schematics of tunable sources via nonlinear crystals pumped by various lasers. We note that rapid-tuning (beam steering) may be achieved via OPO pumped by tunable lasrs, in which angle-tuning is not required. Rapid tuing of mid-IR ranges (1.5-3 um) may be performed in BBO and KTP crystals pumped by tunable lasers such as Alexandrite, Ti:sapphire and dye lasers.[8-9]

As shown in Fig. 2, tunable diode-pumped lasers (400-1600 nm) may be achieved by SHG or SFM in high-efficiency crystals of KTP and $KNbO_3$. The noncritical phase matching conditions of this type of application are presented elsewhare.[5-10] Greater details on tunable lasers and other applications of nonlinear crystals such as photorefractive effects (phase conjugation) and E-O modulation are shown in Ref. [11].

Table 1—Nonlinear Crystals Transparent From 0.2-5 μm

Material	Transparency Range (nm)	Phase-Matching Range (Type)	Damage Threshold (GW/cm^2)	Relative Figure Of Merit (1)	Reference
KDP	200-1500	517-1500 (I)	0.20	1.0	5
KDP	200-1500	732-1500 (II)	0.20	1.8	5
D-CDA	270-1600	1034-1600 (I)	0.50	1.7	5
Urea	210-1400	473-1400 (I)	1.50	6.1	6
Urea	210-1400	600-1400 (II)	1.50	10.6	6
BBO	198-3300	400-3300 (I)	5.00	26.0	1
BBO	198-3300	526-3300 (II)	5.00	15.0	1
LAP	220-1950	440-1950 (I)	10.00	40.0	7
LiIO$_3$	300-5500	570-5500 (I)	0.50	50.0	5
m-NA	500-2000	1000-2000 (I)	0.20	60.0	9
MgO:LiNbO$_3$	400-5000	800-5000 (I)	0.05	105.0	8
KTP	350-4500	1000-2500 (II)	1.00	215.0	11
POM	414-2000	830-2000 (I)	2.00	350.0	10
MAP	472-2000	900-2000 (I)	3.00	1600.0	12
KNbO$_3$	410-5000	840-1065 (I)	0.35	1460/1755 (2)	13
COANP	480-2000	960-2000	—	4690.0	14
DAN	430-2000	860-2000	—	5090.0	15
Structure-modulated LiNbO$_3$	400-5000	800-5000	0.05	2460N^2 (3)	16

(1) The relative figures of merit are based on the value of KDP in Type I phase-matching for SHG at 1.06 μm. All are at room temperature except D-CDA, MgO:LiNbO$_3$ and KNbO$_3$, where high-temperature noncritical phase-matching is employed. The absolute value for KDP is calculated in Reference 2 as 0.025 pm/V.

(2) For KNbO$_3$ crystal, room-temperature 90° phase-matching at 860 nm uses d32, while phase-matching at 986 nm uses d31. Numerically, d32 = 1.33 × d31 = 20.4 pm/V.

(3) Structure-modulated LiNbO$_3$ is grown in quasi-phase-matching conditions, where N is the number of periodic layers of the structure-grown crystal.

Table 2—Nonlinear Crystals Transparent From 0.5-20 μm

Material	Transparency Range (μm)	Absorption (cm^{-1} at 10.6 μm)	Damage Threshold (MW/cm^2)	Relative Figure Of Merit (1)	Reference
AgGaS$_2$	0.5-13	0.090	15	1.0	17
CdSe	0.75-20	0.016	50	1.6	5
AgGaSe$_2$	0.71-18	0.050	12	6.3	18
TAS	1.26-17	0.040	16	6.5	19
CdGeAs$_2$	2.4-18	0.230	40	9.2	5
ZnGeP$_2$	0.74-12	0.900	3	14.0	5
Te	3.8-32	0.960	45	270.0	5

(1) The relative figure of merit is based on the value of 14.0 pm/V for SHG at 10.6 μm in AgGaS$_2$.

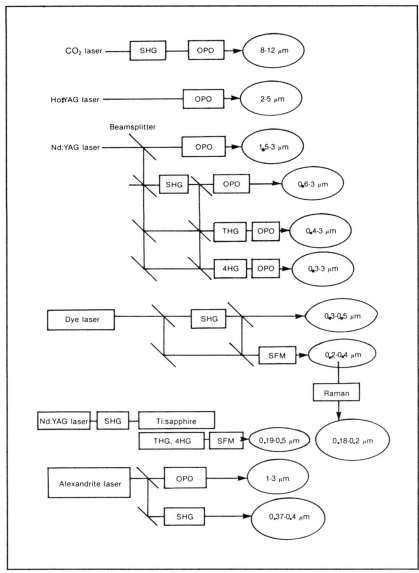

Figure 1. Schematics of tunable sources via nonlinear crystals pumped by various commercial lasers. The appropriate crystals for these processes are characterized in Tables 1 and 2.

70

REFERENCES

[1] Lin, J. T., in Technical Digest, CLEO'87, paper TuH4. (1987)

[2] Lin, J. T., in Proc. of LASER'86 (STS press, VA), pp.262-269 (1986).

[3] Lin, J. T. and Chen, C., in Laser & Optronics, Nov. (1987), pp.59-63.

[4] Burghardt, B. et al, in Proc. of LASER'87, (STS press, VA, 1988), in press.

[5] Lin, J. T., in Proc. of LASER'87, (STS press, VA, 1988), in press.

[6] Lenth, W. et al, Proc. SPIE, Vol. 898, 61 (1988).

[7] Anthon, D. W. et al, Proc. SPIE Vol. 898, 68, (1988).

[8] Lin, J. T., in Proc. of LASER'87 (STS press, VA), in press.

[9] Lin, J. T., Yao, J. Q. and Liu, K. C., Proc. of Topical Meeting on Nonlinear Properties of Optical Materials, (Troy, NY, Aug. 22-25, 1988), in press.

[10] Lin, J. T., in Technical Digest of this conference.

[11] Lin, J. T., Technical Digest of First LEOS Annual Meeting (Santa Clara, CA, Nov.
2-4, 1988), invited paper.

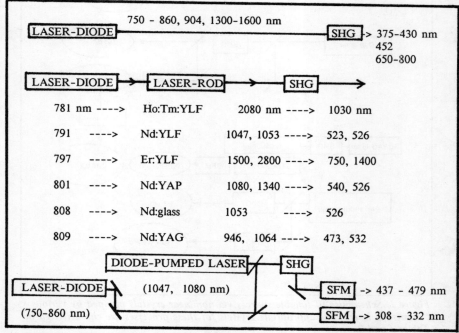

Figure 2. Tunable coherent sources in diode-pumped systems using
various frequency conversion techniques.

PARAMETRIC SCATTERING IN β-BaB$_2$O$_4$ CRYSTAL INDUCED BY PICOSECOND PULSES

Chen Bosu, Wang Peilin, Guo Chiang,
Yue Guming, Song Zhihong

Anhui Institute of Optics and Fine Mechanics,
Academia Sinica, P.R. China

Abstract
Tunable parametric fluorescent emission over the range from
0.97 to 1.29 μm has been obtained in β-BaB$_2$O$_4$ crystal pumped
by a frequency-doubled and mode - locked Nd:YAG laser.

Parametric scattering or parametric fluorescence is a
nonlinear optical process. When a powerful pump light ex-
cites a nonlinear crystal, under energy and momentum matching conditions, it will spontaneously emit radiations at
frequencies w_s and w_i[1]. By rotating the crystal, different pairs of signal and idler photons will be obtained.
The tuning characteristics of parametric scattering are
similar to that of optical parametric oscillator.

Here we report generation of tunable parametric scattering in β-BaB$_2$O$_4$ crystal pumped by second harmonic of a
mode-locked Nd:YAG laser. As we know this is the first report. β-BaB$_2$O$_4$ is a new crystal which possesses the advantages of large nonlinear susceptibility(6 deff KDP), highly
damage resistant (10GW/cm^2), working at room temperature
and not easy to be deliquescence. For reasons given above,
β-BaB$_2$O$_4$ will be an attractive optical parametric crystal.

Our experimental set up is shown in Fig.1. A passive
mode-locked Nd:YAG laser provided the pulse trains at 1.064
μm consisted of about 6 pulses separated by 7 nsec. After
a single stage amplifier, each pulse typically had the peak
power of 200 MW and duration of 35 psec. This was converted
into the second harmonic in an orientation phase-matched β-
BaB$_2$O$_4$ crystal at about 50% efficiency. After a beam splitter, the green pulses at 0.532 μm were focused into a 4mm-
long β-BaB$_2$O$_4$ parametric crystal by a lens of 2 m focal

length. β-BaB$_2$O$_4$ crystal was placed on a rotary platform with precision of about 0.05. The incident pump light was polarized in the c-axis plane, the output signal and idler light were polarized perpendicularly to the c-axis plane, and the observed direction was at the incident forward direction.

After passing through the β-BaB$_2$O$_4$, the pump beam intensity was strongly attenuated by a cutoff filter and a polariod. The emitted signal and idler radiation were focused into a WDS-3 monochromator. The signal emerging from the monochromator was detected by a RJ-7100 energy meter or a GDB 239 photomultiplier. The parametric scattering pulses at degenerate point were displayed by a PIN photodiode and 7834 oscilloscope (see Fig.2). The bandwidth of parametric fluorescence was about 7Å, and the maximum energy was about 0.6 μj.

If we took off the parametric crystal or turned it away from the phase-matching angle, there was no any signal detected, this verified what we detected was generated by an optical parametric process. We have calculated the theoretical curve in terms of Sellmeier's equations given by K. Kato.[2]

By rotating β-BaB$_2$O$_4$ and detecting different pairs of signal and idler radiation, the tuning curve of parametric scattering was obtained (see Fig.3.).

The experimental data are basically in accord with the theory, but this agreement is not sufficient. We attribute this diviation to the unstability of crystal rotation.

Since the limit of wave range of grating and cutoff filter, parametric tuning range is limitted between 0.97 and 1.29μm. If improve experimental conditions and add another β-BaB$_2$O$_4$ crystal, the tuning range and output of optical parametric amplifier will be increase considerably.

[1] Yariv, A., "Quantum Electronics", 2d edition, 460-465 (1975)
[2] Kato, K., IEEE J. Quant. Electron., QE-22 , No.7, 1013 (1986)

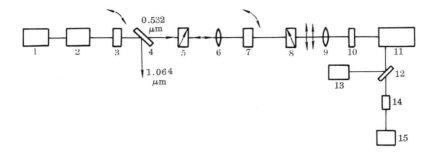

Fig.1

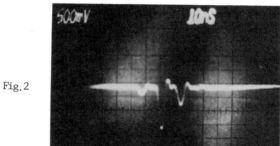

Fig.2

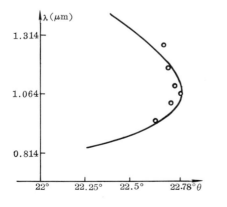

Fig.3

GENERATION OF PICOSECOND PULSES AT 193 nm BY FREQUENCY MIXING IN β-BaB$_2$O$_4$

P. Lokai, B. Burghardt, D. Basting, W. Mückenheim

Lambda Physik, P.O. Box 2663, D-3400 Göttingen, FRG

ABSTRACT: By frequency mixing of a picosecond excimer laser (248 nm) and a pulsed dye laser in a Beta-Barium-borate crystal an efficient method to generate picosecond pulses at 193 nm has been made accessible.

The nonlinear material β-BaB$_2$O$_4$ (BBO) developed by Chen et al. /1/ has proved very efficient for frequency conversion in the UV /1-3/. Phasematching for type-I second harmonic generation (SHG) is possible down to 204.8 nm, and also with respect to most other relevant properties like large nonlinear coefficient, high damage threshold, low deliquescence, and ease of fabrication, this negative uniaxial crystal is superior to the known nonlinear materials such as urea, lithiumformiate or potassium pentaborate. Since the transparency range of BBO stretches below 190 nm, it should be possible to obtain the important excimer wavelength 193 nm.

A minimum wavelength of 197.4 nm has been achieved by mixing the fundamental output of a dye laser beam with its second harmonic using type-I sum frequency generation (SFG) /4/. By cooling the crystal it was possible to satisfy the phasematching condition for this simple method of type-I SFG even beyond this value /5/. At a crystal temperature of 95 K the wavelength 195.3 nm could be generated. But from the wavelength-versus-temperature relation it can be estimated that this method is restricted to wavelenghts above 194.4 nm.

The Sellmeier coefficients given by Chen et al. /1/ as well as those given by Kato /2/ indicate, that shorter wavelengths can be generated by employing fundamental wavelengths with a larger difference. This conjecture has been verified by our experiment /6/.

The fundamental wavelengths were supplied by a pulsed dye laser (λ_2 = 865 nm) and a picosecond excimer laser operating with KrF at λ_1 = 248.4 nm. The dye laser (Lambda Physik, FL 3002) emitted pulses of about 10 ns duration and 0.2 cm^{-1} bandwidth. The power density in the crystal was 2.5 MW/cm^2. The picosecond excimer laser (Lambda Physik, PSL 4000) delivered 30-ps pulses which were attenuated to a maximum power density of 20 MW/cm^2. Its bandwidth was 16 cm^{-1} . Assuming the same pulsewidth for the SFG pulses (which could not be measured with our streak camera, but can be assumed to be not longer than 30 ps), an SFG efficiency

$$\eta = P(\omega_1 + \omega_2) / \sqrt{P(\omega_1) P(\omega_2)}$$

of 7 % can be calculated from the 0.5 MW/cm^2 SFG power density generated, using a BBO crystal of 8 mm length. The pulse shapes, all taken with a fast photodiode, are shown in Fig. 1.

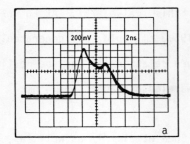 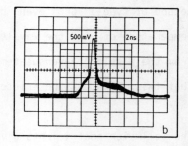

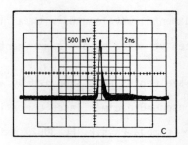

Fig. la-c: Pulse shapes of (a) the dye laser, (b) the picosecond excimer laser with that of the dye laser superimposed (changed scale), and (c) the SFG at 193 nm.

References:

/1/ C. Chen, B. Wu, A. Jiang, G. You: Sci. Sinica B 28 (1985) 235

/2/ K. Kato: IEEE J. Quantum Electron. QE-22 (1986) 1013

/3/ K. Miyazaki, H. Sakai, T. Sato: Opt. Lett. 11 (1986) 797

/4/ W. L. Glab, J. P. Hessler: Appl. Opt. 26 (1987) 3181

/5/ P. Lokai, B. Burghardt, W. Mückenheim: Appl. Phys. B 45 (1988) 245

/6/ W. Mückenheim, P. Lokai, B. Burghardt, D. Basting: Appl. Phys. B 45 (1988) 259

MIXING FREQUENCY GENERATION OF 271.0-291.5 nm IN β-BaB$_2$O$_4$

Lu Shiping, Yuan Yiqian, Tang Yonggui, Xu Wenjie, Wu Cunkai

Laboratory of Laser Spectroscopy, Anhui Institute of Optics and Fine Mechanics, Academia Sinica, P.O.Box 25, Hefei,P.R.C

The nonlinear optical properties of crystal β-BaB$_2$O$_4$ have been studied by C.Chen and K.Kato. We have calculated two theoretical curves for the type-1 phase-matching of second-harmonic generation (SHG) of Rh6G dye laser and sum frequency generation (SFG) between the dye laser and its pump source at 532 nm.

We have obtained the output of continuously tunable UV laser over the range of 281.0 - 291.5 nm (SHG), the range of 271.0 - 281.4 nm (SFG) in the same crystal.

1 Theoretical Calculation

The Sellmeier's equations of β-BaB$_2$O$_4$ reported by C.Chen and K.Kato in ref. [1] [2].

Based on the two groups refractive indexes equations mentioned literature the collinear type-1 phase matching condition:

$$K^e_3 = K^o_1 + K^o_2, \qquad W^e_3 = W^o_1 + W^o_2$$

and the dependence of phase matching angle on the indexes ellipsoid equation:

$$\Theta_m = \sin^{-1} \frac{[n^e_3(\Theta_m)]^{-2} - (n^o_3)^{-2}}{(n^e_3)^{-2} - (n^o_3)^{-2}}{}^{\frac{1}{2}}$$

We have calculated the tunable curves as shown in Fig.1, which reflected the dependence of the SFG wavelength λ_3 on phase matching angle Θ_m.

2 Experimental system

The experimental crystal is cut at 31.3° for type-1 phase matching with size of 7 $\times$ 6 x 5 (mm)^3.

The experimental optical arrangement for sum-frequency generation is shown in Fig.2. In outline: the 532 nm radiation with 200 mJ from a Nd:YAG SHG laser Splitted into two beams by a splitter. One of them (include 50 % energy) drived a home-made Rh6G dye laser oscillator-amplifier system, operating in the range 552–598 nm. The dye laser pulse with output energy 20mJ at central wavelength at 565 nm was introduced into the BBO crystal passed through a dichroic mirror. The other green beam was reflected by M_2 into the dichroic mirror passed through a halfwave plate, an optical delay-line before the BBO crystal. On a high precision turn-table, which can be hand-operate with the accuracy of 0.018°/graduation or was controlled by the monoboard computer with the accuracy of 0.0018°/step. The wavelengths were analyzed with w-44 model grating monochromator, 9584QB model PMT and oscilloscope. The UV output power was monitored by a calibrated LEp-1 model joulementer.

3. Experimental results

Using the experimental system, we have measured the incident angles of SFG and SHG. The experimental tuning curves for type-1 phase matching of SFG have been obtained by calculation from the external angle to the internal angle for the BBO crystal and shown with "x" in fig.1. The experimental results agree with the calculated curve using K.Kato's Sellmeier equations. The deflection is only o.1° between the experimental curve and theoretically calculated curve. The tuning range 271.0–281.4 nm was obtained. In the same crystal the tuning range 281.0–291.5 nm of Rh6G laser also have been obtained. The UV output energy exceeds 200 μJ. The corresponding peak power is about 10kW, and the conversion efficiency is 3–5%. Due to the laser damage threshold of some optical elements are low, so that the input energy of BBO is limited.However, using the UV radiation we have performed the experiments of the Laser Enhanced Ionization Spectroscopy in flame for five metal elements: Zr, Mn, Ti, Li and Y, and obtained good results.

This project was supported by the Fund of Natural Science of China.

References

[1]. Chen, C., Wu, B., Jing, A. and You, G., Sci. Sinica (Ser.B), <u>28</u>, 235-243 (1985)

[2]. Kato, K., IEEE J.Quantum Electron., <u>QE-22</u>, 1013-1014 (1986)

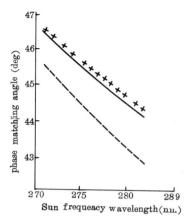

Fig.1 The tunable curves for type 1 SFG of dye-laser with 532nm
in BBO. The deshed and solid line are theoretical curves
using Sellmeier equation of Chen, C. and Kato, K., respecti
vely. *:experimental points

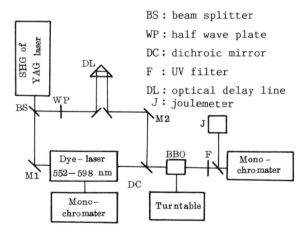

BS : beam splitter
WP : half wave plate
DC : dichroic mirror
F : UV filter
DL : optical delay line
J : joulemeter

Fig 2 The experimental optical arrangement for sum frequency
generation

LOW TEMPERATURE ABSORPTION STEPS NEAR ULTRAVIOLET INTRINSIC EDGE IN BETA BARIUM METABORATE

Zhang Guangyin, Yang Yanrong, Zhang Chunping

Department of Physics, Nankai University, Tianjin, P.R. China

Crystalline beta barium metaborate (β-BaB$_2$O$_4$) is one of the best SHG (second-harmonic generation) materials used in the ultraviolet (UV) wavelength region[1,2]. Therefore we measured the UV absorption spectra of β-BaB$_2$O$_4$ crystals in the temperature range from 300 K to 78 K. The results show that while the temperature drops to 10°C three new absorption steps near the UV intrinsic edge which in the vicinity of 190nm at room temperature appear, and the absorption increases rapidly as the temperature is decreased until they reach saturation. The 1[#] specimen is 3.15mm thick and its faces are perpendicular to the optical axis of the crystal. The 2[#] specimen, whose faces are parallel to c axis of the crystal, is 1.82mm thick. Fig.1 shows the transmition spectra of the 1 specimen with polarization perpendicular to the optical axis (i.e. E$\perp$C) at different temperatures. From Fig.1, one can see that there are three new absorption over-lapping steps, which we marked A.B.C. respectively. At 10°C (283 K), the A, B and C absorption bands begin to apper. As the temperature decreases below 10°C, the strength of these new bands increase rapidly and at -9°C (264 K), their absorption coefficients reach over 70 percent of their maximum value so that light passing through a 1 cm thick specimen reduces to 60% of its original intensity at 270nm. From Fig.2, we can see that the three absorption bands have the same dependence on temperature. All the solid curves, which are for perpendicular polarization (E $\perp$C), present an abrupt increase in about the same range from 283 K down to 255 K, and are relatively flat below 255 K indicating a saturation of absorption.

The experimental results show that the absorption spectra in the parallel polarization (E ∥ C) at low temperature present similar absorption steps but are about twice as strong as in E⊥C polarization seen in Fig.2. The abrupt rise is shifted to a lower temperature range between 265 K and 235 K which indicates the dependence of the active energy of the bands on polarization. The most plausible reason for the low-temperature absorption steps is the presence of a donor energy state induced by certain impurity centers contained in the crystal. These impurity centers are ionized at room temperature, but at low temperature, they capture electrons from the conduction band to become neutral impurity centers which are able to bind excited electron-hole pairs to from bound excitons which may be produced by photo-excited electron in valence band. Because the intrinic absorption edge of β-BaB$_2$O$_4$ is the indirect edge [3], the bound excitons in the crystal may be excited by photons with assistance of phonons. These are two reasons why the absorption bands of the indirect bound exciton (IBE) are expanded to emerge in the shape of steps. The amount of photo-excition E responsible for the intensity of the absorption is determined by the number of neutral impurity centers, and so increase rapidly with a drop in temperature. According to the experimental fact that the three absorption steps have a similar temperature change, it is assummed that they are related to the same kind of donor impurity centers, and therefore, they correspond to IBE transitions of the electrons from three different valence subbands up to the donor level. The polarization characteristics of the new absorption steps observed in the experiment might be explained by the "layer molecular structure" model [4] which suggests that the interreactive force between two molecular layers is even weaker that that between atoms in one layer. According to the low-temperature absorption spectra of the specimens measured by us, the transmittance of the specimen with 1 cm thick at 222nm fall below 22 percent just at 0°C. So, in order to assure the application value of β-BaB_2O_4 crystal, the impurities possibly contained in a crystal ought to be removed as far as possible.

REFERENCE

[1] C. Chen, B. Wu, G. You, A. Jian, in Dig. Tech. Papers, XIII IQEC, paper MCC5, (1984).
[2] J.M. Halbout, S. Blit, W. Donaldson, C.L. Tang, IEEE J. Quantum Electron, EQ-15, 1176, (1976).
[3] Yanyong Yang, Guanyin Zhang, Chunping Zhang, Optica Acta Sinica, 6, 1105, (1986).
[4] Guangyin Zhang, Yanyong Yang, Wu Bochang, Optica Acta Sinica, 5, 548, (1985). (In Chinese)

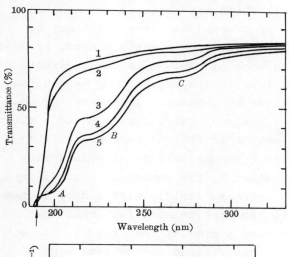

Fig. 1. The UV transmisson spcctra for the 1 -BaB_2O_4 specimen of 3.15 mm in $\vec{E} \perp \vec{C}$ at different tempetatures: (1) RT, (2) 283K, (3) 264K, (4) 188K and (5) 164K. The arrow indicates the position of the absorption edge.

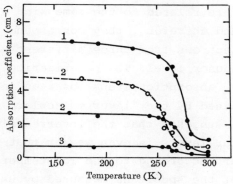

Fig. 2. The dependence of the absorption coefficients of -BaB_2O_4 on temperature at three fixed wavelengths: (1) =198nm (2) =222nm and (3) = 270nm ".": $\vec{E} \perp \vec{C}$ '' '': $\vec{E}//\vec{C}$.

THE GROWTH AND PROPERTIES OF BaTiO$_3$ CRYSTALS

Wu Xing,   Zhu Youg,
Zhang Daofan,   Jian Yandao
Yang Huaguang

Institute of Physics, Chinese Academy of
Science. P. O. Box 603, Beijing, China.

ABSTRACT

BaTiO$_3$ crystal is very useful for photorefractive applications but the crystal growth is difficult in good optical quality. Based on a detail analysis on the growth condition and a special design of the equipment the BaTiO$_3$ crystals in good optical quality and in big size were grown successfully. The main physical properties of these crystals which grown in our lab. were measured and are presented in this paper.

1. Introduction

BaTiO$_3$ crystal is familiar material to physicists. It was the first ferroelectric perovskite to be discovered. Recently, the interests in BaTiO$_3$ crystal rised once again due to the discovery of photorefractive phenomena. This crystal was considered to be the best one for photorefractive applications because it has a large electrooptical coefficient and some other special properties [1]. However, the growth of BaTiO$_3$ crystals in good optical quality is very difficult.

Phase diagram [2] shows that the crystals of the desired modification cubic of the pure BaTio$_3$ can not be grown directly from the melt because of the existance of a hexagonal phase just below the melting point . This hexagonal phase is nonferroelectric and remain metastable phase even down to room temperature. This case induces that the cubic pure BaTiO$_3$ crystal can not be grown directly from melt by Czochralski technique.

Flux growth technique is one of the ways for avoiding the hexagonal phase, many works have been done by this way with KF, BaCl$_2$, BaB$_2$O$_4$ and so on as solvent, but the crystals are not good enough in their optical quality and the size of them is too small to use for photorefractive applications.

On principle, Top Seed Solution Growth (TSSG) technique which was developed by Lize [3] and G.Godefroy [4] has given a possibility to grow the pure $BaTiO_3$ crystals in good optical quality and big size. But the crystal quality still was not good enough. It is necessary to improve the $BaTiO_3$ crystal growth for meeting the urgent needs of photorefractive experiments.

2. The Growth of $BaTiO_3$ Crystals with High Optical Quality

Some more efforts on this object have been done in our laboratory recently. TSSG technique with overplus TiO_2 has been selected in our work also, In order to progress the growing process of $BaTiO_3$ to get the crystal in good optical quality. We analysed the growing technique and condition based on some experiment effects and made special equipment as well as get very good results in crystal growth. All of them will be shown in following.

A.   Temperature distribution: Comparing with usual flux growth, the main advantage of TSSG tech. is that the spontaneous nucleuses could be avoided and the growing process happened only on the seed which was put in melt manually. This means that we should create a suitable growth condition so that the seed could be in an advantageous position on the growth condition compared with that at the other part of the melt. Otherwise, the spontaneous nucleuses would present and impede the growth on the seed. Besides, it is very important to keep the supersaturation on the growing interface to be suitable and stable for avoiding crystal defects. All of these facts depend on the temperature gradient on the growing interface. Therefore, a suitable arrangement of temperature gradient on the growth interface is very important for growing a good $BaTiO_3$ crystal.

B.   Select a suitable seeding temperature: It is very important to select a right seeding temperature. If it was too high, the seed would be dissolved. In opposite case, the growth rate would be too fast to keep the stable growth on the growing interface and would induce crystal in a bad quality. In order to determine the right seeding temperature, one should measure the saturate temperature of the melt accurately. But it is very difficult to do this measurement during the cry-

stal growth. For solving this problem, a special equipment with a precision balance for BaTiO$_3$ crystal growth was constructed in our laboratory. By this equipment, we can know the real growth rate synchronously during the crystal growth. It is very useful to control all of
the crystal growing process.

C.   Convection of the high temperature melt: The overplus TiO$_2$ in melt
as solvent will be rejected from growing interface and accumulated in
the liquid immediatly ahead the interface during crystal growth. This
accumulation of solvent induces that the concentration of TiO$_2$ in this
layer ahead growing interface becomes appreciably higher than that in
the bulk of the melt, then, results in the growing interface to become
unstable and presents a lot of defects in crystal. Diffusion process
can remove these accumulative TiO$_2$ into melt bulk but it is a very slow
process. The other fact should be paid attention is that the density of
TiO$_2$ and BaTiO$_3$ are quite different. It induced the composition of the
melt to be unhomogeneity and weakened the thermal convection in the
crucible, it is very inadvisable to BaTiO$_3$ crystal growth. In this
case, selecting an suitable seed rotation pattern and using the crucible acceleration rotation technique would be very useful to overcome
this difficult.

D.   Stability of the growing interface: Because the available stable
growth rate is quite small due to the much more solvent in the melt,
any small temperature disturbance and mechanic vibration would induce
unstable growth and a lot of crystal defects. In this case, a high precision temperature program controller and mechanism for pulling and
rotation are necessary. For example, the pulling rate is about 1 mm
per day and it is not so easy to have a mechanism worked at such slow
rate on very homogeneous.

3. Properties of BaTiO$_3$ Crystal

A.   Domains and poling: In room temperature the macroscopic domain
structure of a BaTiO$_3$ crystal includes two different types of domain
boundaries, that is 90° and 180° domain boundaries. the samples have

been mechanically poled to eliminate $90°$ domains and electrically poled to control $180°$ domains. If there are some inclusions in the crystal, one can not completely obtain a single domain sample because these
inclusions would impede the poling process.

B.   Optical properties: Optical transmission measurements between
0.38 μm and 10 μm obtained with a CARY2390 spectrometer (from 0.38 to
2.5 μm) and PE983G spectrometer (from 2.0 to 10 μm). The refractive
indices were measured with a minimum deviation method. For the measurement of the two principal refractive index, we used a prism. The refractive index data N_o and N_e are shown in Table 1. for different wavelength in $18.5°$C.

C.   Dielectric and electro-optical properties: The type ROK10-300
cryostat and a silicon oil bath with a type DWT761 precise temperature
controller were respectively used to change the temperature from 10 to
500 K. The low frequency dielectric constants of $BaTiO_3$ crystals were
measured between 130 and 440 K along the c-axis and a-axis with a HP
4274A LCR meter. Temperature dependence of diecectric constants is
shown in Fig.1. The Curie temperature is $132 \pm 1°$C. The DC conductivit-
ies of the crystals are 1×10^{-13} to 5×10^{-12} $(\Omega \cdot cm)^{-1}$ at $23°$C. Electro-optical properties were measured with modulating method described
by Günter [5] at a modulating frequency f = 10 kHz. One gets $r_a = r_{33}$
$- (n_o/n_e)^3 r_{13} = (83 \pm 12) \times 10^{-12}$ m/v which is near ref. [6] and $r_{42} =$
$(1640 \pm 235) \times 10^{-12}$ m/v. These crystals show very good results on
self-pump phase conjugation and two waves coupling effects. These
experiments will be published in other paper.

The authors would like to thank Zhou Tang, Hu Boqing, Wang Jianhong
for the refractive indices and optical transmission measurements,
Cheng Ximin for polishing the crystals, as well as Chen Yingping for
the $BaTiO_3$ crystals growth.

References
[1] Brody,P.S., Efron,U. and etc, Applied Optic, 26, 220-224 (1987)
[2] Rase,D.E. and Rastum. Roy., J.Am.Ceram.Soc, 38, (3),111 (1955)
[3] Belruss,V., Kalnais,J., Lize,A. and Folweiler,R.C., Mat. Res. Bull,

6, 899-906 (1971)

[4] Godefroy,G., Lompre,P., Dufas, C. and Arend,H., Mat.Res. Bull, 12, 165-170 (1977)

[5] Günter,P., Electro-Optics/Laser 76, edited by Jerrard,H.G., (IPC Science and Technology Press, England 1976, 121-130)

[6] Klein.M.B.,Electro-Optic and Photorefractive Materials Edited by Günter,P., (Springer-Verlag, 1987, 280)

TABLE 1: Barium titanate crystal properties
Molecular formula: $BaTiO_3$

Phase	4	3	2	1	
State	F	F	F	P	
Crystal system	rhombo-hedral	ortho-rhombic	tetragonol	cubic	hexagonal
Space	$R3m-C_{3v}^5$	$Amm2-C_{2v}^{14}$	$P4mm-C_{4v}^1$	$Pm3m-O_h^1$	$C63/mmc-D_{6h}^4$

Transition temp.	−80	13		132	1460 °C
Lattice constants	a = 3.9928 Å,		c = 4.0383 Å.		(26°C)
Density		= 6.06 g/cm³			(26°C)

Colour   transparent, light brown.
Melting point   1612°C
Solubility: do not dissolve in water and organic solution,
           but dissolve in HCl

Refractive indices	4880 Å	5145 Å	6328 Å
N_o	2.5200	2.4912	2.4160
N_e	2.4478	2.4247	2.3630

Transmission region   0.4 - 6.3 μm

Dielectric constants	(23°C)	ε_c	ε_a
1 kHz		140	4600
10 MHz		70	1300

DC electrical conductivity (23°C) $1\times10^{-3}-5\times10^{-12}(\Omega.cm)^{-1}$

Electro-optic (26°C) 10 kHz
coefficients $\gamma_a = 83\pm12)\times10^{-12}$ m/v; $\gamma_{42}=(1640\pm245)\times10^{-12}$ m/v

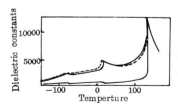

Fig 1 The dielectric constants of $BaTiO_3$ crystals as a function of temperature at f=1KHz. —— Heating, ‒‒‒Cooling.

HIGH-ORDER PHENOMENA ACCOMPANIED WITH
SELF-PUMPED PHASE CONJUGATION IN BaTiO

Zhiguo Zhang, Gang Hu, Xing Wu, Yong Zhu, Yandao Jiang
and Peixian Ye

Institute of Physics, Chinese Academy of Sciences,
P.O. Box 603 Beijing, China

ABSTRACT

We report here for the first time some phenomena of high-order degenerate wave mixing in BaTiO when self-pumped phase conjugation occurs. Applications of these phenomena are discussed.

In recent years much attention has been paid to photorefractive crystals because of their nonlinear response at very low cw laser power and their special interaction mechanism. Many interesting effects, such as two-wave coupling and self-pumped phase conjugation, have been observed, while some high-order effects due to the very low power response have not yet been studied extensively. In this paper we report some phenomena of high-order degenerate wave mixing observed in a BaTiO crystal when self-pumped phase conjugation occurs.

In order to observe the high-order effects, we used a BaTiO crystal grown in the Institute of Physics, Academia Sinica. An Ar laser operating at 514.5 nm was used to generate beams 1,2 and 7, as showm in Fig. 1. B1 and B2 were coherent with each other, and B7 was made incoherent with the others by a suitable relative time delay. Beams 1 and 2 were incident at external angle of 70 and 116 with respect to the c-axis, respectively. The powers of beams 1,2and 7 were 4.4 mW, 2.2 mW and 0.2 mW, respectively. All of the incident beams were e-polarized rays. Initially, we turned on only the incident beam B1 (Fig.1) and the self-pumped phase conjugate beam, B3, was observed after 150 sec. At the same time B3', a reflected beam of B3 from the internal surface of the crystal also appeared. Then because B3' interacted with B1 in the crystal, an index grating called an A-grating would be formed. At this time, if another incident beam B7 with the same frequency entered the crystal along the direction opposite to B1, it would be diffracted according to the Bragg condition on this index grating along the direction which coincided with the reflected beam (BB) of B1 from the external surface of the crystal. Fig.2 shows our observation of the intensity of B8 versus time after B3' has built up. Here B7 was turned on and B1 off at t and t, respectively. The abrupt increase of the intensity of B8 at t demonstrates that there exists the Bragg diffraction of B7 on the A-grating, which is superposed on the reflected beam of B1 in the same direction. Obviously, the occurrence of this diffracted beam can be understood as a result of high-order degenerate wave mixing of beams 1 and 7. In addition, when a third incident beam B2 enters the crystal with the direction shown in Fig.1, the phase conjugate beam, B4, will occur in the direction opposite to B2, as a diffracted beam of B7 on the index B-grating formed by the interaction between B1 and B2 in the crystal. However, it should be noted that as B3 has appeared, it is also diffracted on this B-grating in the same direction as B4 and superposed on the latter. Fig.3 shows our observation. Here all the beams 1, 2 and 7 were turned on at t = 0, and we verified that when only B2 was turned on, it did not give rise to a self-pumped phase conjugate beam.The fact that accompaning the build-up of the self-pumped phase conjugate beam of B1 there is a synchronous increment of the diffracted beam B4 demonstrates that B4 is formed not only from the Bragg diffraction of B7 but also from that of B3 on the same B-grating. Actually, if there exist only two incident beams, B1 and B2, the phase conjugate beam of B2 also appears in the same direction as B4 because of the generation of B3. This phenomenon can also be understood as high-order degenerate wave mixing of beams 1 and 2.

Finally, applications may be found for these phenomena. For example, we can construct a nonexternally pumped device which can generate two or more reflected phase conjugate signals simultaneously for two or more incident beams, respectively, but requires only one of them have self-pumped phase conjugation.

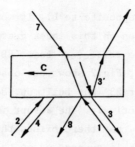

Fig. 1

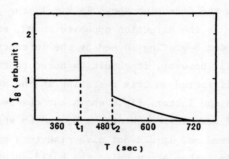

Fig. 2

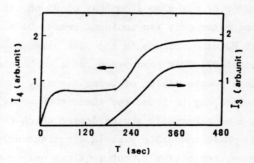

Fig. 3

GROWTH AND LASER DAMAGE ESTIMATION OF POTASSIUM DIHYDROGEN PHOSPHATE CRYSTALS FOR LASER FUSION

Takatomo SASAKI, Atsushi YOKOTANI, Kunio YOSHIDA,
Tatsuhiko YAMANAKA, Sadao NAKAI and Chiyoe YAMANAKA

Institute of Laser Engineering, Osaka University
2-6, Yamada-Oka, Suita, OSAKA 565 JAPAN

(Abstract)

A 40cm X 40 cm KDP crystal for laser fusion was successfully grown by three vessel system. The improvement of the bulk laser damage threshold could be done by reducing organic carbon in growing solution.

1. INTRODUCTION

A very large and damage resistant nonlinear crystal, potassium dihydrogen phosphate (KDP) is necessary as a frequency converter to obtain a short wavelength laser for laser fusion. For this purpose, for example, a 40cm x 40cm sectional and 1m high KDP single crystal is needed to grow. Besides, the bulk laser damage threshold more $10J/cm^2$ against 1ns pluse at 1μm is necessary. In this paper, at first, we present a crystallizer for growth of such a huge crystal using a constant-temperature, constant-supersaturation technique, which successfully worked. Next we describe the results of the experiments to improve the bulk damage threshold by reducing the residual organic impurity from aqueous solution. We confirmed that the main cause of the bulk laser damage of KDP crystals is due to the organic impurities incorporated from the solution.

2. CRYSTAL GROWTH USING THREE-VESSEL SYSTEM

When we adopt a conventional temperature reduction method, a huge growth tank of which volume is approximately $4m^3$ is necessary in order to grow a 40cm x 40cm cross section, 100cm in height KDP crystal. Such a vessel is too bulky and a stirring system of aqueous solution becomes more complicated to obtain temperature uniformity of the solution. The three vessel system using a constant-temperature, constant supersaturation technique is smaller and easier to handle because the additional KDP powder materials can be put in the course of the crystal growth. But there has been no report of the growth of the large scale crystals by this method because of the problem of choking the interconnected pipes between vessels by the spurious crystals during the long growth term. We successfully grew 40cm x 40cm crystals with the three vessel system by setting the temperature of the crystallizer lower than the room temperature, which prevented the spurious crystals from growing in the interconnected pipes.

Fig. 1 shows the growth history in the c-axis and the supersaturation of the solution in the crystalizer. We could obtain the final growth rate of 3.4 mm/day in c-axis. The relation between the flow rate and supersaturation were calculated as a parameter of a growth rate and were compared with the experiment.

3. ESTIMATION OF LASER DAMAGE THRESHOLD

Previously we reported that incorporation of organic material such as microbes or their carcasses which increased in the KDP solution during the growth was a severe problem to obtain damage resistant KDP crystals for use in high power fusion lasers[1] Here we present an effect of reducing organic carbon (OC) concentration from KDP solution on the damage threshold of the crystals.

Typically KDP solution prepared by combination of usual deionized water and the guaranteed reagent contained OC of approximately 10ppm, and the damage threshold of crystals grown from such solution was 6-9J/cm^2(1ns, 1.05µm). We reduced OC from KDP solution by ph-oto-chemical reaction with the combination of UV light and oxidants and grew the crystals.

The measured damage threshold of the crystals grown in the solution which contained OC of 1ppm and bellow 0.1ppm were 15-18J/cm^2 and 20-22J/cm^2, respectively and showed very stable values, sample from sample. Fig. 2 illustrates the histogram of the bulk laser damage threshold grown in the organic-reduced solution.

There is a report that the dislocation existing in the crystal play a main role of the bulk damage threshold[2]. But we think the main cause of the damage is due to residual OC in the crystals rather than dislocation.

REFERENCES

1) A.Yokotani,T.Sasaki,K.Yoshida,T.Yamanaka and C.Yamanaka, Appl. Phys. Lett. 48,1030 (1986)
2) V.V.Azarov,L.V.Atroshchenko,Yu.K.Danileiko,M.I.Kolybaeva,A.V. Sidorin and B.I.Zakharkin, Sov.J.Quantum Electron. 15,89(1985)

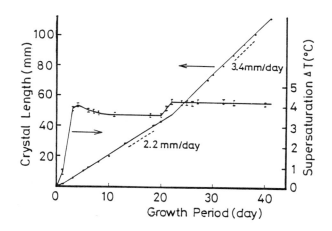

Fig. 1
Growth length histry of c-axis and the supersaturation of the solution.
(KDP crystal: 400 x 400mm)

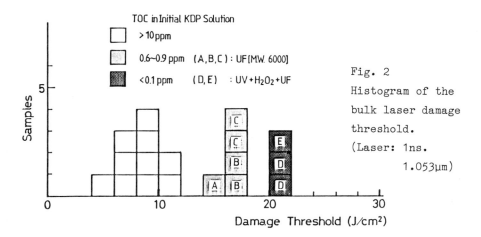

Fig. 2
Histogram of the bulk laser damage threshold.
(Laser: 1ns. 1.053μm)

NONCRITICALLY PHASE-MATCHED KTP FOR
DIODE-PUMPED LASERS (400-700 nm)

J. T. LIN

Center for Research in Electro-Optics & Lasers (CREOL)
University of Central Florida, Orlando, FL 32816 USA

ABSTRACT

Various noncritical phase-matching(NPM) conditions of KTP for visible diode-pumped lasers are presented. Extremely large angular acceptance widths and effective nonlinear coefficients (6.1-7.6 pm/V) are found.

INTRODUCTION

Green light sources at 532 nm (or 527 nm) are commercially available from the intracavity doubling of diode-pumped Nd:YAG (or Nd:YLF) using KTP as a doubling crystal. However, visible diode-lasers (450-500 nm) with reasonable powers (2-10 mW) are not yet available due to the lack of suitable nonlinear crystals (used as a doubler) or the lack of diode sources emmiting in the 905-1000 nm spectral regimes.

The new crystal of $KNbO_3$ has been used for frequency doubling of the direct output from a laser diode and the diode-pumped systems. This crystal however requires precise temperature control and the room temperature noncritical phase-matching(NPM) conditions are not readily achievable. Furthermore, the angular temperature and spectral acceptance widths of $KNbO_3$ cyrstal are rather narrow for practical applications. NPM using KTP for sum frequency mixing (SFM) of 808 nm and 1064 nm (in diode-pumped Nd:YAG) has been recently reported. However, the general rule of SFM using KTP as the NPM crystal other than this reported case has not yet been explored. In this paper we shall present the new application of KTP for the generation of visible sources in diode-pumped lasers where the NPM condition may be achieved at various configurations. The effective nonlinear coeefficients and angular acceptance width at NPM are analyzed.

PHASE MATCHING CURVES

Based upon the available accurate refractive indexes of KTP, we have calulated all the six possible polarization configurations for type II operation (which is always more efficient than type I). The polarization directions of the mixing beams λ_1, λ_2, and λ_3 may be (ac,a), (ac,c), (bc,b), (bc,c), (ab,b) or (ab,a), where a, b, c correspond to the x, y, z axies with $n_x < n_y < n_z$. Our computer simulation indicates that only the configurations of (ac,a) and (bc,b) provide us the NPM operation within the useful spectral ranges, namely, for the pumping diode spectra of 750-860 nm, and 1-1.6 microns, and the diode-pumped output wavelengths of 1.047-1.6 microns.

The major features of Fig. 1 are summarized as follows:

(1)   Visible light at 459 nm may be efficiently generated from the intracavity SFM of the pumping diode(808 nm) and the diode-pumped Nd:YAG (1064) where the pumping wavelength (808 nm) perfectly matches the peak absorption of Nd:YAG and the NPM configuration of (ac,a), see curve B. Our calculation also shows an extremely wide angular acceptance widths compared with that of the SHG of 1064 nm: about 6 and 1.7 time in θ and ϕ angle, where θ and ϕ are the angle respect to the z and x-axis. [See Ref. 3].

(2)   Intracavity SHG of diode-pumped Nd:YAP (1080 nm output) may be achieved at the near NPM operation, referred to curve A (bc,b).

(3)   For the cases where the pumping wavelengths are not matching the absorption peak of the lasing materials and simultaneously meet the NPM conditions, we shall still be able to generate visible sources by the external SFM from the commercially available diode sources (750-860 nm) and the diode-pumped outputs under the NPM conditions. Examples indicated by Fig. 1 are: 467 nm generated from SFM of 842 nm and diode-pumped Nd:YLF (1047 nm); 464 nm from SFM of 830 nm and 1053 nm (Nd:glass); (using curve B), 452 nm from SFM of 778 nm and 1080 nm (Nd:YAP); UV source of 378 nm from 532 nm (double-YAG) and 1313 nm (Nd:YLF); 439 nm from 1318 and 659 nm (i.e. triple-YAG at 1318 nm); (using curve A).

We should note that Fig. 1 also provides the guidance for the NPM conditions for optical parametric oscillation (OPO), the reverse process of SFM. For example, 532 nm-pumped OPO with output of 1040 and 1090 nm, and 1080 nm-pumped OPO with output of 1650 and 3126 nm, where the NPM condition is achievable in both cases.

The effective nonlinear coefficients(deff) for curves A and B are calculated to be 6.1 and 7.6 (pm/V), respectively, in KTP under the NPM operation vs. 0.43 pm/V of d_{36} of KDP.

REFERENCES

[1] Lin, J. T., in Technical Digest, CLEO'87, paper TuH4. (1987).

[2] Lin, J. T. and Chen, C., in Laser & Optonics, pp. 59-63, Nov. (1987).

[3] Lenth, W. et al, Proc. SPIE 898, 61 (1988).

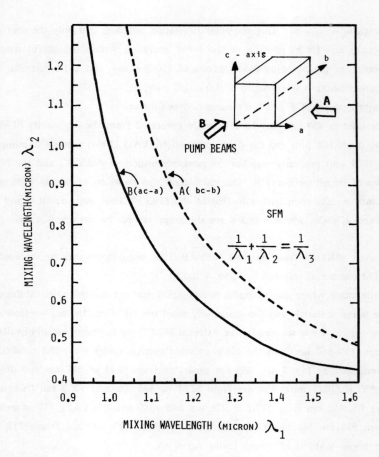

Fig.1 Noncritically phase-matched SFM in type II KTP crystal at various configurations: IIA[bcb], IIB[aca] with output signal polarizations along b and a-axis, respectively.

POTASSIUM TITANYL PHOSPHATE (KTP) :
PROPERITES AND NEW APPLICATIONS

J. D. Bierlein, E. I. duPont de Nemours & Company, Inc., Central Research & Development Department*, Experimental Station, Wilmington, DE 19898

ABSTRACT

$KTiOPO_4$ (KTP) is a nonlinear optical material widely used for second harmonic generation of Nd:YAG. This paper reviews basic KTP properties, describes recent advances and discusses new applications.

KTP is a superior nonlinear optical material that is widely being used commercially for second harmonic generation (SHG) of the Nd:YAG 1.06 micron laser[1]. It is not hydroscopic, has large nonlinear optic coefficients and high damage thresholds, has small beam walkoff, and has large thermal and angular bandwidths. In addition KTP has large electrooptic coefficients and low dielectric constants which make it potentially useful for electrooptic applications.[2]

Some of the nonlinear optical properties of KTP are shown in Table I where they are compared to other materials. Although a few specific characteristics of other materials are better, as Table I shows, KTP has

TABLE I

Nonlinear Optical Materials

	Phase Matchable Wavelength (μ)	d^2/n^3 (re:KDP)	Optical Damage Threshold (GW/cm^2)	Acceptance Angle $(mrad\text{-}cm)$	Temp Range $(°C\text{-}cm)$
$KTiOPO_4$	1.0 - 2.5	62	1 - 10	26/70	30
$\beta\text{-}BaB_2O_4$	0.4 - 3.3	13	1 - 10	1	37
$KNbO_3$	0.4 - 5	270	0.35	45	0.3
$LiNbO_3$	0.4 - 5	23	0.05	5 - 25	0.6
$LiIO_3$	0.3 - 5.5	40	0.5	0.8	50
KH_2PO_4	0.2 - 1.5	1	0.2 - 10	3 - 6	6 - 25

a combination of properties that make it unique for nonlinear optical applications. Table II lists some electrooptic and dielectric properties of KTP and compares them to other materials for an optical waveguide

TABLE II

Electrooptic Waveguide Materials

	r (pm/V)	n	ε_{eff} ($\sqrt{\varepsilon_{11}\varepsilon_{33}}$)	$n^3 r/\varepsilon_{eff}$ (pm/V)
KTP	35	1.86	13	17.3
$KNbO_3$	25	2.17	30	9.2
$LiNbO_3$	29	2.20	37	8.3
$Ba_2NaNb_5O_{15}$	56	2.22	86	7.1
SBN(25-75)	56-1340	2.22	119-3400	5.1 - .14
GaAs	1.2	3.6	14	4.0
$BaTiO_3$	28	2.36	373	1.0

modulator application. Here ε_{eff} is an effective dielectric constant and the last column is a figure-of-merit. Since $n^3 r/\varepsilon_{eff}$ is nearly double that of any of the other inorganic materials listed, KTP appears to be superior for electrooptical waveguide applications as well.

KTP belongs to the orthorhombic mm2 point group (space group $Pna2_1$) and the crystal structure consists of corner sharing titanium octrahedra chains connected by phosphate tetrahedra.[3] There are two chains per unit cell whose direction alternates between (011) and (0$\bar{1}$1). Alternating long and short Ti-O bonds occur along these chains that result in a net z-directed polarization and is the major contributor to KTP's large nonlinear optic and electrooptic coefficients. The potassium ion sits in a large coordination number site and is weakly bonded to both the Ti-octrahedra and P-tetrahedra. Channels exist along the z-axis ((001) direction) whereby potassium can diffuse with a diffusion coefficient several orders of magnitude greater than in the x-y plane.

Although orthorhombic in structure, KTP shows a very small optical birefringence in the x-y plane which results in very wide phase matching acceptance angles. Its relative low refractive indicies (1.76 to 1.86 at 0.63 µ) together with the use of Type II coupling results in its very wide phase matching temperature range. It also has a high Curie temperature (938 °C)[4] which results in nearly temperature independent nonlinear optic and electrooptic properties.

Because KTP decomposes on melting, normal melt processes are not possible for growing this material. However, early work showed that KTP could be grown using both hydrothermal and flux techniques and the hydrothermal route was subsequently developed into a commercial process. Unfortunately, because of equipment restrictions due to the high temperatures and pressures involved in the hydrothermal technique, crystal size is limited. Recently, routes to a lower temperature hydrothermal process have been demonstrated that may result in larger crystals.[5] Also, several laboratories worldwide, including several in China, have made significant progress in improving the flux growth process.[6-9] Using these improved processes crystals of fairly large size are now becoming available

Although the linear, nonlinear and electrooptical coefficients of KTP crystals grown by the flux and hydrothermal techniques are similar, differences have been observed in some of the dielectric properties and in optical damage characteristics. The low frequency dielectric properties which are dominated by potassium ion conductivity vary over four orders of magnitude depending on growth process with some hydrothermal material possessing low ionic conductivity and flux grown material very high conductivities. The specific origin of this variation is not well known but is probably related to impurity ions and/or to growth defects. Optical damage also varies with growth process and for different samples using the same process. Although only limited data is available, the general trend is that hydrothermally grown material shows higher damage thresholds. Damage thresholds have been increased significantly by efficient heat sinking and by operating at elevated temperatures (eg. 100 °C).[10]

In addition to the progress in crystal growth, advances have also been made in modifying KTP for other applications. In particular, since

KTP forms solid solutions with its Rb, Cs (partial), and Tl isomorphs exchanging K with the heavier Rb, Cs, and Tl ions near the crystal surface raises the surface refractive index thus forming optical waveguides without adversly affecting the nonlinear and electrooptic properties.[11] The optical waveguides formed show refractive index increases of up to 0.2 (for Tl exchange) and optical losses less than 0.4 db/cm. Using metal masks to form ion exchanged channel waveguides a Mach-Zehnder optical waveguide modulator has been fabricated with a bandwidth in excess of 16 GHz.[12] Efficient waveguide second harmonic generation has also recently been demonstrated.[13]

With the availability of larger crystals and the new optical waveguide fabrication technology, several new applications for KTP are emerging. In addition to the standard uses for frequency doubling in scientific and industrial lasers, KTP is currently being used commercially in an intracavity frequency doubling geometry for generating in excess of 20 watts of green for general laser surgery,[14] for intra- and extracavity doubling of laser diode pumped Nd:YAG to generate green,[15] for intracavity mixing of 0.81 micron laser diode with 1.06 micron YAG to give blue,[16] and for intra- and extracavity doubling of the Nd:YAG 1.3 micron line to produce red. Also, using optical waveguides, several integrated optical devices such as modulators, switches, etc. are being developed and promising possibilities exist for various waveguide nonlinear optical applications as well.

Finally, KTP may also be useful in optical parametric applications. Using the second harmonic of Nd:YLF as the pump, a tuning range of from 600 nm to 4.3 µm and single pass conversion efficiencies of 10% have recently been demonstrated.[17] Generally, parametric sum and different frequency mixing is phase matchable over the entire KTP transparency range.

In summary, KTP is a unique, nonlinear optic and electrooptic material that is being used widely for SHG of Nd:YAG. With recent advances in crystal growth and optical waveguide fabrication these types of applications are continuing to expand and entirely new applications are being developed.

REFERENCES

1) Zumsteg, F.C.; Bierlein, J.D. & Gier, T.E., J. Appl. Phys., 47, 4980 (1976).

2) Bierlein, J.D. & Arweiler, C.B., Appl. Phys. Lett., 49, 917 (1986).

3) Tordjman, I.; Masse, R; & Guitel, J.C., Z. Kristallogr, 139, 103 (1974).

4) Yanovskii, V.K.; Varonkova, V.I.; Leonov, A.P. & Stefanovich, S.Y., Sov. Phys. Solid State, 27, 1508 (1985).

5) Laudise, R.A.; Cava, R.J. & Caporaso, A.J., J. Crystal Growth, 74, 275 (1986).

6) Jacco, J.C.; Loiacono, G.M.; Jaso, M.; Mizell, G. & Greenberg, B., J. Crystal Growth, 70, 484 (1984).

7) Bordui, P.F. & Jacco, J.C., J. Crystal Growth, 82, 351 (1987).

8) Defan, Cai & Zhegtang, Yang, J. Crystal Growth, 79, 974 (1986).

9) Dezhong, Shen & Chaoen, Huang, Prog. Crystal Growth and Charact., 11, 269 (1985).

10) Vanherzeele, H.A., to be published in Applied Optics.

11) Bierlein, J.D.; Ferretti, A.; Brixner, L.H. & Hsu, W.Y., Appl. Phys. Lett. 50, 1216 (1987).

12) Chouinard, M.P.; et al., to be presented at ECOC, Brighton, UK, Sept. 1988.

13) Bierlein, J.D., presented at MRS Int. Mtg. on Adv. Mat., Tokyo, Japan, June 1988.

14) Huth, B.G. & Kuizenga, D., Lasers & Optronics, p.59, Oct. 1987.

15) For example, see Baer, T.M., Proc. Soc. Photo-Opt. Instrum. Eng., 610, 45 (1986).

16) Risk, W.P.; Baumert, J.-C.; Bjorklund, G.C.; Schellenberg, F.M. and Lensh, W., Appl. Phys. Lett., 52, 85 (1988).

17) Vanherzeele, H.A.; Bierlein, J.D. & Zumsteg, F.C., to be published in Applied Optics.

* Contribution Number: 4836

A KIND OF NEW DEFECT IN KTP CRYSTAL
AND ITS SHG ENHANCED EFFECT

Liu Yaogang, Wang Jiyang, Shao Zhongsho, Wei Jinqian

Shi Luping, Zhang Jiguo

(Institute of Crystal Meterials, Shandong University)

We have found a kind of new defect in high SHG efficiency crystal KTP. On the edge of the defect, large scale of enhanced SHG effect is observed.

The growth straitions are always presented on (100) face of KTP crystal, which are the macrosteps resulted from the two dimensional nucleation mechanism of this face. Generally, at the middle of the face of (100), in corresponding to this convex line, a line similar to the "ghost" line in KDP crystals is existed in the inner part of the crystal. It shows as a dark band on the macro-camera photo, and looked like a kind of dispersion straitional defect after been amplified 126 times by supermicroscopic method. Acoording to the results of our observation, it is different from some well known defects such as growth section brounaries, grain boundaries so that it would be a new kind of defect.

A crystal plate (10x12x6mm) consists of such defect mentioned above is cut according to SHG direction of KTP. Then the plate is mounted in Nd:YAG laser beam, strong 0.53μ green laser is produced. On corresponding to the defect, there exists a dark line in the spot of green laser. By observing carefully we find that on both edges of the dark line, the green laser beams produced are obviously stronger than that of the other parts. For the further determination of the effect of this defect on SHG performance, we mounted the sample in the beam of 514 QT continuous Nd:YAG laser and measured $0,53\mu$ green laser output with small laser spot on each point of the region (~0.5mm). The output of 0.53μ laser is received by electrooptical amplifier and further amplifiered with Boxcar, then recorded by X-Y recorder. KTP plate mounted on the sample holder being removed in the direction which is perpendicular to SHG direction. We take the 0.53μ laser intensity as the signals in y axis and the distances removed as the signals in X axis. We have observed that the intensities for the region consisting of defect are weak, but the intensities for the edge region are obviously enhanced, the ratio of maximum intensity to normal region intensity is about 10. The enhanced SHG effects in both edge sides of the defect are asymmetric. The results of phasematching angle experiments confirmed this asymmetry. Many KTP samples have been tested. This defect does exist in all the samples observed. We can regard this as one of the intrinsic defects of KTP crystal. The enhanced SHG effect is observed in most of the samples observed.

The enhanced SHG effect results probably from the modulated structures produced in the regions near the defect. Also, it maybe resulted from larger distorations of Ti-O octahedrons in KTP structure close to the defect. Further investigations are under proceeding.

NUCLEATION AND GROWTH OF THE NON-LINEAR OPTICAL CRYSTAL POTASSIUM PENTABORATE TETRAHYDRATE

Chen Wan-chun, Xie An-yan and Ma wen-yi
(Institute of Physics, Chinese Academy of Sciences)

ABSTRACT
A serices experiments for the dependence of crystal growth parameters on pH were carried out. A solubility of KB_5 crystal was determined and a optimum growth conditions were presented for growing large KB_5 crystals.

The potassium pentaborate tetrahydrate crystal is a nonlinear optical crystal which has been applied to tunable laser devices in the vacuum ultraviolet region. Radiation at wavelengths as short as 185 nm have been generated in KB_5 crystal of a cube with 10-mm sides utilizing the sum frequency by mixing the outputs of two tunable lasers, one operation in the IR, the other in the UV which were pumped by the fundamental output of a ruby laser. Efficiencies for the upconversion of UV input radiation to sum frequency radiation of 8 - 12 % were realized at input power densities of 500 MW/cm^2. Several microwatts of tunable cw radiation near 194 nm in a linewidth of less than 2 MHz was produced by sum-frequency mixing the radiation from a frequency-doubled argon-ion-laser with the radiation from a ring dye laser in a 30-mmx5-mmx5-mm brewster-cut KB_5 crystal, it would be used to measure the absolute wave number for $^{202}Hg^+$. Tunable picosecond pulses in the ultraviolet region down to 196.7 nm with more than 20 KW peak power were attained by the sum frequency mixing of fourth or third harmonic pulses of a mode-locked YAG laser with tunable pulses produced by $LiNbO_3$ parametric oscillator pumped by second harmonic pulses of the YAG laser by using a 7mmx7mmx7mm KB_5 crystal.

Since the metastable region of KB_5 crystal in the saturated solution is quite narrow, it is difficulty to control the spontaneous nucleation, dentritic growth, twin crystal growth, veil and clouds during growing. In order to investigate a optimum growing conditions,we have determined the solubility of KB_5 in H_3BO_3-H_2O-KOH system for studying the thermodynamical stabilty, measured the value of R, the ratio of an amount of acid to base added in a solution for setting the experiments and studied the the dependence of crystal growth parameters on pH of the solution.

1.   The solubility of KB_5 crystal in the H_3BO_3-H_2O-KOH system

The solution was prepared by desolving H_3BO_3 , KOH, and $2K_2CO_3 \cdot 3H_2O$ into ion-free water and maintained at $70^{\circ}C$ for 3 days,then cooling in rate of $0.25^{\circ}C/day$.The solubility was determined by the weighing method. The results shown that:1)the solubility of KB_5 increases with pH of the solution; 2)the solution is more stable when pH is up;3)the metastable region is quite narrow;4)it is not possible to have crystal nucleation when pH is more than the critical value of 8.88.

2.   The dependence of R on the pH of the solution

R was calculated from the experimental data. The pH was measured by E632 Digital-pH-meter with the accuracy of 0.01 unit (Metrohm Switzerland). The results indicated that: 1) R increases with decreasing of the concentration of the solution at pH<8.0;   2) R increases with exprimental temperture at pH<8.0;   3) the relationship between log R and pH is linear when pH is less than 6.30; 4) the R from the H_3BO_3-H_2O-KOH is more than that one from H_3BO_3-H_2O-K_2CO_3 solution system at the same temperature, concentration and pH.

3.   Optimum conditions for growing the crystal of KB_5

The crystal growth experiments were carried out by the slow cooling method. The optimum conditions are: 1) Systems: using H_3BO_3-H_2O-K_2CO_3 system; 2) Initial temperature: 50^0-55^0 C; 3) pH of the solution: 6.50<pH<7.0; 4) Seeds: twinning-free and oriented growth;   5) solution: it would be stable with purifying and recrystallization treatment.

We have been used the results to improve a technique for crystal growth. A large, macro-perfect KB_5 crystals were grown for vacuum ultraviolet later devices in our lab.

QUASI-PERIODIC OSCILLATIONS IN PHOTOINDUCED CONICAL LIGHT SCATTERING FROM LiNbO$_3$:Fe CRYSTALS

Si-min Liu, Guang-yin Zhang, Jin-long Wang, Xiao-yan Ma

Department of Physics, Nankai University, Tianjin, China

Yuan-fen Fu

Department of Physics, Tianjin Institute of Science and Technology, China

Photoinduced conical light scatterings in photorefractive materials have been extensively reported in the past few years, and their conical structures were always stationary. This paper reports for the first time a new kind of photoinduced conical light scattering with quasi-periodic oscillation.

The sample of LiNbO$_3$:Fe (0.08 wt % Fe) is illuminated by a focused beam from a 10-mW He-Ne laser (λ=6328 A) along (or near) the $-c$ axis of crystal. Two sets of conical light scatterings oscillating synchronously in space and time with quasi-period, are observed in the forward and backward directions. The center lines of the forward and backward scattered light cones are antiparallel to the reflected and incident beams, respectively. When the incident beam propagates exactly along the c axis, the center points of rings observed on the screens located in the forward and backward directions, coincide with the transmitted and reflected beam spots, respectively (Fig.1). When the incident beam propagates deflecting the $-c$ axis, the center points of scattered light cones will depart from the transmitted and reflected beam spots, respectively (Fig.2). These rings swell wave-likely from small to bigger with a quasi-period repetition (about 0.1 - 2 Hz). When the diameter of a ring reaches the maximum value (the corresponding maximum half cone angle is 3.5°), the ring disappears. In general, there exist 5 or 7 rings of different simultaneously. As long as the illumination remains, rings of different diameter simultaneously. As long as the illumination remains this process will repeat. Particularly, in the oscillating of the two scattered beams, the transmitted beam spot has no obvious change in shape maintaining a small circle; and the transmitted beam do not form the oscillating rings.

First of all, the conical light scatterings could not be explained by self-focusing or self-diffraction induced by the incident light, and it also could not be explained by the thermal effect induced by absorption of the incident light, because the transmitted beam spot has no obvious change and the transmitted beam does not form oscillating light cone.

We tentatively consider that this phenomenon is a special photoinduced four-wave mixing caused by photorefractive effect. When a focused beam illuminates the crystal along (or near) the -c axis, a photorefractive grating, whose wave-vector parallelling to the optical axis, is written into the crystal because of interaction between incident beam and reflected beam at the rear face. This phase grating induces energy transfer from the incident beam to the reflected beam. At the same time, because of inhomogeneity of the crystal, the incident and reflected beams, with their scattered beams respectively, will write noise phase gratings. Some energy can be transfered into scattered beams by these gratings. On proper condition, two scattered beams antiparallelling to the incident and reflected beams, respectively, will read the phase grating recorded by the incident and reflected beams and can produce two conjugate beams of the incident and reflected beams. These form two sets of four-wave mixings.

The photoelectrons excited from Fe^{2+} by the incident wave will shift along the +c axis in the asymmetric polarization field of the ionic centers. More and more photoelectrons will accumulate in the +c terminal of the crystal with time lasting. When the density of photoelectrons exceeds certain extent, they will diffuse in the direction perpendicular to the optical axis. Simultaneously, the positive charge (Fe^{3+}) in the -c terminal of the crystal attract these photoelectrons. Convention of electron flow occurs near the incident and reflected beams in the crystal. In this case, the photorefractive phase grating is not the usual one depicted by a grating vector parallelling to the optical axis, but a moving one depicted by one set of conical vectors. Due to the moving of the grating, a quasi-periodic changing phase shift in space and time exists between the phase grating and the interference field of

the incident and reflected beams. This produces the synchroneous oscillating of two sets of conical light scatterings from the photoinduced
four-wave mixing.

Fig 1 Oscillating rings in the backward when the
incident beam propagates axactly
along the c axis

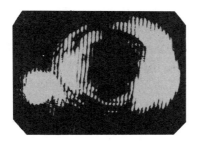

Fig 2 Oscillating rings in the backward when the
incident beam propagates deflecting the c axis

LASER EXCITED PHOTOREFLECTANCE OF $Ga_xIn_{1-x}As/InP$ MULTIPLE QUANTUM WELLS

X.M.Fang, S.C.Shen

Laboratory for Infrared Physics, Shanghai Institute of
Technical Physics, Academia Sinica

P.Helgesen

PHYSIK-DEPARTMENT E16 der Technischen Universität
München, Fed. Rep. Germany

ABSTRACT

We report the low temperature photoreflectance of interband excitonic transition for undoped $Ca_{0.47}In_{0.53}As/InP$ multiple quantum wells. The n=1 and n=2 heavy- and light-hole interband excitonic transitions and Δn=2 forbidden transition for the sample with 80 Å well and 300 Å barrier are investigated. The transition from an impurity level to a subban is also observed.

High quality semiconductor injection laser can be fabricated from $Ga_{0.47}In_{0.53}As/INP$ superlattices and multiple quantum wells which allow the laser emission wavelength to be shifted continuously from 1.65 μm to 1/3 μm by the change of the well thickness.

The The photoreflectance technique has been described in the literature [1]. The 632.8 nm lines of He-Ne laser is used as the modulated beam and the signal is detected by a low temperature Ge photodiode. The samples used in our experiments were grown by MOCVD. The MQW's structure consists of 15 layers of $Ga_{0.47}In_{0.53}As$ and InP grown on 0.3 μm thick undoped InP buffer layer on top of semi-insulating InP substrate and clad with 1.0 μm thick InP. The GaInAs wells and InP barriers are 80 Å And 300 Å thick,respectively.

We observed at least 6 structures corresponding to interband excitonic transitions in the spectral range between 800 MeV and 1100 MeV at 80 K (shown in Fig.1). The energy positions and the line widths of the spectral structures are determined by fitting the experimental results to Aspnes theory [2], including the excitonic effect on the line shapes. The PR intensity of excitonic transition is proportional to the optical matrix. Thus the lowest energy peak shown in Fig.1 does not correspond to the n=1 heavy-hole interband excitonic transition although its energy is lower than that of the higher energy peak. Shown in Fig.2 is

the photo-inductivity spectroscopy of the same sample at various temperatures[4]. From Fig.2 we can see that the lowest energy peak is related to an impurity because of its temperature behaviour. At 80 K the
energy position of the lowest energy peak in Fig.2 is close to that in
Fig.1. Therefore we identify the lowest energy peak to the transition
related to an impurity.

To identify the other spectral structures shown in Fig.1 we have calculated the energies of the confined electron and hole subband states and
possible excitonic transitions in finite square well by use of the envelope function approximation[3]. Shown in Table I is the results of
the calcualtions and experiments. Thus we have observed the n=1 and n=
2 heavy- and light-hole allowed interband excitonic transitions and
Δn=2 parity forbidden interband transition. The positions of the structures shift systematically to lower energy when temperature increases
from 80 K to 280 K with a temperature coefficient of about -3.0×10^{-1}
MeV/K.

REFERENCES

(1) J.L.Shay, Phys. Rev., B3, 803 (1970).

(2) D.E.Aspnes, Handbook on Semiconductor. vol.2.

(3) C.Bastard, Phys. Rev., B25, 7584 (1982).

(4) F.Koch, private communication.

Table I:

spectral structures	energy (MeV)		
	experiment	theory	line width
11HH	863	859	5
11LH	875	879	10
22HH	952	953	14
221H	1027	1026	15
13HH	930	927	18

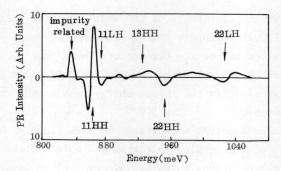

Fig. 1

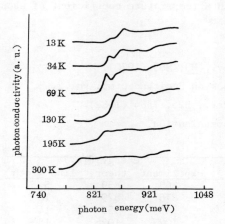

Fig. 2

Growth, Spectroscopic Properties and Applications of Doped $LiNbO_3$ Crystals

Liu Jian-cheng

865 Chang-ning Road, Shanghai Institute of Ceramics, Academia Sinica

ABSTRACT

$LiNbO_3$ with different Li/Nb and $LiNbO_3$:TR(TR=Nd,Er+Mg,Ho,Sm,Eu, Cr,Mn,Co,Fe) were grown. Absorbtion, fluorescence and luminescence were determined and analyzed. Parameters of Coulomb B, C and crystal field D_q for Cr^{3+} under O_h were calculated. Applications of holography and waveguide were investigated.

The pure $LiNbO_3$ with different Li/Nb ratios 0.88, 0.92, 0.945, 1.00, 1.04, 1.083 and $LiNbO_3$:TR(TR=Nb,Er+Mg,Ho,Sm,Eu, and Cr, Mn, Co,Fe) with congruent ratio Li/Nb=0.945 and dopant in the form of oxide with the content of 0.1-1.5 mole% were grown along the C-axis by the Czochralski Technique using resistance heated platinum crucibles at 4 mm/h pulling and 24 r/min. The crystal compositions measured by chemical quantitative analysis showed that Nb_2O_5 in crystal has more concentration then in melt, but Li_2O in crystal has less concentration then in melt and that the distritution coefficient of rare earth ions are a bit large, $K_{Eu}=0.51$, $K_{Tb}=0.48$, $K_{Sm}=0.6$. Accurate values of the lattice parameters a and c of six different Li/Nb ratio obtained from their X-ray powder diffraction data by the treatment of the least square method show that these lattice parameters increase with the decrease of Li/Nb ratio, and that the most likely explanation to the nonstoichiometry in the Li/Nb ratio of $LiNbO_3$ crystals is the partial substitution of lithium atoms by niobium atoms. $LiNbO_3$:Cr,Mn,Co, Fe crystals were treated by heating reduction in Li_2CO_3, oxidation in air and γ-irradiation, the valent state change between $Mn^{2+} \rightleftharpoons Mn^{3+}$ were revealed.

The optical absorption spectra were taken with a UV-5270 spectrophotometer (made in USA) and BGO spectrophotometer in 300 to 800 and 800 to 2500 nm region, the absorption spectra measured by $LiNbO_3$ crystals doped with rare earth ions show that they consist of several bands characteristic of the rare earth ions in crystaks and detail analysis were also given. The absorption spectrum due to cobalt consists of five bands which correspond to the transition from the ground state $^4F_1$ to $^4F_2$(7143 cm^{-1}), $^4A_2$(16474 cm^{-1}), and $^4F_1$(19305, 18349 cm^{-1}). The absorption spectrum due to Co^{3+} must appear a 756 nm peak, but the peak of the sample at 750nm is not too clear, so that cobalt entered the $LiNbO_3$ lattice in the Co^{3+} is less than in the Co^{2+}. The absorption pattern of LiNbO :Mn is the most simple. We only considered the ground state energy splittin so that the ^5D spectrum term of $Mn^{3+}(d^4)$ were split into $^5E(t_2^3e)$ and $^5T_2(t_2^2e^2)$. and obtained a absorption peak ($^5E \rightarrow ^5T_2$). The ^6S spectrum term of $Mn^{2+}(d^5)$ were not split under O_h field, and were not obtained any absorption peaks. The tail part of $LiNbO_3$:Mn were t reated by heating reduction in Li CO_3, the valent state change from Mn^{3+} to Mn^{2+} were revealed, the green crystals turned into yellow crystals and the absorption peak disappeared. When reduced samples were again treated by heating oxidizing in air, the colour occured change from yellow to green, and the absorption peak appeared again. The splitting energy of crystal field in d orbit is D_q=1742 cm^{-1} . Comparison between experimental and theoretical values of spectrum terms for doped with transition metal $LiNbO_3$ crystals were given. We have calculated the magnitues of the parameters of Coulomb interaction B and C and the parameter of the internal crystal field, D_q, for $LiNbO_3$:Cr^{3+} under O_h crystal-field by the secular equations of $3d^3$ electrons, the results obtained are B=609 cm^{-1}, C=3087 cm^{-1} and D_q=1515 cm^{-1}, which lead to the best agreement between experimental and theoretical values.

The fluorescence spectra (corresponding to the $^4F_{3/2} \rightarrow ^4I_{11/2}$, $^4I_{13/2}$, $^4I_{9/2}$ transition for Nd^{3+}; $^4I_{13/2} \rightarrow ^4I_{15/2}$ for Er^{3+}+Mg^{2+}; $^5I_7 \rightarrow ^5I_8$ for Ho^{3+}; $^5D_0 \rightarrow ^7F_2$ for Eu^{3+}, and for Sm) were measured with a spectrophotometer. The room temperature excitation and photoluminescence spectra due to chromium and iron impurities have been observed and explained.

$LiNbO_3$:Fe and $Ba_xSr_{1-x}Nb_2O_6$:Ce are volume phase optical storage media, thus volume phase holographic storage by means of the photorefractive effect becomes possible. The holographic properties and applications of these crystals were investigated for a diffraction efficiency of 10% exposure levels of 200 and 20 mj/cm^2 are required respectively. The maximum diffraction efficiency is more than 80%.

LiNbO$_3$ is finding its wide application in integrated optics as a basic substrate material for optical waveguided devices. In view of this a study has been made on the depth profile of in diffused Ti concentration in LiNbO$_3$ waveguide. It was found that the depth profiles as analysed by EPM, clearly depend on the thickness of the sputtered Ti film. For samples with thin Ti films, depth profiles have the shape of Causs function, while thick film samples have their depth profiles of exponential function.

PHOTOREFRACTIVE AND PHOTOVOLTAIC EFFECT IN DOPED LiNbO$_3$*

Wen Jinke,   Zhu Yaping,   Chang Tiejun,
Wang Hong,   Wang Huafu

Department of Physics, Nankai University,
Tianjin, P.R. China

To get further information about the effect of doping to the photorefraction, we studied the photorefractive and photovoltaic effect in Li-rich Lithium niobate (LN) crystal co-doped with Mg and Fe in comparison with LN:Fe, LN:Mg,and Li-rich LN:Mg.

Single crystals of LN:Fe (0.1 wt%), LN:Mg (5 mol%), Li-rich (Li/Nb=50.5/49.5 in the melt) LN:Mg (5 mol%) and Li-rich LN:Mg (5 mol%):Fe (0.1 wt%) were grown by Czochralski method. All samples were annealed under the same condition at 700°C for 5 hours in air. For estimation of photorefractive damage, we observed the distortion of the transmitted laser beam spot, which will smear and elongate along z-axis when the crystal suffers from optical damage. The y faces of the crystal were irradiated by focused Ar laser beam (λ= 488nm) at 26.7kW/cm^2. As expected, LN:Fe is severely suffered from optical damage, while photorefraction in Li-rich LN:Mg is completely suppressed. Unexpected fact is that the distortion of the transmitted beam for the Li-rich LN:Mg:Fe is very slight and alike to that of LN:Mg, even though photorefraction sensitive dopant Fe was added. This fact indicated that doping 0.1 wt% of Fe in Li-rich LN:Mg does not distictively alter its ability to resist photorefraction.

The photorefraction in ferroelectric crystal is attributed to the formation of an internal field through photovoltaic effect. We studied the photovoltaic current J and photoconductivity σ_{ph} of Li-rich LN:Mg:Fe, Li-rich LN:Mg and LN:Fe by taking current-voltage characteristics. An argon laser (488nm) and a beam expander were used to irradi-

---

* Project supported by the National Science Fundation of China.

ate the y-facets and photocurrent J_{ph} in z-direction was
measured. Since $J_{ph} = J + \sigma_{ph}E$, where E is the applied field,
the intersection of the curve with the ordinate gives J,
and σ_{ph} is obtained from the slope of curve. The results
obtained at 140mW/cm^2 are shown in table 1.

Table 1   Photovoltaic current J and photoconductivity
σ_{ph} of doped LN at 140 mW/cm^2

Samples	J (10^{-11}A/cm^2)	σ_{ph} $(10^{-14}\Omega^{-1}\text{cm}^{-1})$
LN:Fe	20.3	*
Li-rich LN:Mg	0.25	2.12
Li-rich LN:Mg:Fe	0.27	0.17

* too low to be determined

The photovoltaic current of Li-rich LN:Mg:Fe and Li-rich
LN:Mg is nearly equal and is two order of magnitude smaller
than that of LN:Fe. The photoconductivity of Li-rich LN:Mg:
Fe is an order of magnitude smaller than that of Li-rich
LN:Mg and is much larger than that of LN:Fe. The above results mean that doping of Fe in Li-rich LN:Mg does not increase the photoelectrons by photoionization of Fe^{2+}, but
lower the mobility of photoelectrons. A plausible explanation is that the occupation sites of Fe ions in Li-rich LN:
Mg:Fe and LN:Fe may be different. Further works are needed
to clarify this problem.

RECENT ADVANCES IN PHOTOREFRACTIVE NONLINEAR OPTICS

(Invited Paper)

Pochi Yeh
Rockwell International Science Center
Thousand Oaks, California 91360

ABSTRACT

There have been several significant new developments in the area of photorefractive nonlinear optics during the past few years. This paper briefly describes some of the important and interesting phenomena and applications.

1.0 INTRODUCTION

The photorefractive effect is a phenomenon in which the local index of refraction is changed by the spatial variation of light intensity. This spatial index variation leads to a distortion of the wavefront and is referred to as "Optical Damage."[1] The photorefractive effect has since been observed in many electro-optic crystals, including $LiNbO_3$, $BaTiO_3$, SBN, BSO, BGO, GaAs, InP, and CdTe. This effect arises from optically generated charge carriers which migrate when the crystal is exposed to a spatially varying pattern of illumination with photons of sufficient energy.[2,3] Migration of charge carriers produces a space-charge separation, which then gives rise to a strong space-charge field. Such a field induces a change in index of refraction via the Pockels effect.[4] Photorefractive materials are, by far, the most efficient media for optical phase conjugation[5,6] and real-time holography using relatively low intensity levels (e.g., 1 W/cm²).

2.0 TWO-WAVE MIXING

When two beams of coherent radiation intersect inside a photorefractive medium, an index grating is formed. This index grating is spatially shifted by $\pi/2$ relative to the intensity pattern. Such a phase shift leads to nonreciprocal energy transfer when these two beams propagate through the index grating. The hologram formed by the two-beam interference inside the photorefractive media can be erased by illuminating the hologram with light. Thus dynamic holography is possible using photorefractive materials.[7,8] Some of the most important and interesting applications are discussed as follows.

Laser Beam Cleanup

Two-wave mixing in photorefractive media exhibits energy transfer without any phase crosstalk.[9,10] This can be understood in terms of the diffraction from the self-induced index grating in the photorefractive crystal. Normally, if a beam that contains phase information $\psi(r,t)$ is diffracted from a fixed grating, the same phase information appears in the diffracted beam. In self-induced index gratings, the phase information $\psi(r,t)$ is impressed onto the grating in such a way that diffraction from this grating will be accompanied by a phase shift $-\psi(r,t)$. Such a self-cancellation of phase information is equivalent to the reconstruction of the reference beam when the hologram is read out by the object beam. Energy transfer without phase crosstalk can be employed to compress both the spatial and the temporal spectra of a light beam.[10] This has been demonstrated experimentally using $BaTiO_3$ and SBN crystals.[9-11]

Photorefractive Resonators

The beam amplification in two-wave mixing can be used to provide parametric gain for unidirectional oscillation in ring resonators. The oscillation has been observed using $BaTiO_3$ crystals.[12] Unlike the conventional gain medium (e.g., He-Ne), the gain bandwidth of photorefractive two-wave mixing is very narrow (a few hertz's for $BaTiO_3$). Despite this fact, the ring resonator can still oscillate over a large range of cavity detuning. This phenomenon was not well understood until a theory of photorefractive phase shift was developed.[13] This theory also predicts that the unidirectional ring resonator will oscillate at a frequency different from the pump frequency by an amount directly proportional to the cavity-length detuning. Furthermore, in a photorefractive material with moderately slow response time τ, the theory postulates a threshold where oscillation will cease if the cavity detuning becomes too large. The theory has been validated experimentally in a $BaTiO_3$ photorefractive ring resonator.[14]

Optical Nonreciprocity

It is known in linear optics that the transmittance as well as the phase shift experienced by a light beam transmitting through a dielectric layered medium is independent of the side of incidence.[15] This is no longer true when photorefractive coupling is present. Such nonreciprocal transmittance was first predicted by considering the coupling between the incident beam and the reflected beam inside a slab of photorefractive medium.[16] The energy exchange due to the coupling leads to an asymmetry in the transmittance. In the extreme case of strong coupling ($\gamma L \gg 1$), the slab acts as a "one-way" window. Nonreciprocal (optical) transmission has been observed in $BaTiO_3$ and $KNbO_3$:Mn crystals.[17,18] In addition, there exists a nonreciprocal phase shift in contra-directional two-wave mixing. Such nonreciprocity may be useful in applications such as the biasing of ring laser gyros.[18,19]

Conical Scattering

When a laser beam is incident on a photorefractive crystal, a cone of light (sometimes several cones) emerges from the crystal. This has been referred to as Photorefractive Conical Scattering. It is known that fanning of light occurs when a laser beam is incident on a photorefractive crystal.[20] Because of the strong two-beam coupling, any scattered light may get amplified and thus lead to fanning. In conical scattering, the noisy hologram formed by the incident light and the fanned light further scatters off the incident beam. The fanning hologram consists of a continuum of grating vectors, but only a selected portion of grating vectors satisfies the Bragg condition for scattering. This leads to a cone of scattered light. Photorefractive conical scattering has been observed in several different kinds of crystals.[21-23]

Cross-Polarization Two-Wave Mixing

Cubic crystals such as GaAs and InP exhibit significantly faster photorefractive response than many of the oxide crystals. In addition, the isotropy and the tensor nature of the electro-optic coefficients allow the possibility of cross-polarization two-wave mixing in which the s component of one beam is coupled to the p component of the other beam and vice versa. A coupled mode theory of photorefractive two-wave mixing in cubic crystals was developed.[24] The theory predicts the existence of cross-polarization two-wave mixing in crystals possessing a point group symmetry of $\bar{4}3m$. Such a prediction was validated experimentally using photorefractive GaAs crystals.[25-27] Cross-polarization two-wave mixing provides extremely high signal-to-noise ratios in many of the applications which employ photorefractive two-wave mixing.

Photorefractive Optical Interconnection

A new method of reconfigurable optical interconnection using photorefractive dynamic holograms was conceived and demonstrated.[28] Reconfigurable optical interconnection using matrix-vector multiplication suffers a significant energy loss due to fanout and absorption at the spatial light modulators. In the new method, the nonreciprocal energy transfer in photorefractive media is employed to avoid the energy loss due to fanout. The result is a reconfigurable optical interconnection with a very high energy efficiency. The interconnection can be reconfigured by using a different SLM pattern. The reconfiguration time is limited by the formation of holograms inside the crystal. Once the hologram which contains the interconnection pattern is formed, such a scheme can provide optical interconnection between an array of lasers and an array of detectors for high data rate transmission.

3.0 OPTICAL PHASE CONJUGATION

Optical phase conjugation has been a subject of great interest because of its potential application in many areas of advanced optics.[4-6] For nonlinear materials with third-order susceptibilities, the operating intensity needed in four-wave mixing is often too high for many applications, especially for information processing. Photorefractive materials are known to be very efficient at low operating intensities. In fact, high phase conjugate reflectivities have been observed in $BaTiO_3$ crystals with very low operating power. In what follows, we will briefly describe some of the most important and interesting recent developments.

Self-Pumped Phase Conjugation

A class of phase conjugators which has received considerable attention recently are the self-pumped phase conjugators.[12,29] In these conjugators, there are no externally supplied counterpropagating pump beams. Thus, no alignment is needed. The reflectivity is relatively high at low laser power. These conjugators are, by far, the most convenient phase conjugate mirrors available. Although several models have been developed for self-pumped phase conjugation,[30-34] the phenomena can be easily understood by using the resonator model.[13,14,35] In this model, the crystal is viewed as an optical cavity which supports a multitude of modes. When a laser beam is incident into the crystal, some of the modes may be excited as a result of the parametric gain due to two-wave mixing. If the incident configuration supports bi-directional ring oscillation inside the crystal, then a phase conjugate beam is generated via the four-wave mixing. The model also explains the frequency shift of these conjugators.[35]

Mutually Pumped Phase Conjugators

Another class of phase conjugators consists of the mutually pumped phase conjugators (MPPC) in which two incident incoherent beams can pump each other to produce a pair of phase conjugate beams inside a photorefractive crystal. The spatial wavefronts of the beams are conjugated and the temporal information is exchanged. The phase conjugation requires the simultaneous presence of both beams. Recently, conjugators were demonstrated experimentally using two incoherent laser beams in $BaTiO_3$.[36,38,39] These phenomena can be explained in terms of either hologram sharing[39,40] or self-oscillations[37] in a four-wave mixing process or resonator model.[41]

Phase Conjugate Michelson Interferometers

We will now consider a Michelson interferometer which is equipped with phase conjugate mirrors. Such an optical setup is known as a phase conjugate Michelson interferometer and has been studied by several workers.[42-44] By virtue of its names, this interferometer exhibits optical time reversal. Consequently, no interference is observed at the output port. The output port is, in fact, totally dark.[44] Such an interferometer is ideal for parallel subtraction of optical images because the two beams arriving at the output port are always out of phase by π. This has been demonstrated experimentally using a $BaTiO_3$ crystal as the phase conjugate mirrors.[45] Using a fiber loop as one of the arms, such an interferometer can be used to sense nonreciprocal phase shifts.[46] A phase conjugate fiber optic gyro has been built and demonstrated for rotation sensing using $BaTiO_3$ crystals.[47,48]

Phase Conjugate Sagnac Interferometers

Using the mutually pumped phase conjugators mentioned earlier, a new type of phase conjugate interferometer was conceived and demonstrated.[49] In the new interferometer, one of the mirrors of a conventional Sagnac ring interferometer is replaced with a MPPC. Such a new interferometer has a dual nature of Michelson and Sagnac interferometry. As far as wavefront information is concerned, the MPPC acts like a retro-reflector and the setup exhibits phase conjugate Michelson interferometry and optical time reversal.[44] As for the temporal information, the MPPC acts like a normal mirror and Sagnac interferometry is obtained. Such a new phase conjugate interferometer can be used to perform parallel image subtraction over a large aperture. With optical fiber loops inserted in the optical path, we have constructed fiber-optic gyros and demonstrated the rotation sensing.

Other Developments Related to Photorefractive Nonlinear Optics

In addition to those described above, there are other significant developments. These include polarization-preserving conjugators,[50] phase shifts of conjugators,[51] optical matrix algebra,[52] fundamental limit of photorefractive speed,[53] nondegenerate two-wave mixing in ruby crystal,[54] and nonlinear Bragg scattering in Kerr media.[55]

REFERENCES

1. Ashkin, A., et al, Appl. Phys. Lett. 9, 72 (1966).
2. Vinetskii, V.L., et al, Sov. Phys. Usp. 22, 742 (1979).
3. Kukhtarev, N.V., et al, Ferroelectrics 22, 961 (1979).
4. Yariv, A., and Yeh, P., "Optical Waves in Crystals," (Wiley, 1984).
5. See, for example, Pepper, D., Sci. Am. 254, 74 (1986).
6. Yariv, A., IEEE J. Quantum Electronics, QE-14, 650 (1978).

122

7. Staebler, D.L., and Amodei, J.J., J. Appl. Phys. 34, 1042 (1972).
8. Vahey, D.W., J. Appl. Phys. 46, 3510 (1975).
9. Chiou, A.E.T., and Yeh, P., Opt. Lett. 10, 621 (1985).
10. Yeh, P., CLEO Technical Digest (1985) p. 274.
11. Chiou, A.E.T., and Yeh, P., Opt. Lett. 11, 461 (1986).
12. White, J.O., et al, Appl. Phys. Lett. 40, 450 (1982).
13. Yeh, P., J. Opt. Soc. Am. B2, 1924 (1985).
14. Ewbank, M.D. and Yeh, P., Opt. Lett. 10, 496 (1985).
15. Knittl, Z., Optics of Thin Films (Wiley, New York, 1976), p. 240.
16. Yeh, P., J. Opt. Soc. Am. 73, 1268-1271 (1983).
17. Zha, M.Z., and Gunter, P., Opt. Lett. 10, 184-186 (1985).
18. Yeh, P., and Khoshnevisan, M., SPIE 487, 102-109 (1984).
19. Yeh, P., Appl. Opt. 23, 2974-2978 (1984).
20. Feinberg, J., J. Opt. Soc. Am. 72, 46 (1982).
21. Odoulov, S., Belabaev, K., and Kiseleva, I., Optics. Lett. 10, 31 (1985).
22. Temple, D.A., and Warde, C., J. Opt. Soc. Am. B3, 337 (1986).
23. Ewbank, M.D., Yeh, P., and Feinberg, J., Opt. Comm. 59, 423 (1986).
24. Yeh, P., J. Opt., Soc. Am. B4, 1382 (1987).
25. Cheng, L.-J. and Yeh, P., Opt. Lett. 13, 50 (1988).
26. Chang, T.Y., Chiou, A.E.T. and Yeh, P., OSA Tech. Digest 17, 55 (1987).
27. Chang, T.Y., Chiou, A.E.T. and Yeh, P., J. Opt. Soc. Am. B5, 1724 (1988).
28. Yeh, P., Chiou, A.E., and Hong, J., Appl. Opt. 27, 2093 (1988).
29. Feinberg, J., Opt. Lett. 7, 486 (1982); Opt. Lett. 8, 480 (1983).
30. Chang, T.Y. and Hellwarth, R.W., Opt. Lett. 10, 408 (1985).
31. Lam, J.F., Appl. Phys. Lett. 46, 909 (1985).
32. MacDonald, K.R., and Feinberg, J., J. Opt. Soc. Am. 73, 548 (1983).
33. Ganthica, D.J., et al, Phys. Rev. Lett. 58, 1644 (1987).
34. Gower, M.C., Opt. Lett. 11, 458 (1986).
35. Ewbank, M.D., and Yeh, P., Proc. SPIE Vol. 613, 59 (1986).
36. Ewbank, M.D., Opt. Lett. 13, 47 (1988).
37. Cronin-Golomb, M., et al, IEEE J. Quantum Electron. QE-20, 12 (1984).
38. Weiss, S., Sternklar, S. and Fischer, B. Opt. Lett. 12, 114 (1987).
39. Eason, R.W., Smout, A.M.C., Opt. Lett. 12, 51 (1987); 12, 498 (1987).
40. Yeh, P., submitted to Opt. Lett. (1988).
41. Yeh, P., Chang, T.Y., and Ewbank, M.D., J. Opt. Soc. Am. B5, 1743 (1988).
42. Ewbank, M., Yeh, P., and Khoshnevisan, M., SPIE Proc. Vol. 464, 2 (1984).
43. Chen, W.H., et al, SPIE Proc. Vol. 739, 105 (1987).
44. Ewbank, M.D., et al, J., Opt. Lett. 10, 282 (1985).
45. See, for example, Chiou, A.E.T., and Yeh, P., Opt. Lett. 11, 306 (1986).
46. Yeh, P., McMichael, I., and Khoshnevisan, M., Appl. Opt. 25, 1029 (1986).
47. McMichael, I., and Yeh, P., Opt. Lett. 11, 686 (1986).
48. McMichael, I., Beckwith, P., and Yeh, P., 12, 1023 (1987).
49. McMichael, I., et al, CLEO Tech. Digest Vol. 7, 134 (1988).
50. McMichael, I., Yeh, P., and Khoshnevisan, M., Opt. Lett. 11, 525 (1986).
51. McMichael, I., and Yeh, P., Opt. Lett. 12, 48 (1987).
52. Yeh, P., and Chiou, A.E.T., Opt. Lett. 12, 138 (1987).
53. Yeh, P., Appl. Opt. 26, 602 (1987).
54. McMichael, I., Yeh, P., and Beckwith, P., Opt. Lett. 13, 500 (1988).
55. Yeh, P., and Koshnevisan, M., J. Opt. Soc. Am. B4, 1954 (1987).

STUDY ON THE DOUBLING-FREQUENCY AND ANTI-PHOTOREFRACTIVE PROPERTY OF HEAVILY MAGNESIUM-DOPED LITHIUM-RICH LITHIUM NIOBATE CRYSTALS

Zhang Hongxi, Xu Yuheng, Xu Chongquan, Xiao Difan

Department of Applied Chemistry, Harbin Institute of Technology, Harbin, P.R. China

Wu Zhongkang

Physics Department of Nankai University, Tianjin, P.R. China

1. Introduction

Lithium niobate ($LiNbO_3$, LN) crystal has been widely studied and used in laser technology because of its unique nonlinear optical property[1]. Doping MgO into congruent LN reduces the photorefraction at moderate power irradiation.[2] A lot of experiments have indicated that composition of LN has an evident influence on its quality. In order to further study the relation between composition and property of LN, a new series of LN crystal-heavily MgO doped Li-rach LN has been developed, and their doubling-frequency and anti-photorefractivity have been studied. The result indicates that this series of LN crystal has more excellent SHG and anti-photorefractive property. And their improved resistance to photorefraction is also attributed to the increased photoconductivity.

2. Samples preparation

The crystals were pulled along C-axis by Czochralski technique. And their quality was checked by polarization microscopy. The composition and homogeneity of the samples are shown in Tab. 1.

For SHG experiment the samples' front and back Y-faces were platted with 1.06 μm and 0.53 μm anti-reflection film respectively.

3. SHG and photorefractive experiment

Extra-cavity SHG experiment has been carried out of the samples in Tab.1. The schematic diagram is shown in Fig.1.

124

And the experiment result is shown in Tab.2. From Tab.2 can
be seen that the Tpm of the samples changes little in comparison with those reported in reference [3] ($118^{\circ}C$) and
[4] ($107^{\circ}C$), but the SHG converting efficiency is greatly
increased.

The anti-photorefractive property of the samples were
studied by direct-observation method and the result indicated that there is no photorefraction observed when the
power density of Ar^+ and pulse Nd:YAG laser is $7.5 \times 10 w/cm^2$
and $100Mw/cm^2$ respectively.

4. Photovoltaic effect

We have measured the photocurrent of sample 2 and 4 with
different applied field by Voltage-Ampier method and the
photoconductivity is calculated and the result is listed in
Tab.3. For reference, the data of a congruent (01) and an
only Mg-doped (02) LN is also presented.

5. Discussion

From our results can be seen that the improved resistance of heavily Mg-doped Li-rich LN to photorefraction is
companied by the increased σ_{ph} and high SHG converting efficiency; and the macroscopic mechanism of anti-photorefractivity is the increased σ_{ph} instead of photocurrent.

The influence of σ_{ph} on photorefraction may be interpreted by the electro-optic equation:

$$\delta(\Delta n) = -\frac{1}{2}(n_e^3 r_{33} - n_o^3 r_{13})E$$

Based on diffusion model, it can be deduced that the increase of σ_{ph} makes photovoltaic field decrease consider-
ablly. If the change of n_e, n_o and r_{13}, r_{33} resulting from
doping are neglected, then from the above equation can be
seen that the decrease of E depresses birefringence variation $\delta(\Delta n)$, which means that the resistance to photorefraction improves.

Project supported by the Chinese National Foundation of Science

REFERENCE

[1] D.A. Bryan, R.R. Rice, Robert Gerson, H.E. Tomaschke,
    K.L. Sweeny, L.E. Halliburton, Opt. Eng., 24, 138(1985).
[2] Zhong Jiguo, Jin Jian, Wu Zhongkang, 11th International Quantum
    Electronics Conference, IEEE Catalog No.80 CH$_{156-0}$, New York
    (1980)631.
[3] Zhong Jiguo, Acta Physics Sinica, 32, 795(1983).
[4] R.L. Byer, Appl. Phys. Lett., 39, 17(1981).

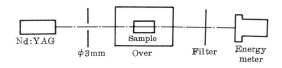

Fig 1 Schematic diagram for doubling frequency experiment

Tab. 1.   Composition and homogenity of the samples

Sample	1	2	3	4
Mg content (mol%)	5	5	5	5
Li/Nb mol ratio	0.97	0.99	1.01	1.02
Birefringence gradient (cm^{-1})	1.5×10^{-5}	1.1×10^{-5}	1.1×10^{-5}	4.8×10^{-5}

Tab. 2.   Result of the SHG experiment

Sample	1	2	3	4
SHG converting efficiency (%)	41.9	45.8	46.2	46.2
T_{pm} ($^{\circ}$C)	110	107	105	108

Tab. 3.   Data of photoconductivity σ_{pn}

Sample	1	2	01	02
σ_{pn} ($\Omega^{-1} cm^{-1}$)	2.5×10^{-13}	2.1×10^{-13}	2.0×10^{-15}	3.0×10^{-14}

A NEW TECHNIQUE FOR INCREASING TWO-WAVE MIXING
GAIN IN PHOTOREFRACTIVE $Bi_{12}SiO_{20}$ CRYSTALS

XU GAN and YE SHOUNIAN

Department of Applied Physics, Beijing University
of Aeronautics and Asteonautics

SUN YINGUAN

Department of Physics, Beijing Normal University

ABSTRACT

Dramatic increase of two-wave mixing gain in BSO by using alternating electric fields is reported. Detailed theoretical analysis is also given.

In photorefractive materials of the photoconductive type such as BSO, dc-electric fields are usually used to increase the beam-coupling efficiency in two mixing (2WM) and four-wave mixing (4WM) processes. Although the amplitude of the index grating in the crystal is increased by the dc-fields, the phase shift between the index grating and the light fringes cannot be maintained as $\pi/2$, resulting in inefficient enhancement of the 2WM gain. One method to overcome this problem is to artificially introduce an additional phase shift to the grating through a frequency shifter (the moving-grating techmique)[1]. In this paper, a new technique using ac-electric fields is presented.

Let us consider the effect of an external electric field on Es, the amplitude of the effective space-charge field grating (the component shifted by $\pi/2$ from the light interference fringes) in 2WM process in a BSO crystal (Fig.1). Assuming that the external field is a) $E=E_0$ (dc-field), b) $E=E_0 \sin(2\pi t/T)$ (ac-sinusoidal field), c) an ac-, square-wave field (amplitude E_0, period T), the maximum values of Es for the three cases can be shown to be[2]

$$(Es)_{max} = E_0/2 \quad \text{at} \quad K_{opt} = (RE_0)^{-1} \quad \text{for case a)}, \tag{1}$$

$$(Es)_{max} = [(P/R)^{\frac{1}{2}}-1] E_0 / [(P/R)^{\frac{1}{2}}+1] \quad \text{at} \quad K_{opt} = (PR)^{-\frac{1}{2}}E_0^{-1}$$
$$\text{for case b)}, \tag{2}$$

$$(Es)_{max} = (P/R)^{\frac{1}{2}}E_0/2 \quad \text{at} \quad K_{opt} = (PR)^{-\frac{1}{2}}E_0^{-1} \quad \text{for case c)}, \tag{3}$$

where $P = \mu\tau$, $R = \varepsilon_r\varepsilon_0/eN_A$; μ, τ - mobility and lifetime of photoelectrons; ε_r - dielectric constant, N_A - trap density, e - charge of electron; $K = 2\pi/\Lambda$, Λ - grating spacing. Expressions (2) and (3) are obtained by assuming the period T of the external fields satisfy $\tau \ll T \ll \tau_g$ (formation time of the index grating) and the distribution of the space-charge field is a time averaged effect of the external field[2]. Fig.2 shows Es as functions of Λ at $E_0 = 20$ kV·cm^{-1} for the three cases. Since the 2WM gain coefficient $\Gamma \simeq$ Es, so that Γ_{max} depends on the factor P/R $(= \mu \tau e N_A / \varepsilon_r \varepsilon_0)$ of the crystal parameters and E_0. P/R $\gg$ 1 usually holds for BSO, therefore Γ_{max} under a sinusoidal field is nearly two times as large as that under a dc-field of the same amplitude, while Γ_{max} can be further increased by using ac-, square-wave fields. Interestingly, we find that the experessions for $(Es)_{max}$ and K_{opt} in (3) are exactly the same as the formulae obtained by using the moving-grating technique[3].

Our experiments were carried out by using an argon ion laser (514.5 nm) in a typical configuration for 2WM in BSO (Fig.1). To satisfy the requirement for T mentioned above, sinusoidal electric fields of 11 kHz frequency were applied to the crystal. The results are shown in Fig.3 and a comparison of Γ_{max} obtained by using different methods is given in Table 1. As can be seen, the measured values of Γ_{max} were indeed doubled under sinusoidal fields compared to the dc-field case. The actually achieved $\Gamma_{max} = 3.3$ cm^{-1} was much larger than the dc-field case and was comparable to that obtained by using the moving grating technique at the same wavelength, because very high ac voltages could be easily applied to the crystal without electric breakdown. The output stability of the signal beam was found significantly improved by the ac-field technique. Analysis shows this is due to the absence of intrinsic oscillation which exists when dc-fields are applied. These results and other advantages disclosed by the experiments indicate that the new method provides a simple but effective technique for efficient 2WM.

REFERENCES

1) Rajbenbach H., Huignard J.P. and Loiseaux B., Optics Comm. 48, 247 (1983).

2) Xu Gan, Ye Shounian and Sun Yinguan, Optics Comm. 66, 155 (1988).

3) Refregier P., Solymar L., Rajbenbach H. and Huignard J.P., J. Appl. Phys. Phys. <u>58</u>, 45(1985)

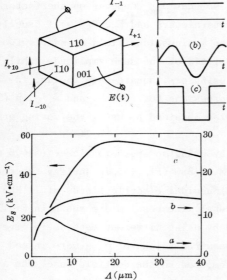

Fig.1 Geometry of 2WM in BSO under a) do-, b) sinusoidal, c) square wave external electric fields.

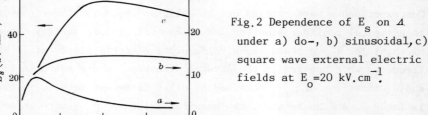

Fig.2 Dependence of E_s on Λ under a) do-, b) sinusoidal, c) square wave external electric fields at E_o=20 kV.cm^{-1}.

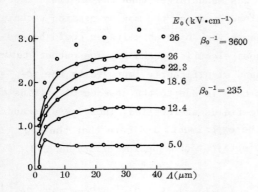

Fig.3 Experimental curves of 2MW-gain coefficient Γ as a function of Λ under sinusoidal electric fields.

I_{+10}=70mW.cm^{-2} β=I_{-10}/I_{+10}

Experimental proof: There Existing Another Mechanism of Photorefractive index in crystal Ce-SBN

Xu Huaifang.

Department of Physics. Shanghai Teacher's University.

Abstract

We have discovered experimently that the change of reflective index of crystal Ce-SBN is of the order 10^{-5}, by laser beam illuminating the whole light-receiving arer of being measured by interferometer.This has also been proved by the evidence that the phase of two-wave-mixing export laser beam is of shift,   approaching to stability in nearly 30 seconds.

Carrying out the two-wave-mixing experiment in photorefractive crystal Ce-SBN , We have found out that the phase of export beam, after being mixed, is changing very slowly during the period of the first 30 seconds, and then approaching to the stability(see Fig1). During the first 10 seconds, the phase is of much less shift, and the last 20 seconds, chang noticeably. The energy coupling process between the two beams is howver nearly completed in 5 to 10 seconds.   Therefore these two phenomenon are controled by different mechanisms. From phase shift direction of exeport beam, we know that the direction is equivalent to the ligth path increase through the erystal. The reasons of export beam phase chang can be explained from following experiment(see Fig2). By using laser beam (480nm, 300mw/cm^2, E light) illuminate whole light-receiving area of the crystal Ce-SBN (size of crystal:10x12x2mm^3; light-receiving area :10x12mm^2) in order to avoid the appearance of light intensity gradient, and by using then interferometer to observe the refractive index change of this crystal (light in interferometer:480nm, 0.3mw, 1mm, 0 light), we have found the refeactive index is increased:$\Delta n \doteq 1.22 \times 10^{-5}$ (the number of fringe shifted in interferometer is 1,this can be calculated from $2\Delta nd = \lambda$, here d=0.2cm), and the time for shifting is about 30 seconds, In the same interferometer, heating the crystal, we have found that the

130

direction of fringe shife is opposite to that produced by illuminating
the crystal. This shows that the illuminating-resulted fringe shift is
not from thermal effect of light. We thus obtain the result: refractive
index chang is directly produced by light illumination, not necssarily
needing light intensity gradient (generally using interference fringes).
Though we have not known clearly so far the mechanism of refractive index chang produced directly by light illumination in crystal Ce-SBN , it
is existence is proved experimently. So we can image that we are carrying out the two-wavemixing experiment in crystal Ce-SBN. There exist two
index gratings (phase gratings), we call the grating, which has phase
shift with interference fringes, as first phase grating; the grating
without phase shift or overlaooed with interference fringes, the second
phase grating. The refective index fluctuation of the second phase grating is about 1×10^{-5} , lower in fluctuation than the first one (the
fluctuation of the first one is about 1×10^{-4}[1]) and more slowly in
formation. These two grating satisfy the Bragg's condition for the two
incident beams. The first grating makes the energy coupled and transfered, the second one makes the export beam phase shifted. When the intensity of these two beams is of several mv/mm^2, then phase shift is
about π.

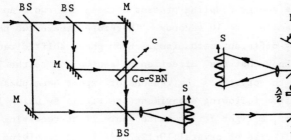

Fig1. The experimental proof of
phase shift of export beam in two
wave maxing in crystal Ce-SBN.

Fig2. The experimental set for
proving the change of refractive
index by uniform illumination in
crystal Ce-SBN.

Reference

1. XU Huaifang, "Chinese Physics Letters" Vol.14 NO.4,pp 260-5 (1987)

EFFECT OF CRYSTAL ANNEALING ON HOLOGRAPHIC RECORDING IN BISMUTH SILICON OXIDE

Pan Shoukui, Ma Jian

Shanghai Institute of Optics and Fine Mechanics,
Academia Sinica, Shanghai, P.R. China

1. Introduction

In this paper we present a report on the effect of crystal annealing and constituent control of starting materials on holographic recording in BSO, which may be helpful to understand photorefractive mechanism of BSO crystal.

2. Experiment method

We used 5N and 4N purity of Bi_2O_3 and SiO_2 as the starting materials and grew some crystals with different concentration of absorption center from the melt with various excess in Bi_2O_3 by induction heating method, which were inclusion and voids free. Then these crystals were annealed at high or low temperature in atmosphere.

3. Experimental results

We learn by Table 1 that the diffraction efficiency of the grown BSO crystal increased with the increase of excess in Bi_2O_3 in the melt. From Fig.1 we know that the transmission of BSO crystals grown from the melt with excess in Bi_2O_3 reduces with the increase of excess in Bi_2O_3 in the melt. From Table 2 we realize that the crystal annealing can raise the diffraction efficiency either at high or low temperature, and the performed at high temperature was available.

4. Discussion

1) On absorption center

Abrahams[1] in 1967, pointed out the Germanium (Ge) occupation probability was $(91.1 \pm 8.4)\%$ in the structure of BGO crystal. This result was also suitable for the isomorphic BSO crystal. Thus the absorption shoulder[2] in absorption spectra of BGO or BSO was interpreted to be due to Ge or Si defects. Recently, R. Oberschmid[3] considered the absorption center in BGO or BSO was M(Ge or Si) vacancy (V_M) or an incorrect occupation of an M site in oxygen tetrahedron by a Bi atom(Bi_M). So BSO crystals grown from the melt with excess in Bi_2O_3 would have more absorption centers than the stoichiometric BSO crystal had, which consistented with the results in Fig.3. The results in Table 1 show the sample having the largest diffraction efficiency is that one having the largest absorption. Therefore, we speculate about the main absorption center in BSO crystal may be the incorrect occupation of an M site in oxygen tetrahedron by a Bi atom (Bi_M), which is ($Bi^{3+}O_4^{2-}$ + h). But we also do not rule out the possibility of V_M as absorption center, because the excess in Bi_2O_3 was equal to the difficiency of GeO_2.

2) On the effect of crystal annealing

In view of the experimental results that the effect of crystal annealing at high temperature was available we supposed that the effect of crystal annealing might be correlated with the formation of the thermal defects related to absorption center V_{Si}. We think the research of thermal deffects in BSO crystal may be helpful to understand the nature of the absorption center.

REFERENCES

[1]   S.C. Abrahams, P.B. Jamiesen, J.L. Bernstein, J. Chem. Phys.,
      47(1967) 4034.
[2]   S.L. Hou, R.B. Lauer, R.E. Aldrich, J. Appl. Phys., 44(1973) 2652.
[3]   R. Oberschmid, Phys. Stat. Sol., (a) 89(1985) 263.

Table 1 Crystal Constituent Diffraction Efficiency (η)
and Refractive Index Change (Δn)

sample No.	Input and diffraction intensity		η	Δ
B_0	Input intensity	4.7 mW	0.29%	6.67×10^{-5}
	Diffraction intensity	0.0136 mW		
B_5	Input intensity	4.7	0.37%	9.59×10^{-5}
	Diffraction intensity	0.0174 mW		
B_{20}	Input intensity	4.7 mW	0.61%	1.11×10^{-4}
	Diffraction intensity	0.0290 mW		

B_0: The melt constituent was stoichiometric (6:1) amounts of Bi_2O_2 and SiO_2

B_5: The melt constituent was 0.5 wt% excess in Bi_2O_3

B_{20}: The melt constituent was 2 wt% excess in Bi_2O_3

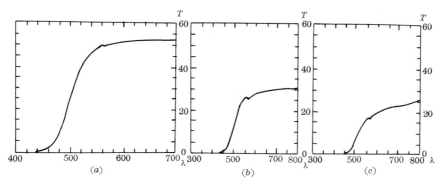

Fig 1 Absorption spectra of BSO crystals grown from the melt with various excess in Bi_2O_3

a) B_0: grown from the melt in stoichoimetric (6:1) amount of Bi_2O_3 and SiO_2

b) B_5: grown from the melt with 0.5wt% excess in Bi_2O_3

c) B_{20}: growm from the melt with 2wt% excess in Bi_2O_3

Table 2 The Effect of Crystal Annealing on Diffraction
Efficiency (η) and Refractive Index Change (Δn)

Sample No. η Δn / Annealing No.	B_0	B_{20}
A_0	6.67×10^{-5} · 0.29%	1.11×10^{-4} · 0.61%
A_1	1.33×10^{-4} · 1.16%	1.34×10^{-4} · 0.87%
A_2	1.18×10^{-4} · 0.91%	1.12×10^{-4} · 0.61%
A_1'	9.43×10^{-5} · 0.38%	9.18×10^{-5} · 0.41%

A_0 : Before annealing

A_1 : Annealing was performed at 750°C 7h.

A_2 : Annealing was performed at 250°C 7h.

A_1': A_1 sample was laid for 72h after annealing.

TWO WAVE COUPLING IN KNbO$_3$ PHOTOREFRACTIVE CRYSTAL

Zhang Heyi, He Xuehua, Chen Erli

Physics Dept. of Peking Univ., Beijing, China

Shen Dezhong, Jiang Daya, Tong Xiaolin

Research Institute of Synthetic Crystals

Hyatt M. Gibbs

Optical Sciences Center of Univ. of Arizona

Four wave mixing was proposed as a means of generating a phase conjugate replica of an optical wave[1]. Photorefractive crystals are the most sensitive material for performing optical phase conjugation. One of the most important photorefractive crystals is reduced KNbO$_3$. It has fast response time. Here we report the study of two wave coupling of reduced KNbO$_3$ crystal. It gives higher gain Γ and higher energy transfer efficiency η than previous paper[2].

An Ar$^+$ laser with it's wave length λ = 5145A used as a light source. The reduced KNbO$_3$ crystal used in this work is prepared by Dr. Shen Dezhong and his co-workers Jiang Daya, Tong Xiaolin et al.

The size of reduced KNbO$_3$ is axbxc = 5.4x5x4mm^3. The absorption coefficient α = 1.5cm^{-1} for λ = 5145A. Fig.1(a) is the intensity ratio β dependence of gain Γ and Fig.2(b) is the similar figure which has published by Gunter[2]. One can find that the gain Γ which showed in Fig.1(a) is higher than Günter's data. The β is defined as

$$\beta = I_{p_o} / I_{s_o}$$

The energy transfer efficiency in two wave coupling is defined as

$$\eta = (P'_p - P''_p) / P'_p$$

P'_p is the laser power of pumping beam P_{p_o} passed through the crystal without two wave coupling that means P signal beam power Ps_o = 0.

136

P''_p is the laser power passed through the crystal with two wave coupling.

Fig.2 showed the fringe dependence of η and $C = P'_p/P''_p$. It is showed that 90% pumping beam energy passed through the crystal transfered to the signal beam by two wave coupling for $\Lambda = 0.75$ μm, for $\Lambda = 1.4$ μm $\eta = 82\%$. This means that only 10% pumping energy is left.

ACKNOWLEDGMENT

The authors acknowledge the helpful discussions with professor Kelvin Wagner and Dr. Dennis.

REFERENCES
[1] Hellwarth, R.W., J.O.S.A., 67, 1 (1977).
[2] P. Gunter, Phys. Rep., 93, 199 (1982).

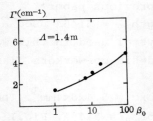

Fig 1 (a) Intensity ratio β_o dependence of gain Γ

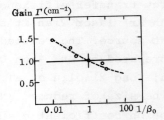

Fig 1 (b) Gain Γ versus intensity ratio β_o of writing beams for $KNbO_3 : Fe^{2+}$ $\Lambda = 2\mu m$ and $I_o = 1W/cm^2$

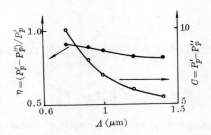

Fig 2 Groting period Λ dependence of η and C

PHOTOREFRACTIVE EFFECTS IN Nd-DOPED FERROELECTRIC $(K_xNa_{1-x})_{0.4}-(Sr_yBa_{1-y})_{0.8}Nb_2O_6$ SINGLE CRYSTAL

Huang Zhengqi, Li Weiliang, and Xu Yuhuan
(Laser and Spectroscopy Lab., and Dielectrics Research Lab.,
Department of Physics, Zhongshan University, Guangzhou, China)
Chen Huanchu
(Institute of Crystal Materials, Shandong University, Jinan, China)

ABSTRACT

The experiments for photorefractive effect has been performed in $(K_{0.2}Na_{0.2})(Sr_{0.6}Ba_{0.17}Nd_{0.02})Nb_2O_6$ crystal. High photorefractive sensitivity in this ferroelectric crystal shows that niobates with tungsten-bronze structure should be a kind of available photorefractive material.

INTRODUCTION

The ferroelectric single crystal series $(K_xNa_{1-x})_{0.4}(Sr_yBa_{1-y})_{0.8}Nb_2O_6$ where $0.50 \leq x \leq 0.75$ and $0.30 \leq y \leq 0.90$ (to be abbreviate to KNSBN) with tetragonal tungsten-bronze type structure have been developed and studied by Xu, et al.[1-7]. These crystals possess very good ferroelectric, pyroelectric, piezoelectric, and linear electro-optical properties. They show strong non-linear optical effect as well.

EXPERIMENTS

KNSBN crystals are transparent for the light with wave-length from 4000Å to 5.6 μm and the indexes of refraction, n_0 and n_e, are 2.3358 and 2.2747 respectively for 5610Å (see Fig.1). But there are four strong absorption peaks about 5200Å, 5800Å, 7400Å, and 8000Å in a Nd-doped KNSBN crystal with chemical formula of $(K_{0.2}Na_{0.2})(Sr_{0.6}Ba_{0.17}Nd_{0.02})Nb_2O_6$.

The experiment for photorefractive effect has been performed in Nd-doped KNSBN crystal. The crystal sample was exposed by a beam of light (λ=5780Å) with variable intensities I_2 from a tunable dye laser, while the rotation angle of the polarization of another beam with intensity $I_1(I_1 \ll I_2)$, which passed the sample, was measured, and then the photoinduced birefringence change was determined. The experimental results in Nd-KNSBN and those in Ce-SBN, for comparison, are shown in fig.2. The photorefractive sensitivities in two crystals are almost equal, but are larger than that in Fe-KNbO3[9].

138

CONCLUSION

Ferroelectric niobates crystals with tungsten-bronze structure possess large photorefractive effect and can be grown more easy than BaTiO$_3$ crystal. For tolerance in tungsten-bronze structure is larger than that in BaTiO$_3$, and therefore the former should be a kind of available photorefractive material.[10]

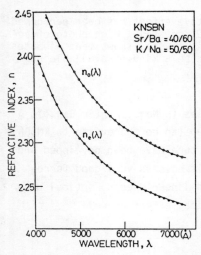

Fig.1 The wavelength dependence of refractive indexes of KNSBN single crystal

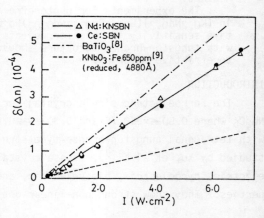

Fig.2 The intensity dependence of the photoinduced birefringence change at 5780Å in Nd:KNSBN, Ce:SBN, and BaTiO$_3$

References
[1] Chen Huanchu and Xu Yuhuan, Physics, 10[12],729,(1931).
[2] Chen Huanchu and Xu Yuhuan, Journal Chinese Silicate Soc. 10[4],406,(1982).
[3] Xu Yuhuan and Chen Huanchu, Acta Physica Sinica, 32[6], 705,(1983).
[4] Xu Yuhuan, Chen Huanchu, and L.E. Cross, Ferroelectrics, 54,123,(1984).
[5] Xu Yuhuan, Chen Huanchu, and L.E. Cross, Ferroelectrics Letters, 2,189,(1984).
[6] Xu Yuhuan, Chen Huanchu, and S.T. Liu, Japanese J. Appl. Phys., 24 (suppl.), 236,(1985).
[7] Xu Yuhuan et al., Chinese Physics, 5[4],870(1985).
[8] Xu Yuhuan and Li Weiliang (unpublished experiments).
[9] Günter,P., Physics Reports, 93,228(1982).
[10] Rodriguez,J., et al., Applied Optics, 26[9],1732(1987)

* This work is supported in part by Ultrafast Laser Spectroscopy Laboratory, Zhongshan University

HIGH PRESSURE RAMAN SPECTRA AND THE EFFECT OF PRESSURE TO THE FERROELASTIC PHASE TRANSITION IN LnP_5O_{15}[*]

G.X.Lan, Q.R.Cai, S.F.Hu, H.F.Wang and G.Y.Hong[**]

Department of Physics, Nankai University, Tianjin, P.R.China

Lanthanide pentaphates(LnP_5O_{14}, Ln=La,Ce,Pr,Nd,Sm,Eu,Gd,Tb) are ferro-elastic crystals which undergo the temperature-induced C_{2h}^5-D_{2h}^7 transition[1,2]. In this work, the PrP_5O_{14} has been investigated by high-pressure Raman spectra. The pressure dependence of observed Raman peaks are shown in Fig.1. These peaks show discontinuous changes at pressure near 75 kbar. Thus it seems probable that a structural phase transition induced by pressure occurs at this pressure.

The high pressure Raman spectra of soft optic modes of all eight kind crystals of LnP_5O_{14} were studied up to 20 kbar. These modes show a large initial increase in frequency with pressure and flattening at higher pressure, while the Grüneisen parameters decrease with pressure. The results of CeP_5O_{14} are shown in Fig.2 by solid lines. The pressure dependence of the soft optic mode frequency and the Grüneisen parameter were calculated from the mean field theory and the results are shown in Fig.2 by dotted lines. It shows that the soft optic modes are lattice vibration involving all atoms of unit cell with strong anharmonicity, which decrease with increasing pressure and the atomic number of Ln^{3+}.

Pressure effect to the ferroelastic phase transition of LaP_5O_{14}, CeP_5O_{14}, PrP_5O_{14} and SmP_5O_{14} has been investigated up to 2 kbar. The mean field theory gives the pressure dependence of T_0 by:

$$(dT_0/dp) = (d\omega/\beta\omega dp)(T_0-T_R)$$

where T_R is the room temperature, $(d\omega/\omega dp)$ is its initial value at room temperature and β and T_0 can be found in ref.[3]. Both experimental and theoretical results of dT_0/dp are listed in table 1.

The ferroelastic phase transition in LnP_5O_{14} are induced by the coupling between the soft optic mode and the acoustic mode, which makes the acoustic mode frequency to zero[2]. We suggest that this coupling is related to the anharmonicity of soft optic modes , which decides the strength

140

of coupling. The anharmonicity of soft optic modes decreases with pressure and lanthanide contraction, and so the coupling decreases and the ferroelastic phase transition temperature rises with pressure.

Reference

[1]   Fox,D.L., Scott,L.F.;Solid State Commun.,18(1976)111.

[2]   Errandones,G.;Phys.Rev(B),21(1980)5221.

[3]   Chen,T., Hong,G.Y.;Acta Physica Sinica,35(1986),1521.

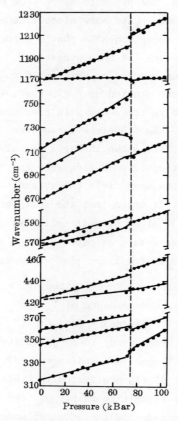

Fig. 1. Pressure dependence of the phonon
frequencies in PrP_5O_{14}

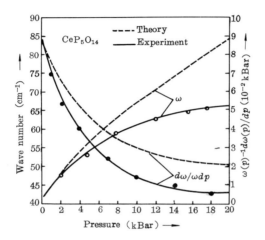

Fig 2 Pressure dependence of ω and $d\omega/\omega dp$ values in CeP_5O_{14}

Table 1.　Experimental and theoretical values of dTc/dp

		LaP_5O_{14}	CeP_5O_{14}	PrP_5O_{14}	SmP_5O_{14}
dTc/dp	Theoretical	31.5	24.9	20.7	16.3
($^\circ$C/kbar)	Experimental	42	41	25	33

TIME-DELAY FOUR-WAVE MIXING WITH INCOHERENT LIGHT IN ABSORPTION BANDS TREATED AS A MULTI-LEVEL SYSTEM

Xin Mi, Ruihua Zhang, Haitan Zhou and Peixian Ye
Institute of Physics, Chinese Academy of Sciences

Abstract

A homogeneous broadening multi-level theory of time-delay four-wave mixing with incoherent light is developed. It is applied successfully to analyse our experiment in the absorption band $^4T_2$ of ruby which is failed to be interpreted with two-level theory.

1. Introduction

Time delay four-wave mixing with incoherent light (TDFWM-IL),developed in the recent years, has attracted much attention. The possibility was proposed that the ultrashort dephasing time (shorter then 1 ps) in condensed matters could be measured by this spectroscopic technique with nanosecond pulse laser or CW laser. Some efforts in this line have been made experimentally. However, the theory used to analyse experiments up to now was based on the two-level model. Here, we develop a homogeneous broadening multi-level theory of TDFWM-IL, and present our experimental results which demostrate that this theory is more suitable than two-level theory for absorption bands that consist of a series of vibration structures with homogeneous broadening.

2. Theory

The basic geometry of TDFWM-IL is the following.Two beams coming from the same broadband laser have a relative delay time τ with respective to each other and a small angle is between them. They interact in the material with a third beam propagating along a direction which satisfies the phase-matching condition. Then, the signal of TDFWM-IL occurs and its intensity $I_S(\tau)$ is a function of τ. For the absorption bands mentioned above, we have developed a homogeneous broadening multi-level model and deduced approximately a formula for the τ-dependence of a signal intensity as the following

$$I_S(\tau) \propto \left[\int_{-\infty}^{\infty} d\omega \frac{g(\omega_i) S_0(\omega,\omega_m)}{(\omega_i-\omega)^2 + 1/T_{2i}^2} \cos\omega\tau\right]^2 + \left[\int_{-\infty}^{\infty} d\omega \frac{g(\omega_i) S_0(\omega,\omega_m)}{(\omega_i-\omega)^2 + 1/T_{2i}^2} \sin\omega\tau\right]^2, \quad (1)$$

where, ω_i and T_{2i}-resonant frequency and dephasing time of the i-th transition which has a weight factor $g(\omega_i)$; ω_m-central frequency of the power spectrum $S_0(\omega,\omega_m)$ of the broadband laser; summation in Eq.(1) is run all of the individual transitions including in the system. The main conclusions of our theory are as follows: (1) The τ-dependence of $I_S(\tau)$ is determined only by the second-order coherence of the laser field, and the fouth-order correction appears as a constant background which can be ignored usually. (2) $I_S(\tau)$ is a symmetric function of . Both of these two conclusions are the same as those of two-level theory. (3) When the levels are broadened enough and become an unresolved continueous band, the nomalized $I_S(\tau)$ can be described with the same formula as in two-level theory, but the Lorentzian profile in the formula has to be replaced by the total absorption spectrum of the band, $F(\omega) = \sum_i g(\omega_i)/[(\omega_i - \omega)^2 + 1/T_{2i}^2]$. Using the formula of two-level theory we can determine not the dephasing time of individual level-transitions but the effective dephasing time which is the inverse of the total absorption width of the band.

3. Experiment and Conclusion

We have studied the TDFWM-IL in the absorption band $^4T_2$ of ruby experimentally using a broadband pulse dye laser with a bandwidth about 50 to 120 Å. The τ-dependence of $I_S(\tau)$ was measured at the temperature range from 80 to 300K and several spectral positions in the band. The main results are in the following : (1) All of the measured $I_S(\tau)$ are symmetric functions of τ, which are consistent with the fact that the band $^4T_2$ consists of a series of vibration structures with large homogeneous broadening at the above temperature range. (2)The dephasing time obtained by using the formula of two-level theory to fit the experiment is independent with temprature at 80 to 300 K. This result is failed to be interpretated by two-level theory because the dephasing time of individual transtions can not be a constant at so large a range of temperature. But if we understand the dephasing time obtained as an effective one, it has to be independent with temperature because the total absorption spectrum of band $^4T_2$ varies not so obviously at that temperature range. This result is consistent with the conclution of our multi-level theory. (3) The effective dephasing time obtained varies with the spectral position what we used in the measurement. This fact can also be understood with our theory because the corresponding wavelength of excitation are in the different position of the total absorption spectrum.

PULSED LASER INDUCED DISLOCATION STRUCTURE
IN LITHIUM FLUORIDE SINGLE CRYSTALS

Zhou Jiang, Qiao Jingwen and Deng Peizhen

Shanghai Institute of Optics and Fine Mechanics, Academia Sinica
P. O. Box 8211, Shanghai, P. R. China

ABSTRACT The dislocation structure and damage forms
due to the focused pulsed laser irradiation on the
(100) surface of LiF single crystals are studied.

INTRODUCTION

As a structure defect of crystal, dislocation affects the mechanical properties of crystal. Therefore, it is beneficial to make
clear the rule of dislocation generation and movement during laser
irradiation for further understanding the physical processes of laser
damage.

EXPERIMENTS AND RESULTS

A TEA CO_2-laser and a Q-switched Nd:YAG laser are employed in
the experiments. The pulse duration for CO_2-laser is 100ns, and 15ns
for YAG laser, the energy of pulse is about 0.1J for CO_2-laser, and
several milli-joules for YAG laser. The power density of irradiation
is controled by a set of attenuators. The dislocations are revealed
by chemical etching method.

When CO_2-laser irradiates on the (100) surface of LiF, and its
incident power density is above 0.8 GW/cm^2, the characteristic dislocation distribution appears around the irradiated site (Fig.1). This
characteristic dislocation structure is a square high dislocation
density zone which consists of <100> directions glide bands, outside
the square zone, there are glide bands along <110> directions. If
the power density remains constant, as increase of irradiated pulse
number, the square zone widens, and microcracks along <100> generate
in the inner. When YAG laser with a power density above 0.3 GW/cm^2

irradiates on the (100) surface of LiF, the dislocation distribute around some irradiated site, but they do not glide, instead of constitute a ring pattern (Fig.2). The dislocation density is the order of 10^7–10^8/cm^2. With increase of irradiated pulse number, net micricracks along <100> directions appear inside the ring. In case of power density above 2.2 GW/cm^2, the breakdown ocuurs, the dislocation distribution around the damage site is an irrigular square.

DISCUSSION AND CONCLUSION

The characteristic dislocation structure induced by CO_2-laser irradiation is mainly caused by thermal elastical stress due to the surface layer heahing. In fact, as the irradiation intensity in the focus has a Gaussian distribution, then the temperature at the surface (in the linear absorption approximation) will obey the same relation. Ignoring the reflection of surface, we can get the expression of temperature at the sample surface,

$$T = \frac{2 \beta E}{c \rho \pi^2 W^2} \exp (-2 r^2/ W^2) \tag{1}$$

where E is the pulse energy, c is the specific heat, ρ is the density, β is the absorption coefficient of LiF single crystal. Expression of the surface stress can got based on the theory of thermal stress and the experimental parameters,

$$\sigma = \sigma_o \exp (-2 r^2/ W^2) \tag{2}$$

where σ_o = 8.9 x 10^3 Mpa, W = 300µm is the Gaussian radius in the focus of CO_2-laser, r is the distance from the axis. In LiF single crystals, the stresses under which the dislocations homogeneously nucleate is $\sigma' \approx G/30 = 1.8$ x 10^3 Mpa (G = 55.12 x 10^3 Mpa is the shear modulus for LiF). Therefore, the dislocations nucleate in a zone which radius is about 270µm, then they move in the {110}<110> slip system and multiply in duration of pulse, finally form the characteristic dislocation distribution observed in the experiments.

The estimations based on the experimental parameters and the theory of thermal stress show that the dislocations can not be induced by YAG laser irradiation in perfect LiF single crystals. The characteristic dislocation structure which has been observed in the experiments is caused by inhomogeneous absorption of radiation due to impurity such as inclusion. This inhomogeneous absorption make crystal burn and evaporate locally, whereby, plasma forms. The observed dislocation distribution is completed by thermal shock due to the plasma torch initiated by material evaporation.

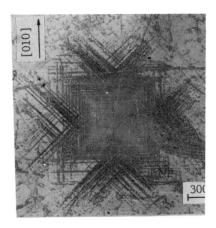 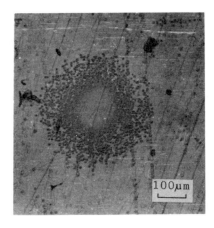

Fig.1 Dislocation structure on LiF surface induced by CO_2-laser.

Fig.2 Dislocation distribution around the site irradiated by YAG laser.

NONCLASSICAL RADIATION FROM SINGLE-ATOM OSCILLATORS

Herbert Walther

Sektion Physik der Universität München and
Max-Planck-Institut für Quantenoptik,
Garching, FRG

1. Abstract

Single atom oscillators emit nonclassical radiation under suitable conditions. Two examples will be discussed: the one-atom
maser and a single ion stored in a quadrupole trap.

2. Introduction

Modern methods of laser spectroscopy allow us to study single atoms
or ions in an unperturbed environment. This has opened up interesting
new experiments, among them the detailed study of radiation-atom coupling. In the following, two experiments of this type are reviewed: the
single-atom maser and the resonance fluorescence of a single stored
ion.

3. The one-atom maser and cavity quantum electrodynamics

The simplest and most fundamental system for studying radiation-
matter coupling is a single two-level atom interacting with a single
mode of an electromagnetic field in a cavity. It received a great deal
of attention shortly after the maser was invented, but at that time,
the problem was of purely academic interest as the matrix elements describing the radiation-atom interaction are so small, so that the field
of a single photon is not sufficient to lead to an atom-field evolution
time shorter than other characteristic times of the system, such as the
excited state lifetime, the time of flight of the atom through the cavity mode damping time. It was therefore not possible to test experimentally the fundamental theories of radiation-matter interaction which
predicted, inter alia, (a) a modification of the spontaneous emission
rate of a single atom in a resonan cavity, (b) oscillatory energy exchange between a single atom and the cavity mode, 3 and (c) the dis-
appearance and quantum revival of optical (Rabi) nutation induced in a
single atom by a resonant field.

The situation has changed drastically in the last few years after
frequency-tunable lasers allowed to excite large populations of highly
excited atomic states characterized by a high main quantum number n of

the valence electron. These Rydberg states are very suitable for observing quantum effects in radiation-atom coupling for three reasons. First, the states are very strongly coupled to the radiation field (the induced transition rates between neighbouring levels scale as n^4); second, transitions are in the millimetre wave region, so that low-order mode cavities can be made large enough to allow rather long interaction times; finally, Rydberg states have relatively long lifetimes with respect to spontaneous decay (for reviews see Refs. 1 and 2.).

The strong coupling of Rydberg states to radiation resonant with transitions to neighbouring levels can be understood in terms of the correspondence principle: with increasing n the classical evolution frequency of the highly excited electron becomes identical with the transition frequency to the neighbouring level; the atom therefore correspnds to a large dipole oscillating with the resonance frequency.

In order to understand the modification of the spontaneous emission rate in an external cavity, we have to remember that in quantum electrodynamics this rate is determined by the density of modes of the electromagnetic field at the atomic transition frequency ω_o which in turn depends on the square of the frequency. If the atom is not in free space, but in a resonant cavity, the continuum of modes is changed into a spectrum of discrete modes of which one may be in resonance with the atom. The spontaneous decay rate of the atom in the cavity γ_c will then be enhanced in relation to that in free space γ_f by a factor given by the ratio of the corresponding mode densities:

$$\gamma_c / \gamma_f = \rho_c(\omega_o)/ \rho_f(\omega_o) = Q\lambda_o^3 /4\pi^2 V_c$$

where V_c is the volume of the cavity and Q is the quality factor of the cavity which expresses the sharpness of the mode. For low-order cavities in the microwave region $V_c \sim \lambda_o^3$ which means that the spontaneous emission rate is increased by roughly a factor of Q. However, if the cavity is mistuned, the decay rate will decrease. In this case, the atom cannot emit a photon, since the cavity is not able to accept it, and therefore the energy has to stay with the atom.

Recently, a number of experiments have been made with Rydberg atoms to demonstrate this enhancement and inhibition of spontaneous decay in external cavities or cavity-like structures (for the most recent experiment see Ref. 3). More subtle effects due to the change of the mode density can also be expected: radiation corrections such as the Lamb shift and the anomalous magnetic dipole moment of the electron are modified with respect to the free space value[4]). Changes are of the order of present experimental accuracy. Roughly speaking, one can say that such effects are determined by a change of virtual transitions and not by real transitions as in the case of spontaneous decay.

In the following, attention is focused on discussing the one-atom maser in which the idealized case of a two level atom interacting with a single mode of a radiation field is realized; the theory of this system was treated by Jaynes and Cummings[5] many years ago. We concentrate on the dynamics of the atom-field interaction predicted by this theory. Some of the features are explicitly a consequence of the quantum nature of the electromagnetic field: the statistical and discrete nature of the photon field leads to new dynamic characteristics such as collapse and revivals in the Rabi nutation.

The experimental setup of the one-atom maser is shown in Fig. 1. Rubidium atoms are excited with frequency-doubled light of a dye laser to the $63P_{3/2}$ Rydberg level and are then injected into a superconducting cavity. On emerging they are monitored using field ionization which can be performed state-selectively by choosing the proper field strength. The cavity is tuned by squeezing it with piezo electric elements. The flux of atoms is very low so that their average number in the cavity at a time is usually much less than unity. In most of the experiments[6,8] the transition $63P_{3/2}$ - $61d_{5/2}$ with a frequency of 21.456 GHz has been investigated. When the cavity is tuned in resonance to this transition, the number of atoms in the upper state does indeed decrease owing to enhanced spontaneous emission with very low atomic-beam flux, the cavity of the single-atom maser contains essentially thermal photons only, whose number is a random quantity obeying Bose-Einstein statistics. At high atomic-beam fluxes, the atoms deposit energy in the cavity and the maser reaches the threshold so that the number of photons stored in the cavity increases and their statistics changes.

For a coherent field the probability distribution is Poissonian resulting in a dephasing of the Rabi oscillations, and therefore the envelope of $P_e(t)$ collapses. After the collapse, $P_e(t)$ starts oscillating again in a very complex way. Such changes recur periodically, the time interval being proportional to $n^{1/2}$. Both collapse and revival in the coherent state are pure quantum features without any classical counterpart[7].

Collapse and revival also occur in the case of a thermal field where the spread in the photon number is far larger than for a coherent state, therefore the collapse time is much shorter. In addition, revivals overlap completely and interfere, producing a very irregular time evolution. On the other hand, a classical thermal field represented by an exponential distribution of the intensity shows collapse, but no revival at all. From this it follows that revivals are pure quantum features of the thermal radiation field, wherease the collapse is less clear-cut as a quantum effect[7].

The above-mentioned effects have been demonstrated experimentally, using a Fizeau velocity selector to vary the interaction time (see Fig. 1). Fugure 2 shows a series of measurements obtained with the single-atom maser[7], where the probability $P_e(t)$ of finding the atom in the upper maser level is plotted against interaction time for increasing atomic flux N. The strong variation of $P_e(t)$ for interaction times between 50 and 80 μs disappears for larger N and a revival shows up for N = 3000 s^{-1} for interaction times larger than 140 μs. The average photon number in the cavity varies between 2.5 and 5, about 2 photons being due to the black-body field in the cavity corresponding to a temperature of 2.5 K.

There are two approaches to the quantum theory of the one-atom maser. Filipowicz et al.[9] use a microscopic approach to describe the device while Lugiato et al.[10] use the standard macroscopic quantum laser theory and arrive at the same steady-state photon number distribution. The special features of the micromaser are not emphasized in standard laser theory because the broadening due to spontaneous decay obscures the Rabi cycling of the atoms. When similar averages in the microscopic theory associated with inhomogeneous broadening are performed, equivalent results are obtained. Both theoretical approaches predict that the distribution of the photons in the maser cavity depends on the interaction time of the atoms in the cavity. It is mostly sub-Poissonian, and with a high Q cavity, $Q \approx 5 \times 10^{10}$, the photon number reaches a steady value i.e. a Fock state[11]. To achieve such a state experimentally, two conditions have to be met. The first concerns the temperature; thermal photons have to be suppressed because they not only induce statistical decay but also result in a superposition of number states. The second is that losses of photons stored in the cavity can be neglected which means that the Q value has to be higher than 10^{10}. Both conditions can now be realized and the predicted distributions demonstrated.

New insight has been obtained into the statistical properties of ordinary masers and lasers where, in contrast to the single-atom maser, Poisson statistics is observed. The reason for this is that in the cavities of macromasers and lasers there is much stronger damping present than in the micromaser; in addition, atomic or molecular transitions usually show stronger damping than the Rydberg states. Furthermore, the selected velocity of the atoms used in connection with the collapse and revival measurements[8] in the micromaser leads to fixed interaction time in the cavity; this also helps to reduce the photon number fluctuations since the photon exchange between the atom and cavity field can be exactly controlled. The smallest fluctuations are achieved when the atoms leave the cavity again in the upper state. Of course, it is necessary that energy be deposited in the cavity in order to maintain maser oscillation, but the losses are very small with a high-Q-cavity and P_e for atoms leaving the cavity can be made very close to unity.

There are other interesting aspects of Rydberg masers which can only be briefly discussed here. Recently, a two-photon maser was realized by Brune et al.[12]. The two-photon transition was chosen such that there is an intermediate level nearly halfway between the upper and lower maser levels, thus enhancing the transition amplitude. In such a device new features not present in one-photon masers can be observed, e.g. delayed start-up time at threshold. Unlike the one-photon maser, which behaves at threshold similarly to a 2nd-order phase transition, the two-photon maser is analogous to a 1st-order phase transition. At this point we should also mention that it was pointed out that the micromaser can be used to investigate aspects of chaos and problems of measurement theory[11,13].

4. Resonance fluorescence of stored ions

Another problem in radiation-matter interaction which has received a lot of interest is resonance fluorescence. Unlike the problem discussed in connection with the one-atom maser, here the atom decays under the influence of the vacuum fluctuations of free space and the emitted photons disappear and cannot be reabsorbed.

Most current work is concerned with theoretically and experimentally determining the spectrum of the fluorescent light radiated by a two-level atom driven by an intense monochromatic field. This is the situation that gives rise to a dynamic Stark effect in which, for sufficiently strong fields, it is found that the spectrum of the scattered light splits into three peaks: a central peak, centred on the driving field frequency with a width $\gamma/2$ ($1/\gamma$ being the Einstein A coefficient) and a height three times that of two symmetrically placed sidebands, each of width $3\gamma/4$ and displaced from the central peak by the Rabi frequency. In addition, there is a delta-function (coherent) contribution, also positioned at the driving frequency. In the limit of strong driving fields, the energy carried by this last contribution is negligible in relation to the three-peak contribution. This result was first predicted by Mollow and has now been confirmed experimentally (for a review see Refs. 14 and 15).

However, it is not only the spectral property of the fluorescent light that has come under investigation. Examination of the intensity correlation of the scattered field in the basic two-level atom has also attracted much attention since fluorescent light exhibits interesting statistical propertied, especially when there is only a single atom at a time interacting with the laser beam. Under these conditions the phenomenon of photon-antibunching can be observed[14,15]. The single-atom condition cannot easily be fulfilled if experiments are made on neutral atoms, whereas the new techniques of laser spectroscopy of single ions in a radio-frequency trap are very suitable for this purpose, as has recently been demonstrated by Diedrich and Walther[16].

The Paul trap used in the experiment is shown in Fig.3. To investigate the photon statistics, the second-order correlation function (intensity correlation) was measured in a Hanbury-Brown and Twiss Experiment. The intensity correlation is proportional to the probability of detecting a second photon at a time after a first one. For thermal and non-coherent light this probability has a maximum for $\tau = 0$ and decreases for larger τ. The behaviour is called bunching. The intensity correlation for coherent light is independent of τ. Quantized fields may show additional behaviour: the intensity correlation can have a minimum at $\tau = 0$, a behaviour known as antibunching. Such a field is produced by the single stored ion in the following way: after a photon is emitted, the trapped ion returns to the ground state; before the next photon can be emitted, the ion has to be excited again. This happens through Rabi nutation in the external laser field. On the average, a time of half a Rabi period has to elapse until another photon can be observed. The probability of two photons being emitted in a short time interval is therefore zero.

The results for the intensity correlation of a single stored ion are shown in Fig. 4. Plotted is $g_I^{(2)}(\tau)$, which is defined by $g_I^{(2)}(\tau)= < I(t)I(t+\tau) > / <I(t)>^2$. Owing to a time delay in one of the signal channels the intensity correlation $g_I^{(2)}(\tau)$ could also be measured for negative . The laser intensity decreases from a to d and therefore the average time interval in which a second photon follows a first one increases.

The fluorescent light of a single stored ion has another interesting property: the fluctuations of the photon number recorded in a small time interval δt is narrower than that expected for a Poisson distribution, i.e. the variance is smaller than the mean value of the photon number and we again find sub-Poissonian statistics. The reason is that the single ion can only emit a single photon at a time and fluctuations only occur because of the finite detection probability.

Another interesting phenomenon recently observed by Diedrich et al.[17] in a Paul trap concerned clouds of 2 to about 50 simultaneously stored Mg^+ ions. Two phases of the ions could be clearly distinguished by their excitation spectra, one corresponding to a cloud-like state, in which the ions move randomly, and the other to a crystalline state, in which they are fixed in a regular array (crystals of 2, 3, 4, and 7 ions are shown in Fig. 5). The latter state is determined by the pseudopotential of the trap and the Coulomb repulsion of the ions. Such crystals are formed when laser cooling reduces the kinetic energy of the ions to a value which is much smaller than the potential energy of the ions in the trap.

The ion crystals in the trap represent a very neat model system and it is of considerable interest to investigate how the conditions depend on the stored ion number: Certain ion configurations can be expected to be more stable than others and will therefore need less laser cooling when they are formed. Furthermore, the vibrational modes of the crystalline structure can also be investigated as well as the dynamics of the crystallization and evaporation process[18].

References

1. Haroche, S., Raimond, J.M. in Advances in Atomic and Molecular Physics, Vol. 20, eds. D. Bates and B. Bederson, 347, Academic Press 1985
2. Gallas, J.A.C., Leuchs, G., Walther, H. Figger, H. in Advances in Atomic and Molecular Physics. Vol. 20, eds. D. Bates and B. Bederson, 413, Academic Press 1985
3. Ihe, W. Anderson, A., Hinds, E.A., Meschede, D., Moi, L., Haroche, S., Phys. Rev. Lett. 58, 666 (1987)
4. Barton, D., Proc. Roy. Soc., London A 410, 147 and 175, (1987)
5. Jaynes E.T., Cummings, F.W., Proc. IEEE 51, 89 (1963)
6. Meschede, D., Walther, H., Muller, G., Phys. Rev. Lett. 54, 551 (1985)
7. Yoo, H.I., Eberly, J. H., Phys. Rev., 118 (1985) 239 and Knight, P.L., Radmore, P.M., Phys. Lett. 90A, 342 (1982)
8. Rempe, G., Walther, H., Klein, N., Phys. Rev. Lett. 58, 353 (1987)
9. Filipowicz, P., Javanainen, J., Meystre, P., Phys. Rev. A 34, 3077 (1986)
10. Lugiato, L.A., Scully, M.O., Walther, H., Phys. Rev. A 36, 740 (1987)
11. Krause, J., Scully, M.O., Walther, H., Phys. Rev. A 36, 4547 (1987)
12. Brune, M., Raimond, J. M., Goy, P., Davidovich, L., Haroche, S., Phys. Rev. Lett. 59, 1988 (1987)
13. Meystre, P., Opt. Lett. 12, 669 (1987) and Meystre, P., Wright, E. M., Phys. Ref. A to be published
14 Cohen-Tannoudji, C., in Frontiers in Laser-Spectroscopy, ed. by R.

156

Balian, S. Haroche, S. Liberman, North Holland, 1977
15. Cresser, J.D., Häger, J., Leuchs, G., Rateike, M., Walther, H. in
    Dissipative Systems in Quantum Optics, ed. by R. Bonifacio,
    Topics in Current Physics, Vol. 27, 21, Springer Verlag 1982
16. Diedrich, F., Walther, H., Phys. Rev. Lett. 58, 203 (1987)
17. Diedrich, F., Peik, E., Chen, J.M., Quint, W., Walther, H.,
    Phys. Rev. Lett. 59, 2931 (1987)
18. Blümel, R., Chen, J.M., Peik, E., Quint, W., Schleich, W., Shen,
    Y.R., Walther, H., Nature 334, 309 (1988).

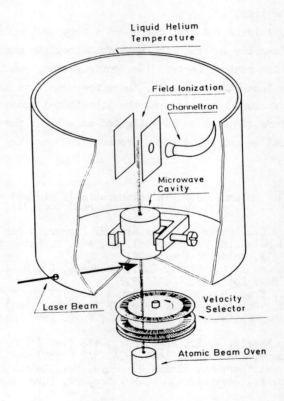

Fig. 1: Scheme of the single-atom maser for measuring
        quantum collapse and revivals (see Ref. 8).

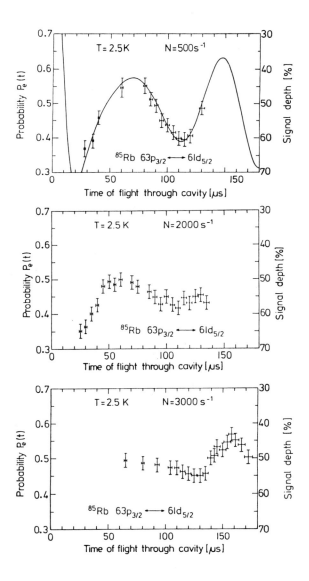

Fig. 2: Quantum collapse and revival in the one-atom maser.
Plotted is the probability $P_e(t)$ of finding the
atom in the upper maser level for different fluxes
N of the atomic beam (see Ref. 8).

158

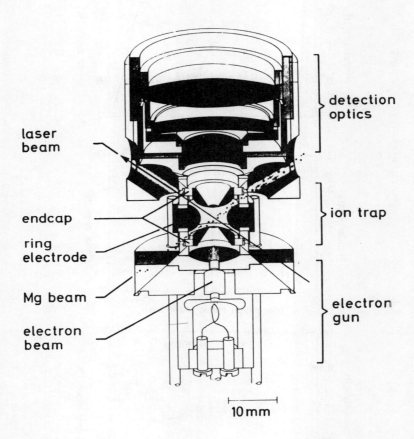

laser beam

detection optics

endcap

ring electrode

Mg beam

electron beam

ion trap

electron gun

10 mm

Fig. 3: Scheme of the Paul trap used for the
experiments (for details see Ref.16).

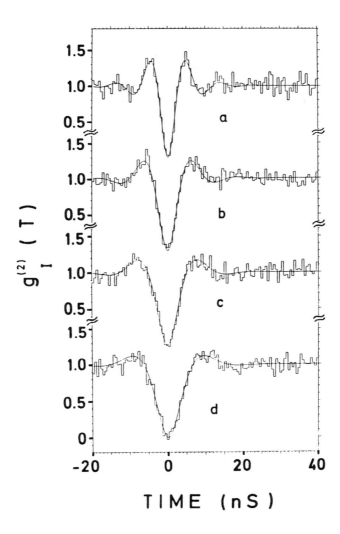

Fig. 4: Results for the intensity correlation. Antibunching for a single ion for different laser intensities, increasing from bottom to top (see Ref. 16 details).

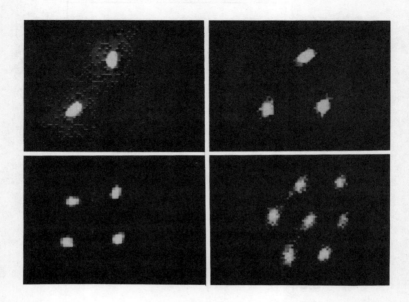

Fig. 5: Ion crystals in a Paul trap, observed
by a photon – counting imaging system
(see Ref. 17 and 18 for details).

Laser Spectroscopic Studies of Molecules in Highly Excited
Vibrational State

Tadao Shimizu, Ken'ichi Nakagawa, Yukari Matsuo, and Takahiro Kuga

Department of Physics, University of Tokyo, Hongo, Bunkyo-ku,
Tokyo 113, Japan

Abstract
      Highly sensitive visible and infrared spectrometers based on
White-type cell have been constructed and successfully employed for
studies of steady state and dynamic properties of molecules in high
vibrational states. Process of localization of molecular vibration
into some particular bond has been investigated as a function of
vibrational quantum number.

1. Introduction

      Highly excited vibrational states of molecules have been
attracting a great interest in recent years. Their evergy level
structures and dynamic properties are directly concerned with
mechanisms of multiphoton dissociation, chemical reaction, relaxation
in lasing medium, and atomic and molecular processes in interstellar
space and in planetary atomosphere. Stationary and dynamic behaviors
of high vibrational states may provide an idea on the localization of
molecular vibration into some particular bond.

      Very weak transitions with changes of several vibrational quanta
are expected to be observed in the visible and near infrared regions.
In the 1930's absorption spectra of H_2O and NH_3 in the visible region
were systematically investigated by Mecke et al.[1-3] and Badger[4],
respectively. Because the observation was made at high pressure,
spectral lines were broad and the rotational structure was rarely
resolved. Rotational analysis of the 645 nm band of NH_3 was atempted
by McBride and Nicholls.[5]

      Recent developments of tunable dye lasers and diode lasers have
allowed us to construct highly sensitive and high resolution
spectrometers in visible and infrared regions. H_2O molecule was
studied by an optoacoustic technique in the frequency range of 16750-
17030 cm^{-1} [6] and the 645 nm band of NH_3 was also studied with

lasers.[7,8] Only a few works with diode lasers have been reported.[9,10] In the present paper we report construction of generally applicable, sensitive and high-resolution laser spectrometers with dye and diode lasers and analysis of observed high-overtone spectra of NH_3 on the basis of a local mode theory.

2. Experimental

When noise and instabilities in light sources are absent, observation of linear absorption enables us to construct the most sensitive and reliable spectrometer. The total amount of photon energy absorbed by molecules can be detected in this method, while only a fraction of absorbed photon energy is utilized in other methods such as fluorescence, polarization, and optoacoustic spectroscopy. Moreover in these cases the signal intensity does not correctly represent the strength of molecular absorption.

Since excess noise in lasers decreases rapidly with an increase in the frequency, an employment of high-frequency modulation and phase sensitive detection attains the high sensitivity in laser spectrometers. the optimum absorption length may be given by

$$l_{opt} = 2/\alpha_0$$

where α_0 is the optical power loss per unit length in the absorption cell. In a realistic case α_0 is an order of 10^{-5} cm^{-1} and it gives $l_{opt} \simeq 2\times10^3$ m. The minimum detectable absorption coefficient[11]

$$\alpha_{min} = 7.7 \ \Delta f/\eta n \alpha_0$$

becomes 5×10^{-13} cm^{-1} at visible region when the post detection bandwidth is $\Delta f = 1$ Hz, the quantum efficiency of the detector is $\eta = 0.8$, and the number of incident photons corresponds to the laser power of 10 mW.

Dye laser spectrometer[11]

With a White type absorption cell of 200 m effective length and 100 kHz Stark modulation, we have achieved the detectable absorption coefficient of 3×10^{-9} cm^{-1}. More than thousand $\Delta v = 5$ overtone transitions of NH_3 in the 15260-15590 cm^{-1} region and several hundreds $\Delta v = 6$ trnasitions in the 18000-18230 cm^{-1} region were measured with very good signal-to-noise ratios[12]. All rotational lines were resolved.

Diode laser spectrometer[13]

A reasonably high sensitivity of 3×10^{-8} cm^{-1} was achieved with the 0.78 μm AlGaAs diode lasers. Two types of absorption cells were employed. A White-type cell with 8 m effective length was used for source modulation detection at 100 kHz. A 60 cm long Stark absorption cell was employed for Stark modulation detection at 100 kHz. Several tens $\Delta v = 4$ overtone transitions of NH_3 were observed with good signal-to-noise ratio. All Stark components of stronger vibration-rotation lines were resolved.

3. Analysis

The $\Delta v = 5$ transitions are almost fully analyzed. A global feature of the spectrum is quite different from but much more simple than that expected from the normal mode theory. The parallel ($\Delta K=0$) and perpendicular ($\Delta K=\pm1$) bands appear closely with each other, and the band origins and the rotational constants in these two bands are almost identical. In the $v=5$ state the anti-symmetric inversion levels lie below the symmetric inversion levels. This order is contrary to the order in the ground and low-lying vibrational states.

All these unusual features of the high-vibrational state are difficult to be explained by the normal mode model, but well interpreted by the present analysis based on the local mode theory. Each of three bonds of NH_3 vibrates in the Morse potential of $U(r) = D[1-\exp(-ar)]^2$. When all bonds are independent, the vibrational energy is given by

$$E(n_1, n_2, n_3)/hc = (v+\tfrac{3}{2}) \omega - \sum_{i=1}^{3}(n_i+\tfrac{1}{2})^2 \omega_x \tag{1}$$

where

$$\omega = \frac{2a\hbar}{hc}(D/2\mu)^{1/2} \tag{2}$$

$$\omega_x = \frac{(ah)2}{2hc} \tag{3}$$

$$v = n_1 + n_2 + n_3 \tag{4}$$

and μ is the reduced mass. Here we introduce the coupling between two equivalent N-H bonds with an coupling constant λ. The matrix element of the total Hamiltonian with C_{3v} symmetry should be

$$\langle n_1' n_2' n_3' | H(C_{3v}; \omega, \omega_x, \lambda)/hc | n_1 n_2 n_3 \rangle$$

$$= [(v+3/2)\omega - \sum_{i=1} (n_i + \tfrac{1}{2})^2 \omega_x] \delta n_1' n_1 \delta n_2' n_2 \delta n_3' n_3$$

$$+ \lambda \sum_{ijk} [(n_i+1)n_j]^{1/2} \delta n_i' n_i+1, \, \delta n_j' n_j-1 \, \delta n_k' n_k \tag{5}$$

Because of the presence of inversion motion, the Hamiltonian should be modified to have D_{3h} symmetry. The energy levels split into symmetric and antisymmetric levels due to the interaction between the left- and right- handed systems. However the matrix can be divided into two C_{3v} sub-matrices, each of which has modified vibrational frequencies , the unharmonicity constants ω_x, and the coupling parameters λ of $\omega^s = \omega - \omega^{inv}$, $\omega^a = \omega + \omega^{inv}$. $\omega_x^s = \omega_x - \omega_x^{inv}$, $\omega_x^a = \omega_x + \omega_x^{inv}$, $\lambda_s = \lambda - \lambda^{inv}$, and $\lambda_a = \lambda + \lambda^{inv}$. The vibrational energy can be obtained by diagonalizing C_{3v}-sub matrix. The energy levels are calculated as a function of ω_x/λ. The limit $\omega_x/\lambda \to \infty$ represents the pure local mode character, while the limit $\omega_x/\lambda \to$ 0 pure normal mode character. The value of the parameter where the calculation fits to the observed NH_3 level is 2.1. This means the case is fairly close to the local mode limit. The A_1- and E-symmetry levels of the (500) state locate quite closely with each other. The transition probability to the (410) state is calculated to be only 3% of that to the (500) state. Those to other states are much weaker. This makes the spectrum quite simple.

The transitions to the v=2, v=4, and v=6 are observed and partly analyzed. Systematic dependence of the rotational constant on the vibrational quantum number is obtained. All observed results are in good agreement with the analyzed results. Careful studies of line shape functions of the overtone transitions show that even in the high vibrational state the collision induced transitons between inversion levels give dominant contributions to the broadening parameters.[14]

The method of double resonance probing inversion transitions in the high vibrational state has been developed.[15] Inversion frequencies are more accurately determined by more than two digits than those reported previously.[16]

References
1) R. Mecke: Z. Phys. 81 313 (1933)
2) W. Baumann and R. Mecke: Z. Phys. 81 445 (1933)
3) K. Freudenberg and R. Mecke: Z. Phys. 81 465 (1933)
4) R.M. Badger: Phys. Rev. 35 1038 (1930)
5) J.O.P. McBride and R.W. Nicholls: Can. J. Phys. 50 93 (1972)
6) A.B. Antipov, A.D. Bykov, V.A. Kapitanov, V.P. Lopasov, Yu.S. Makushkin, V.I. Tolmachev, O.N. Ulenikov, and V.E. Zuev: J. Mol. Spectrosc. 89 449 (1981)
7) K.K. Lehmann and JS.L. Coy: J. Chem. Phys. 81 3744 (1984)
8) S.L. Coy and K.K. Lehman: J. Chem. Phys. 79 5239 (1986)
9) H. Sasada: Optics Letters9 448 (1984)
10) H. Sasada: J. Chem. Phys. 88 767 (1988)
11) T. Kuga, T. Shimizu, and Y. Ueda,: Jpn. J. Appl. Phys. 25 1084 (1986)
12) T. Kuga, T. shimizu, and Y. Ueda: Jpn. J. Appl. Phys. 24 L147 (1985)
13) K. Nakagawa and T. Shimizu: Jpn. J. Appl. Phys. 26 L1697 (1987)
14) Y. Matsuo, K. Nakagawa, T. Kuga, and T. Shimizu: J. Chem. Phys. 86 1878 (1987)
15) Y. Matsuo, Y. Endo, E. Hirota, and T. Shimizu: J. Chem. Phys. 87 4395 (1987)
16) Y. Matsuo, Y. Endo, E. Hirota, and T. Shimizu: J. Chem. Phys. 88 2852 (1988)

INVESTIGATION OF THE STARK EFFECT IN XENON AUTOIONIZING RYDBERG SERIES WITH THE USE OF COHERENT TUNABLE XUV RADIATION

W.E. Ernst*, T.P. Softley[+], and R.N. Zare
Dep. of Chemistry, Stanford University, Stanford CA 94305, USA

The Rydberg spectra of the rare gases have been a matter of interest for both experimentalists and theorists for many years. The Rydberg series converging to the upper (j=1/2) fine-structure level of the ionic core autoionize. Among the heavy rare gases, xenon Rydberg states have been recently measured at high accuracy with classical high resolution vacuum ultraviolet (VUV) absorption spectroscopy/1/. The Stark effect was only investigated for the bound Xe Rydberg states /2/. Here we report the first high resolution Stark effect study of the autoionizing series by single photon excitation from the ground state and present calculations explaining the observations.

Figure 1 shows a schematic diagram of the experimental apparatus. The light from a pulsed Nd:YAG pumped dye laser operated with Rh 590 is frequency doubled in a KDP crystal. The UV pulse is focused into a pulsed free jet of argon or xenon serving as a gaseous nonlinear medium. The XUV is separated from the ~10^6 more intense UV fundamental by using two dichroic beam splitters/3/. One of the XUV reflecting surfaces has a geometry for focusing the XUV into the sample, which is a pulsed Xe beam injected between a pair of metal plates serving as ion repeller and extractor plates of a time-of-flight (TOF) mass spectrometer and also as Stark plates in the electric field studies. Xe atoms excited by the XUV into Rydberg states between the two ionization limits autoionize and the ions are detected in the TOF set up. The electric field strength was varied between 27 and 2362 V/cm. The elec-

*Present address: Institut für Molekülphysik, Freie Universität Berlin, Arnimallee 14, D-1000 Berlin 33, W. Germany
[+]Present address: University Chemical Laboratory, Lensfield Road, Cambridge, CB2 1EW, England

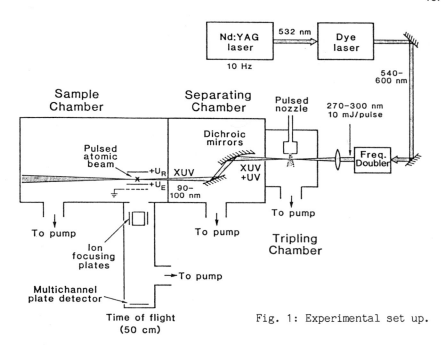

Fig. 1: Experimental set up.

tric field could be chosen parallel or perpendicular to the laser polarization, allowing for excitation with the selection rules ΔM=0 or ΔM=±1. At low electric field strength (e.g. 27 V/cm) the spectrum shows only the narrow ns and broad nd resonances and is virtually identical to the zero field absorption spectrum /1/. Two new series of lines can be observed for |M|=1 at field strengths between 200 and 1000 V/cm. One set of the additional transitions is the np series, the other one the J=2 component of the nd series, both of which are forbidden in zero-field. As the electric field mixes wave functions of opposite parity, the transitions become allowed. Above 1000 V/cm the mixing of ℓ states with negligible quantum defect (ℓ=3 to N-1) causes the appearance of hydrogen-like Stark manifolds. Figure 2 shows a part of the measured Stark spectrum for 1575 V/cm and excitation into states with M=0 (upper trace) and |M|=1 (lower trace). At this field strength of the n=15 manifold covers the region between the 17 d and 19 p lines. Along the wavelength axis the zero-field positions of lines are given which are

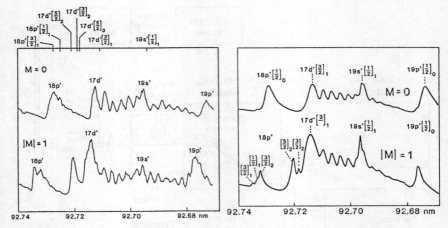

Fig. 2: Part of the measured (left) and calculated (right) Stark
spectrum for E=1575 V/cm and M=0 and |M|=1.

mixed by the Stark field.

In order to explain the observations for different polarizations (M=0
and |M|=1) as well as the field dependence, we performed calculations
of the Stark effect using a simplified theoretical approach. The major
steps of our treatment are: 1. Diagonalization of the Stark energy
matrix for the <u>discrete</u> Rydberg states (matrix of about rank 500), 2.
coupling of the mixed eigenstates to the continuum through no more than
two channels. Bound-continuum and continuum-continuum Stark interactions as well as the autoionization through many more channels are
neglected in this approximation. In Fig. 2 we show a simulation of the
spectrum on the basis of our approach for the same electric field
strengths and wavelength range as in Fig. 2. Our simple model obviously
allows to identify most of the observed features and to understand the
primary autoionization and intensity mechanisms.

REFERENCES

1. K. Yoshino and D.E. Freeman, J.Opt.Soc.Am.B<u>2</u>, 1268(1985).
2. R.D. Knight and Liang-guo Wang, Phys.Rev. A<u>32</u>, 896(1985).
3. T.P. Softley, W.E. Ernst, L.M. Tashiro, and R.N. Zare,
   Chem.Phys. <u>116</u>, 299(1987).

LASER SPECTROSCOPY OF AUTOIONISING 5DNF J=4,5
RYDBERG SERIES OF BA I

W.Hogervorst

Faculteit Natuurkunde en Sterrenkunde,Vrije Universiteit,Amsterdam,The Netherlands

The 5dnf autoionising series with J=4,5 in Ba converging to the $5d_{3/2}$ - and $5d_{5/2}$ - ionisation limits (at resp. 46.909 and 47.710 cm^{-1}) have been studied in a laser-atomic-beam experiment by single-photon excitation from levels of the $5d^2$ configuration, of which the metastable $5d^2\,^1G_4$ level (at 24.696 cm^{-1}) was localised recently [1]. The transitions were induced with light from a frequency stabilised CW ring dye laser or a Nd:YAG pumped pulsed dye laser. The autoionisation process was monitored by either detecting the detached electrons with an electron multiplier after energy selection or by collecting ions with a quadrupole mass filter. In the case of electron detection auto-ionisation into the different end states of Ba$^+$ ($6s_{1/2}$ and $5d_{3/2}$) could be distinguished [2], whereas in the ion-detection setup isotope selective observation of hyperfine structure was possible [3].

In the CW laser experiment the J=4,5 $5d_{3/2}$nf (n=16 -50) and $5d_{5/2}$nf (n=10,11,30, 40 and 50) levels were excited from the $5d^2\,^1G_4$ level. Also some 5dng and 5dnh states were observed. In the $5d_{3/2}$nf multiplet one level with J=5 and two levels with J=4 and J=3 are expected, which can only ionise to 6sεl continua. In the $5d_{5/2}$nf multiplet two levels with J=3,4 and 5 may be excited, which ionise to 6sεl and, for n >11, also to $5d_{3/2}$εl continua. J=3 signals were not observed which may be ascribed to small transition probability from $5d^2\,^1G_4$ as well as to line broadening due to configuration interaction with $5d_{3/2}$np$_{3/2}$ J=3 levels with large autoionisation linewidths.

Remarkable feature of the J=4,5 $5d_{3/2}$nf series is the overall narrow linewidth Γ. Many transitions show linewidths down to a Doppler limited value of 10 MHz. Mean values of the product $(n^*)^3\,\Gamma$ (n^* is effective principal quantum number) are in the order of 500 GHz (HWHM), a factor of 10 to 100 lower than the corresponding values for $5d_{3/2}$nd series [4] also autoionising to the 6s ion state. This large difference may be understood qualitatively by noting that in 5dnl series 5dnf is the first with a non-core penetrating orbit. In the $5d_{3/2}$nf series stabilisation and line broadening effects occur in the vicinity of perturbing states of the $5d_{5/2}$nf and $5d_{5/2}$np configurations.

The narrow linewidths in these autoionising series allow for the observation of hyperfine structure and isotope effects (using the ion-detection setup) as shown in Fig.1 for the $5d_{3/2}39f_{7/2}$ J=5 level. The upper spectrum was obtained by observing the emitted

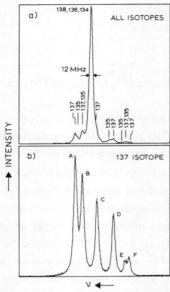

Fig.1. Hyperfine structure of $5d_{3/2}39f_{7/2}$ J=5 level in Ba

a. Signal of all isotopes

b. ^{137}Ba isotope signal only; the transitions indicated are

A: F=11/2→13/2, B: F=9/2→11/2

C: F=7/2→9/2, D: F=5/2→7/2,

E: F=9/2→9/2, F: F=11/2→11/2 and F=7/2→7/2.

electrons, showing contributions from all isotopes. The lower spectrum was taken with the mass filter tuned precisely for mass 137, showing all transitions with a change in the hyperfine quantum number of $\Delta F=1$ and most transitions with $\Delta F=0$. By slightly detuning of the mass filter the reappearance of the single ^{138}Ba- or ^{136}Ba- peak could be observed, thus enabling a determination of isotope shifts.

Pulsed dye laser experiments were performed to study the 5dnf J=4,5 states with lower n values thus extending and completing the data collected in the CW experiments. Also excitations from metastable $5d^2\,^3F$ states to the same 5dnf levels were investigated to confirm earlier assignments [2]. With broadband excitation (0.2 cm^{-1}) and the large scanlength of pulsed dye lasers a better overview of the series was obtained. As an example in the upper part of Fig.2 part of a spectrum taken with a pulsed blue dye laser of the $5d^2\,^1G_4 \to 5d_{5/2}nf$ J=4,5 states above the $5d_{3/2}$ ionisation limit (n>11), including two multiplets of broad lines from excitations of $5d^2\,^3F$ to lower 5dnf states, is shown.

The accurate level energies, hyperfine structure data and autoionisation linewidths of the CW and pulsed laser experiments provided the starting point for detailed MQDT analyses of the 5dnf J=4,5 series. MQDT analyses were performed in the so-called shifted R-matrix formalism [5]. Especially when continuum channels have to be included in the analyses this formalism is easier to apply. In addition it is an advantage that for many of the parameters analytical expressions may be derived in contrast to standard MQDT. Results of the fitting show that the measured level energies of the $5d_{3/2}$nf J=4,5

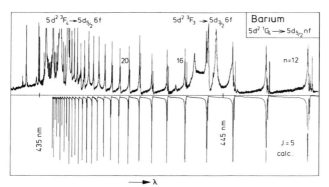

Fig.2. Excitation spectrum of $5d^2 {}^1G_4 \to 5d_{5/2}nf$ J=5 and 4 levels in Ba above the $5d_{3/2}$ ionisation limit (n ≥12). Upper part: spectrum from pulsed dye laser experiment; two multiplets of broad lines from excited from $5d^2 {}^3F$ levels can be observed too. Lower part: calculated J=5 MQDT spectrum (negative).

series are described very well from n=50 down to n=8. For n=4-7 the energy dependence of the quantum defects seems to be non-linear. From the linewidths of the J=5 states below the $5d_{3/2}$ ionisation limit the interaction between the two $5d_{5/2}nf$ J=5 channels could be deduced. This interaction strength was subsequently used to calculate the J=5 excitation spectrum above the $5d_{3/2}$ limit, given in the lower part of Fig.2 and showing excellent agreement with experiment. In the two $5d_{3/2}nf$ J=4 series wavefunctions and series interaction could be determined with the help of the hyperfine structure data. The behaviour of the linewidths in these J=4 series below the $5d_{3/2}$ limit turned out to be completely dominated by interaction with the $5d_{5/2}np_{3/2}$ channel which, however, appears to be weak. The autoionisation linewidths of the 5dnp states are so much larger than those of the 5dnf J=4 channels that a small admixture still produces large effects. From the analysis of the J=5 and 4 series the interaction with continuum channels could be determined. The conclusion of the analyses of autoionising series such as the 5dnf, with small quantum defects and weak interactions, is that highly accurate experimental data are essential.

1. Vassen,W., Bente,E.A.J.M. and Hogervorst,W.,J. Phys.B20, 2383   (1987).

2. Bente,E.A.J.M. and Hogervorst,W., Phys. Rev. A36, 4081 (1987).

3. De Graaff, R.J., Bente, E.A.J.M., Hogervorst,W. and Wännstrom, A., Phys. Rev. A37 (1988), to be published.

4. Neukammer,J.,Rinneberg,H.,Jönsson,G., Cooke,W.E., Hieronymus,H., König,A., Vietzke,K. and Spinger-Bolk,H., Phys. Rev. Lett. 55, 1979 (1985).

5. Cooke,W.E. and Cromer,C.L., Phys. Rev. A37, 2725 (1985).

RESONANCE PHOTOIONIZATION SPECTROSCOPY OF ATOMS : AUTOIONIZATION AND HIGHLY EXCITED STATES OF Kr and U.

Hu Qiquan, Yin Lifeng, Lin Lin,
Lin Fucheng*

Shanghai Institute of Optics and Fine Mechanics,
Academia Sinica, P.R. China

Up to now most photoionization experiments of atomic vapors have been performed with an atomic beam device or a heat pipe tube. In order to obtain a high concentration of metastable noble gas atoms and metallic atomic vapor, we developed a hollow cathode discharge lamp (HCD) with a rectangular pulsed power supply. It was found out that the HCD lamp can not only be used as a spectrum light source, it can also be used as a simple source of metallic atomic vapor and a signal detector for the study of the resonance photoionization process of atoms. In this paper we report an experimental investigation of the autoionization and highly excited states of Kr and U atoms in a HCD lamp by using resonantly stepwise excited photoionization.

The basic experimental setup is shown in Fig.1. The Kr-U HCD lamp is 100mm long with a $3 \times 7 mm^2$ rectangular slot hole, and is filled with 1 torr of krypton gas as a buffer. The through-hole cathode structure protects the photoelectric effects of the cathode by the laser illumination. The Kr-U HCD lamp is driven by a rectangular pulse in order to avoid the discharge interference of the photoionization detection. The large current ensures sufficient discharge sputtering and exciting to produce a high density atomic vapor and metastable noble gas atoms.

Under nondischarging conditions we use the HCD lamp itself to detect the photoionization signal of the Kr and U atoms. A suitable DC voltage is used to prevent sustained discharge of the HCD lamp and to provide sufficient sensitivity for the photoionization detection.

* This work was partially participated with Zhang Yanping, Zhang Guiyan Jing Chunyang, Cui Junwen, Su Haizhang and Zhong Lihong.

By measuring the ionization signals, we obtained 4 auto-ionization states of Kr atom. In our experiment the $1s_5$ metastable state of Kr atom is obtained by HCD exciting. The $2p_3$ excited state of Kr atom is obtained by exciting the $1s_5$ state Kr with a laser wavelength of 557.2nm. The two transitions of the Kr atom $2p_3$ -- autoionization states are $5p\ '[3/2]_2$--$9s[1/2]_1$ and $5p\ '[1/2]_1$--$7d[3/2]_1$, the wavelength is 581.1 and 585.1nm respectively. This result is in agreement with the published data. The other two are newly observed lines, the wavelengths are607.7 and 616.1 nm respectively.

We also obtained the resonant excitation photoionization spectrum of the f^3 dsp electron configuration $^7L_6$ 17362cm^{-1} energy level, and listed a portion of the experimentally measured U energy level wave numbers near 34000 cm^{-1}. The $^7L_6$ energy level of the U atom is obtained by exciting an odd parity $^5L_6^o$ ground-state atom with a laser wavelength of 575.8nm. The parity of the energy levels listed is also odd. However, a number of the experimentally observed energy levels cannot be found in the published data. Measurments also show a large variation in the ionization spectral linewidth. This indicates the possibility of having two-photon resonance excitation in the experiment, whereby the U atom is excited from the f^3 dsp configuration directly to the f^3dp^2 autoionization states. The ionization cross section for the autoionization is at least one order of magnitude greater than the ordinary ionization. Therefore, the autoionization states of the U atom may be studied by identifying the photoionization signal and the energy levels in the ionization spectrum.

Reference

[1] "Laser Program Annual Report —1979. Lawrence Livermore National Laboratory", UCRL 50021-79, 9-15.

[2] E.Miron and R.David et al.; J. O. S. A., 1979, 69 ,No. 2(Feb), 256.

174

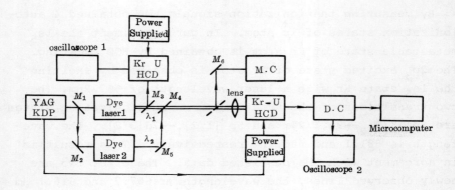

Fig. 1 Experimental set-up

STARK SPECTRA OF STRONTIUM AND CALCIUM ATOMS

Zhang Sen, Qiu Jizhen, Wang Gang, Hu Sufen, Zhong Jianwen

Department of Physics, Zhejiang University, Hangzhou

The Stark structures of levels of Sr and Ca atoms with m=0 in electric field from 0 to 8 kV/cm were obtained by RIS method, using the $\pi\pi$ polarization lasers.[1]

The Sr or Ca atom was excited stepwise and ionized by two dye lasers, corresponding to the following processes, respectively:

$$Sr: 5s^2\ {}^1S_o \xrightarrow{\lambda_1=4609\text{\AA}} 5s5p\ {}^1P_1 \xrightarrow{\lambda_2} Rydberg\ sates \xrightarrow{\lambda_1 or\ \lambda_2} Sr^+ + e^-;$$

$$Ca: 4s^2\ {}^1S_o \xrightarrow{\lambda'_1=4228A} 4s4p\ {}^1P_1 \xrightarrow{\lambda'_2} Rydberg\ states \xrightarrow{\lambda'_1 or\ \lambda'_2} Ca^+ + e^-.$$

The wavelength λ_2 scanned from 4290Å to 4220Å for Sr atom, obtained the Stark spectrum in the vicinity of n=12 and n=13. The λ_2 scanned in 4100Å~3980Å for Ca atom, the Stark spectrum of n=10~13 was obtained. Figures 1 and 2 show the Stark map of Sr atom and the Stark spectra of Ca atom, respectively. In the low field of $F<1/3n^5$, the states of $4<l<n-1$ of Sr or Ca atom, quantum defect of which obey the $\delta n^{-3}\ll 3/2n^2 F$[2], constitute the lhydrogen like linear incomplete Stark manifold, as can be seen from Fig.1 or 2. The adjacent Stark levels in the manifold are separated by $\Delta W=3nF$[3]. The ${}^1S_o, {}^1P_1, {}^1D_2, {}^1F_3$ states have non-negligible quantum defects, they exhibit the second order Stark effect. In the strong field of $F>1/3n^5$, the separation of levels at the edge of manifold become confusion, display the anticrossing structure of energy levels, which because of asymmetry of Coulomb field of core. Such anticrossing structures between the zero-field non-degenerate levels and the Stark levels of hydrogenlike atom is main difference with respect to hydrogenic behavior, which provide the plentiful information about the configuration interaction.

The anticrossing of levels of alkali atoms is dependent on the quantum defect, which had been calculated and measured by M. L. Zimmerman et al.[3], they obtained that large fractional quantum defect δ_l produced large levels repulsion, the energy separation at first anticrossing obeys a simple $\Delta W=1.9\bar{\delta}_\ell n^{-4}$. But that level probably crosses into Stark maniffold, when its quantum defect differs from the neighbouring states by cross to o.5. For Sr atom, $\bar{\delta}_1=0.31$ and $\bar{\delta}_2=0.13$, the $^1P_1$ and 1_{D_2} states are repelled by Stark manifold when they close to the manifold. The ^1S_o state lies close to midway between the $^1P_1$ states because $\bar{\delta}_o-\bar{\delta}_1=0.5$, the repulsions to ^1S_o state from two adjacent $^1P_1$ states are approximately cancelled each other, this level crosses into the Stark manifold when the field is increased, which represent that the interaction between the ^1S_o state and the Stark manifold is negligible. The F state, quantum defect of which is small, gradually joins the manifold as the field is increased. For Ca atom, the ^1S_o and $^1D_2$ states are repelled by Stark manifold, corresponding the $\bar{\delta}_o=0.34$ and $\bar{\delta}_2\sim0.2$. The $^1P_1$ state, $\bar{\delta}_o-\bar{\delta}_1=0.5$, is cross into the Stark manifold when the field increases. The F state joins the manifold because the $\bar{\delta}_3$ is small. Preceding fact shown that regularity of Stark structures of Rydberg states of Sr and Ca atoms bear analogy to alkali atom.

As the field is increased, the valence $3d^2\,^3P_o$ and $3d^2\,^3P_2$ states, which can be excited by optical method for Ca atom,[4] shift toward the low and high energy, respectively, as can be seen from Fig.2. Such obvious shift represent that the valence states are repelled by the n=12 Stark manifold, this repulsion provides clear of the singlet triglet configuration mixing[5].

The strong interference was observed, which leads to the lines of Stark levels in the neighbourhood of $3d^2\,^3P_o$ valence state notably weaken, and the interference enhances as the field increases. Such interference, varies as field, is intimately related to Stark mixing caused by electric field.

[1]   Zhang Sen, Qiu Jizhen et al., Acta Physica Sinica, 37, 983 (1988).

[2]   Chardonnet, C., Delande, D., Gay, J.C., Opt. Comm. 51, 249 (1984).

[3]   Zimmerman, M.L., Littman, M.G. et al., Phys. Rev. A20, 2251 (1979).

[4]   Armstrong,A.J., Esherick,P., Wynne,J.J., Phys. Rev. A15,180(1977).

[5]   Zimmerman,M.L., Ducas,T.W. et al., Phys. B: At. Mol.Phys., 11, L11 (1978).

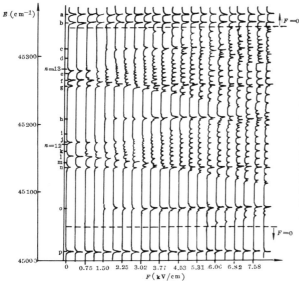

Fig.1.Stark map of Sr atom in the vicinity of n=12 and n=13 (m_1= 0).
a-$5s16d_1{}^{13}D_2$, b-$5s16d_3{}^{13}D_2$
c-$5s16p_1{}^{13}P_1$, d-$5s16p_1{}^{13}P$
e-$5s15d_1{}^{13}D_2$, f-$5s15d_1{}^{13}D_2$
g-$5s16s_3{}^{1}S_o$, h-$5s15p_3{}^{1}P_1$
i-$5s15p_1{}^{1}P$, j-$5s14d_3{}^{1}D_2$
k-$5s12f_1{}^{1}F_3$, l-$5s12f_1{}^{1}F$
m-$5s14d_1{}^{1}D_3$, n-$5s15s_1{}^{1}S$
o-$5s14p_1{}^{1}P_1^2$, p-$5s13d_1{}^{1}D_2^0$

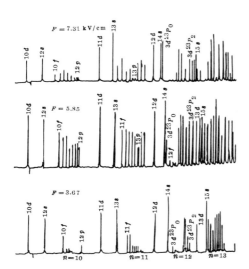

Fig.2. Stark spectra of Ca atom (m_1=0, F=3.67, 5.85, 7 31 kV/cm). The affirmable lines corresponding to the zerofield non-degenerate states are noted by the symbols of excited elec tron in the zero-field.

OBSERVATION OF BIDIRECTIONAL STIMULATED RADIATION AT 330nm, 364nm and 718nm WITH 660nm LASER PUMPING IN SODIUM VAPOUR

Zhang Ping, Hou Fuxing, Xia Zongju

Department of Physics, Beijing University
Beijing, P.R. China

Chen Tianjie

Center of Condensed Matter and Radiation CCAST
(World Laboratory) Department of Physics,
Beijing University

By virtue of the tunable pulsed dye laser with pulse-width and linewidth of 10ns and $0.5cm^{-1}$ respectively, we observed 330nm, 364nm and 718nm bidirectional stimulated radiation in sodium vapor with 660nm two-photon pumping. The sodium is kept in a heatpipe with quartz windows at each end. The dye laser beam is focused by a 0.3m focal-length lens into the heatpipe cell, the beam waist in the focal region is estimated to be $7.1 \times 10^{-4} cm^2$, the working volume of the sodium vapor is about 7cm. A 0.5m grating monochrometer is used to analyse the spectrum and the emission signals are detected by a PMT, then directly fed into a Model 162 Boxcar Averager, finally a strip chart recorder is used. According to the selection rule as well as polarization characteristic, if the initial state is $^1\Sigma_g^+$, the upper-level of the two-photon transition could be $^1\Sigma_g^+$ or $^1\pi_g$. In order to determine the upper-level, we did the polarization experiments. The temperature was maintained at $430^{\circ}C$, with input energy of the laser is 22mj, we got the ratio of σ_{cc} to σ_{11}: $\sigma_{cc}/\sigma_{11} < 1.5$ for all excitation lines. According to Bray's[6] theory, the upper-level is $^1\Sigma_g^+$ in our case. Scanning the laser frequency from 6590A to 6610A, we saw as many as 13 excitation lines added on a butterfly background with the total width about 20A.

We observed a spectrum with the peak centred at 330.0+ 0.2nm showing some structures and having a width about 1nm, which is much wider than the Doppler-linewidth (0.09nm).

This indicates that the UV emission at 330nm can not be a pure atomic process, there must be some molecular effects. Generally speaking, 330.2nm signal responds to Na 4p-3s transition in appearance. Analyses show that the generation of 330nm emission in our case likes neither totally to the SERS and FWM as J.K. Chen[1] and S.G. Dinev[2] indicated, nor totally to Na 4p-3s transition as A.G. Still[3] reported, but a combined processes of atomic and molecular reaction. There exsits apparent differences between our results with those mentioned above. First, our pumping wavelength is 660nm which corresponds to Na 4p-3s two-photon transition, but in terms of dipole transition selection rule, it is forbidden for two-photon absorption. Secondly, our excitation spectrum takes the shape of butterfly profile, also distincts from theirs. Our analyses show that the 330.0+0.2nm signals are mainly due to a molecular two-photon absorption, population transform from Na_2 $(4)^1\Sigma_g^+$ state to Na_2 $C^1\pi_u$ state and then the directly transition from $C^1\pi_u$ to $X^1\Sigma_g^+$. Let us give some details in explaining the generation of these signals. For Na_2 in our two-photon pumping energy region, there exists two degenerate electronic state $(4)^1\Sigma_g^+$ and $C^1\pi_u$. As D.C Stark effect and dipole-quadrupole interaction are taken into account, the pertubation is $1cm^{-1}$ in magnitude which is the same order as the difference of the two adjacent rotational levels of Na_2 in these electronic states, the resulted wavefunction after the pertubations is the superposition of the un-pertubated ones, i.g. $(4)^1\Sigma_g^+$ and $C^1\pi_u$. That is why we did observe the $C^1\pi_u - X^1\Sigma_g^+$ transition which is the origin of our 330nm and 364nm stimulated radiation spectrum.

The 330.2nm signal is generated by the following processes. $Na_2[X^1\Sigma_g^+(0,40]+h\nu_{6597.3A} \rightarrow Na_2[A^1\Sigma_u^+(5,41)]$, $Na_2[A^1\Sigma_u^+. (5,41)]+h\nu_{6597.3A} \rightarrow Na_2[(4)^1\Sigma_g^+(22,42)]+ Na_2[C^1\pi_u(7,41)]$, $Na_2[C^1\pi_u(7,41)] \rightarrow Na_2[X^1\Sigma_g^+(0,42)]+h\nu_{330.2nm}$. The other 12 lines can be explained in much the same way. In our expriments, we also observed some more intense lines near 364nm and 718nm with the resonant two-photon excitation (λ = 6604.5A), the temperature is kept at 430°C, input laser energy is 20mj and 5mj, corresponding to the forward and backward directions respectively. 364nm radiation spectrum is composed of the following six intense lines: 3556A, 3580A, 3606A, 3624A, 3641A and 3663A while the 718nm radiation spectrum has the peaks at 7105A, 7250A, 7326A, 7394A, 7473A, 7551A and 7703A, according to the analyses based on the appropriate spectroscopic information reported before [4,5], we are convinced that the 364nm and 718nm band belong to $C^1\pi_u - X^1\Sigma_g^+$ and $A^1\Sigma_u^+ - X^1\Sigma_g^+$ stimulated emission, it is confirmed by the polarization experiments, too.

In summary, the Na_2 in the ground state $X^1\Sigma_g^+$(v=0=1) absorb one photon reaching to the very near resonant intermediate state of $A^1\Sigma_u^+$ (v=5-6), one portion of them come down to the ground state of $X^1\Sigma_g^+$(v=8-15) at longer internuclear distance, while the other portion at shorter distance absorb another photon transiting to $(4)^1\Sigma_g^+$(v=22,24) state. To a first approximation, we take the D.C Stark effect caused by the ionization and dipole-quadrupole interaction in atom-molecule system into account, as a result of pertubation, some molecules transform from $(4)^1\Sigma_g^+$(v= 22-24) to $C^1\pi_u$(v=7-9). That is the reason that we did observe 330nm and 364nm radiation simultaneously. The 330 nm occurs most likely at shorter inter-nulear distance, however, the 364nm at longer one.

REFERENCE:
[1] J.K. Chen, Appl. Phys. B, 33, 155(1984).
[2] S.G. Dinev, Appl. Phys., B, 39, 65(1986).
[3] A.G. Still, Chemical Physics Letters, 142, No.1,2, (1987).
          J. Phys., B: At Mol Phys, 19, 2735(1986).
[4] W.J. Stevens, J. Chem. Phys., 66, No.3,4, 1477(1977).
[5] K.K. Verma, Journal of Molecular Spectroscopy, 91, 375(1982).
[6] R.G. Bray, Molecular Physics, 31, No.4, 1199(1976).

STUDY OF MOLECULAR RYDBERG STATES
AND THEIR DISCRIMINATIONS IN Na2

Hui-Rong Xia, Jian-Wen Xu and I-Shen Cheng

Department of Physics, East China Normal University
Shanghai 200062, People's Republic of China

The great disparities in equal-frequency two-photon excitation spectra and two-photon-excited fluorescence spectra between the dipole appreximation allowed and spin-forbidden two-photon transitions of sodium dimers have been found by multiplex nonlinear laser spectroscopy.

Study of molecular Rydberg states is an attractive topic recently. Based on the development of the available comprehensive identification methods by using multiplex nonlinear laser spectroscopic techniques, we have systematically studied near-resonantly enhanced equal-frequency two-photon transitions of Na2 for both of the dipole allowed absorption $X\,{}^1\Sigma_g^+ -- (n)\,{}^1\Lambda_g$ enhanced by $A\,{}^1\Sigma_u^+$ state[1-3], and spin-forbidden absorption $X\,{}^1\Sigma_g^+ --(n)\,{}^3\Lambda_g$ enhanced by $A\,{}^1\Sigma_u^+ -b\,{}^3\Pi_{nu}$ mixing. The great disparitity in their coarse structures, fine structurs, hyperfine structures, line shapes and two-photon-excited fluerescence spectra have been observed, which can be described and explained as follows :

1) in vibrational (coarse) structures of excitation spectra: The band density of the latter (spin-forbidden two-photon transitions) was much heavier than the former (dipole allowed one). The phenomeno is illustrated by Fig.1. Evolving Fig.1 to all of the consequent term curves, we have furthermore found that the former has the evolution directions of the V"- and V-progressions towards blue and the V'-progressions towards red, whereas the latter has the opposite evolution diréctions for all of its respective band progressions;

2) in rotational (fine) structures of excitation spectra : The former possesses five spectral branches, involving six near-resonant enhancing centers;Whereas the latter has the doubling number of spectral branches and about one order heavier two-photon lines than the former due to the avoided crossings of the spin-orbital couplings, which form the enhancing zones with averagied rotational constants around the

182

perturbation centers;

3) in hyperfine stuctures of excitation spectra : Each line of the singlet-singlet two-photon transitions was observed to show a narrow resonant peak in the Doppler-free profiles; whereas a line of singlet-triplet two-photon transitions was observed to present multiple Doppler free peaks, corresponding to the magnetic hyperfine splittings of the upper triplet states. The hyperfine constants were thereby obtained;

4) in Doppler-free line shape dependences: The comparison of the line shape variations on offsets Δ_i are shown in Fig.2. While the line shapes in Fig.2(a) are essential symmetric, which agree to the analyses in three-level system, the shapes in Fig.2(b) present remarkable line shapes peculiar to a two-photon transition in a four-level system with a pair of near-resonant mixing levels, which,depending on the relative values of the offsets, contribute in some cases the destructive inter-ference terms to the total two-photon excitation probability to result in asymmetric line shape with reduced Doppler-free peak(s).

5) in two-photon-excited fluorescence spectra: Fig.3 shows the observed cascade fluorescence spectra from the upper singlet (a) and triplet (b) states following respective two-photon transitions. While (a) is composed of two pair of doublets with one component overlaped, (b) provides much complicated spectrum, which allowed one to obtain perturbation shift value of the intermediate coupling levels.

These results have been used to discriminate the Rydberg states of sodium dimers.

* Project supported by the National Science Fund of China

References

(1) H.R. Xia, G.Y. Yan, and A.L. Schawlow: Opt. Commun. 39, 153 (1981)
(2) G.P. Morgan, H.R. Xia, and A.L. Schawlow: J. Opt. Soc. Am. 72, 315 (1982)
(3) J.G. Cai, H.R. Xia, and I.S. Cheng: ACTA OPTICA SINICA 6, 212 (1986)
(4) H.R. Xia and Z.G. Wang: in Laser, Spectroscopy and New Ideas, ed. by M.D. Levenson and W.M. Yen, Springer Ser. Opt. Sci., Vol.54, pp.174-182 (1987)
(5) H.R.Xia,L.S.Ma, J.W.Xu, M.Yuan, I.S.Cheng, Paper presented on Int. Conf. on Laser, Nov. 15-19, Xiamen (1987)

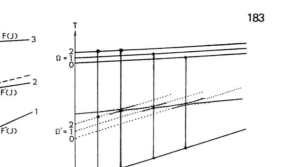

Fig.1 Rotational term curves for illustration of the near-resonant enhanced band centers of molecular two-photon absorptions. (a) $X^1\Sigma_g^+ - (n)^1\Lambda_g$ transitions enhanced by $A^1\Sigma_u^+$; (b) $X^1\Sigma_g^+ - (n)^3\Lambda_g$ transitions enhanced by intermediate $A^1\Sigma_u^+ \sim b^3\Pi_u$ singlet(solid) and triplet (dotted) level mixing (curved lines)

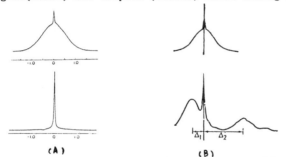

Fig.2 Comparison of the line shape dependences on offset of two-photon transitions. (A) $X^1\Sigma_g^+ - (6)^1\Sigma_g^+$; (B) $X^1\Sigma_g^+ - (3)^3\Pi_g$. upper: $\Delta \ll \Delta\nu_D$; lower: $\Delta \sim \Delta\nu_D/2$

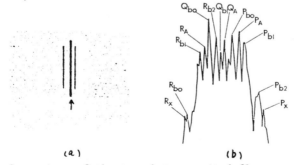

Fig.3 Comparison of the two-photon-excited fluorescence spectra following the Q-branch two-photon excitations around the excitating laser wavelengths from the upper levels of the Q-branch two-photon transitions, where the upper level is (a) singlet state; (b) triplet state

THE MEASUREMENT OF THE HIGH EXCITED SPECTRA OF SAMARIUM BY USING STEPWISE LASER EXCITATION METHOD

Hu Sufen   Qiu Jizhen   Zhang Sen   Wang Gang

Liang Yi   Chen Xing

Department of Physics, Zhejiang University

Hangzhou, People's Republic of China

ABSTRACT

We report the method to select the high excited states from the complex photoionizing spectra. Thirty energy levels of Sm near the ionization limit have been measured.

Photoionizing spectra of Sm near the ionization limit have been observed by using stepwise laser excitation and ionization methods.

The Sm atoms are selectively excited by means of two dye lasers pumped by the harmonics of a Nd:YAG laser. The first dye laser, at wavelength of $\lambda_1 = 6534.0$Å, is used to excite resonantly Sm atoms in atomic beam from thermally populated state $4f^6 6s^2\ ^7F_2$ to $4f^6 6s6p\ ^7G_2$. The second dye laser is scanned from 6710Å to 6970Å to excite Sm atoms from $4f^6 6s6p\ ^7G_2$ to high excited states with J=0,2,4, odd parity with two photons.

Then the Sm atoms are either autoionized or photoionized by another photon of λ_2. The Photoionizing signals are detected by an electron multiplier, the output of which is fed into a boxcar averager and recorded on x-y recorder.

The wavelength is determined by using the interference spectra of a Fabry-Perot etalon as a wavelength scale and the absolute calibration is achieved with the optogalvanic spectra of Ne atoms.

In this experiment several unwanted background peaks can occur when λ_2 is scanned. Those peaks are caused by $\lambda_2 + 2\lambda_2$, $2\lambda_2 + 2\lambda_2$, $\lambda_1 + \lambda_2 + 2\lambda_2$ etc. excited from the ground or thermally populated levels. The background peaks caused by $\lambda_2 + 2\lambda_2$, $2\lambda_2 + 2\lambda_2$ etc. can be identified by blocking λ_1 and repeating the λ_2 scan. The background peaks caused by $\lambda_1 + \lambda_2 + 2\lambda_2$ can be identified by using λ_1 and sending λ_2 with reduced intensity to avoid λ_2 two photons process. The ionization processes are accomplished by utilizing additional 2nd harmonics of the Nd:YAG laser. All of above unwanted background peaks then can be eliminated from photoionizing spectra.   Twelve energy levels of Sm which are located in the 44805.6–45683.6cm^{-1} energy interval have been thus measured.

The measurements have also been performed with different λ_1 and λ_2. In this case the λ_1 is fixed on 6511.0Å to excite Sm atoms from thermally populated state $4f^6 6s^2\ {}^7F_1$ to $4f^6 6s6p\ {}^7G_1$ and λ_2 is scanned from 6506–6950Å to excite Sm atoms from $4^6 6s6p\ {}^7G_1$ to high excited states with J=1,3 odd parity with two photons. Also, the unwanted background peaks have been rejected from photoionizing spectra. Eighteen energy levels of Sm which are located in the 44442.4–$46140.2\ \text{cm}^{-1}$ energy interval have been measured.

In this experiment, we have offered a method to select the high excited states from complex photoionizing spectra.

PRODUCT ANALYSIS IN THE REACTION OF THE TWO-PHOTCN EXCITED Xe(5p^{5}6p) STATES WITH FREONS

Xu Jie

Shanghai Institute of Optics and Fine Mechanics,
Academia Sinica, Shanghai, P.R. China

D.W. Setser

Department of Chemistry,
Kansas State University, U.S.A.

The reaction products of $Xe(2p_5)Xe(2p_6)$ atoms with CF_3Cl CF_2Cl_2, CF_2HCl and CF_2ClBr have been studied in a conventional laser-induced fluorescence cell using two-photon laser excitation technique.

A Nd-YAG-SHG laser was used to pump a dye laser. The laser output was frequency doubled and mixed to obtain about 250nm UV beam for the excitation of $Xe(2p_5)Xe(2p_6)$ states. The fluorescence intensity was monitored via averaging of 3000 laser shots using a digitizer coupled to computer[1].

The pressure dependence of the fluorescence decay from $Xe(2p_5)Xe(2p_6)$ atoms in reagent gases was used to measure the quenching rate constants.

The reaction of Xe(6p) atoms and Freons generated strong XeCl(B-X) emission and weak XeCl(C-A) emission. The time resolved spectrum of products in 200-320nm has been measured. The integrated time was divided into four parts. Fig.1 shows the XeCl(B-X) emission spectra in 0-32ns integration time from $Xe(2p_5)$ atoms in CF_3Cl CF_2Cl_2 and CF_2HCl.

The rate constant ($K_{xecl}*$) for XeCl(B,C) formation was measured by comparing the integrated XeCl(B-X) emission intensity from these reactions to that from Xe/Cl_2 reference reaction. The branching fraction of product ($\Gamma_{xecl}*$) was determined by ratio of $K_{xecl}*$ and quenching rate constant (K_Q). Table 1 lists the rate constants and branching fractions for XeX* formation. Comparing to reaction of Xe(6s, $^3P_2$) atoms an order of magnitude enhancement of XeCl(B,C) formation rate constants was observed[2]. This result can be qualitatively explained by the easier access to the $V(Xe^+, RCl^-)$ ion pair curve for the higher-energy Xe(6p) entrance channel potential.

Table 1: Rate Constants and Branching Fraction for XeX* Formation

Reaction	$K_Q^{a)}$	$K_{XeCl}^{a)}{}^*$	$\Gamma_{XeCl}{}^*$
$Xe(2p_5) + CF_3Cl$	6.1 ± 0.2	0.97 ± 0.2	0.16 ± 0.04
CF_2Cl_2	13.9 ± 0.3	2.3 ± 0.3	0.16 ± 0.04
CF_2HCl	6.8 ± 0.2	0.85 ± 0.2	0.13 ± 0.04
CF_2ClBr	10.8 ± 0.2	2.5 ± 0.3	0.23 ± 0.04
$Xe(2p_6) + CF_3Cl$	11.2 ± 0.2	2.8 ± 0.3	0.25 ± 0.04
CF_2Cl_2	12.4 ± 0.3	1.5 ± 0.3	0.12 ± 0.04
CF_2HCl	14.7 ± 0.3	2.94 ± 0.3	0.20 ± 0.04
CF_2ClBr	10.9 ± 0.2	1.83 ± 0.2	0.17 ± 0.04

a) In 10^{-10} cm^3 molecule^{-1} s^{-1}

REFERENCES
[1] J. Xu, J.K.Ku, D.W. Setser, Chem. Phys. Letters, 132(1986)427.
[2] J.H. Kolts, J.E. Velazco, D.W. Setser, J. Chem. Phys., 71(1970) 1247.

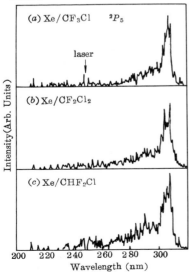

Fig. 1. XeCl (B-X) emission spectra. The monochromator was scanned at 0.5 nm/data point with 1mm slit width.

PHOTOIONIZATION SPECTRA OF Ca AND Sr ATOMS ABOVE THE CLASSICAL FIELD-IONIZATION THRESHOLD

Qiu Ji-zhen   Zhang Sen   Wang Gang

Department of Physics, Zhejiang University
The People's Republic of China

ABSTRACT

We have observed the photoionization spectra if Ca and Sr atoms in the energy region of $E > E_c$ with two-photon excitation. The period of resonances near E_o is proportional to $F^{3/4}$.

According to the classical sadlepoint model, the photo- ionization spectra should be continua when the atomic energy $E > E_c$. But, in 1978 Freeman et al. observed resonances structure of photoionization cross section for Rb in an electric field in the energy region of $E > E_c$, even $E > E_o$[1]. (E_o - zero field limit; E_c - classical field - ionization limit). Rau has shown that the spacing of such strong-field-mixing resonances follows well defined law:

$$dE/dn = (22.5 \text{ cm}^{-1}) (F/4335 \text{ V/cm})^{3/4} \text{ for H atom.}[2]$$

We have observed the photoionization spectra of Ca and Sr atoms above E_c with two-photon excitation. In the region of $E > E_c$, we can see sharp, resonancelike structure which are corresponding to quasi-stable Stark states. And we find that the spacing observed near E_o are in good agreement with the $F^{3/4}$ law. The relative modulation M increases with the increased field F. (see Table I) And both in Ca and Sr, the resonances are closely related with polarization of laser beam.

Our experimental approach for Ca case is briefly the following: the atomic beam spurted from the atomic oven at temperature $T = 750 °C$. The emission atomic beam, laser beam and DC electric field were perpendicular each other. The Ca atom was excited from the ground state $4s^2 \, {}^1S_o$ to above classical field-ionization threshold by two-photon process. Due to the linear polarizer in the laser path , the orientation of the polarization of which was parallel with electric field, the photoionization spectrum of m = 0 final state was obtained from the selection rule $\Delta m = 0$. Three DC electric fields were chosen in the experiment to observe the $\pi - \pi$ excitation of Ca atom. The values were 2.57, 4.71 and

6.86 kV/cm respectively.

In addition, changing the orientation of the polarizer perpendicular
to the electric field, we have observed photo-ionozation spectre above
the classical electric field threshold of $\sigma-\sigma$ excitation of Ca atom.
Here, the spectrum was mixed with m = 0 and m = 2. The method in Sr is
almost the same as in Ca.

References:
(1) R.R.Freeman et al.; Phys. Rev. Lett.,41 (1978)1463
(2) A.R.P.Rau;J.Phys.,B12, L193 (1979)
Table I. The spacing dE/dn and the relative modulation M of the resonance structure with $\pi-\pi$ polarization in Ca and Sr

	Electric field F (kV/cm)	Observed spacing (cm^{-1})	Calculated dE/dn (cm^{-1})	Observed modulation M (%)
Ca	2.57	15.6	15.2	7
	4.71	23.2	24.0	11
	6.86	32.2	31.8	18
Sr	3.27	18.8	18.2	12
	5.05	26.5	25.3	14

EFFECT OF MEDIUM BACKGROUND ON THE HYDROGEN SPECTRUM

Zhu Shitong
Center of Theoretical Physics, CCAST(world laboratory) and
Shanghai Institute of Optics and Fine Mechanics,Academia Sinica,
Shanghai, P.R.C.

Shen Wenda and Ji Peiyong
Department of Physics, Shanghai University of Science and
Technology, Shanghai, P.R.C.

Lin Fucheng

Shanghai Institute of Optics and Fine Mechanics, Academia Sinica
Shanghai, P.R.C.

The general theory of relativity and other metric theories predict that gravitation is manifested as a curvature of space-time. Recent research showed that the energy levels of an atom will be shifted when the atom is placed in a region of curved space-time[1]. The energies of the various levels will be altered in different ways, so that the effect of curvature can be distinguished from other effects such as gravitational , Doppler shifts. The magnitude of the energy level perturbations produced in a freely falling atom by space-time curvature increases as n^4 (n is the principle quantum number). Theoretically, a shift in wavelength of one part in 10^9 requires a radius of curvature of $D \leqslant 30$ km. Obviously, in the usual laboratory condition, such curvature radius of space-time cannot be reached. In this paper,we first study the effect of medium background on atomic spectrum with the help of optical metric in the frame of general theory of relativity[2]. The results show that the effect of medium background on atomic spectrum is similar to that of curved space-time in the presence of gravitational field. Physically , this corresponds to that the local inhomogeneity of medium causes the shift of atomic energy levels. We consider a flat or curved space-time manifold W with the metric $g_{\alpha,\beta}$ and a moving medium with the four-velocity u^α, where $u_\alpha u^\alpha = -1$, and the signat are $(-,+,+,+)$ is adopted.If the medium is isotropic, transparent, and dispensionless, its electric and magnetic properties are characteristized by its scalar permittivity $\mathcal{E}(x)$ and permeability $\mathcal{U}(x)$. The "Optical metric" in the moving medium is

$$\bar{g}_{\alpha\beta} = g_{\alpha\beta} + (1 - 1/\varepsilon\mu)\, u_\alpha u_\beta$$

For simplicity, we assume that gravitational field is absent ($g_{\alpha\beta} = \eta_{\alpha\beta} = (-1, 1, 1, 1)$) and the medium is one-dimension inhomogeneous plasma in rest. $u^\alpha = (1,0,0,0)$ and $\mu = 1$.

Then the optical metric of the medium is

$$ds^2 = \bar{g}_{\alpha\beta}\, dx^\alpha dx^\beta = -1/\varepsilon\, dt^2 + dx^2 + dy^2 + dz^2$$

Generally speaking, the space-time is not flat in the "optical metric" sense.

In the orthonormal cartan frames. Dirac equation in the curved space-time is[1]

$$(\,\underline{\gamma}^\mu \nabla_\mu + m\,)\psi = 0 \tag{3}$$

where $\underline{\gamma}^\mu$ are related by Dirac γ^α in the flat space-time

$$\underline{\gamma}^0 = \varepsilon\, \gamma^0, \qquad \underline{\gamma}^\prime = \gamma^\prime$$

$$\underline{\gamma}^2 = \gamma^2, \qquad \underline{\gamma}^3 = \gamma^3 \tag{4}$$

After calculation. we finds that the Hamitonian of eq. (3) is

$$H = H_0 + H_1$$

with

$$H_0 = -i\alpha^i \partial_i + m\beta - e^2 r^{-1}$$

$$H_1 = i(1+g_{oo})\alpha^i \partial_i - (1+g_{oo})\beta m + i g_{oo}\, \alpha^1 + \frac{dg^{oo}}{dr}$$

In Fermi normal coordinates,

$$g_{00} = -1$$

After calculation, we find that the curvature tensors are

$$R_{0101} = \frac{3}{4}\varepsilon^2 - \frac{1}{2\varepsilon}\varepsilon''$$

$$R_{0202} = R_{0303} = -\frac{\sigma^2 \varepsilon^\prime}{2r}$$

$$R_{1212} = R_{1313} = -\frac{(\sigma^2\, \varepsilon^{-1})}{2r}\varepsilon^\prime \tag{5}$$

$$R_{oo} = \frac{3}{4}\varepsilon^{-2}\varepsilon'^2 - \frac{1}{2\varepsilon}\varepsilon'' - \frac{\sigma^2\varepsilon^\prime}{r}$$

$$R = \frac{2\varepsilon^\prime}{r\varepsilon} - \frac{3}{2}\varepsilon^{-2}\varepsilon'^2 + \frac{1}{\varepsilon}\varepsilon''$$

The energy shift of the one electron atom caused by the background medium is

$$E^{(1)} = A\ R_{00} + B\ R + \sum_{i=1}^{3} C^{ii} R_{oioi} \tag{6}$$

Here σ is an energy parameter. A, B and C^{ii} are constant given by ref[1] For highly excited atoms

$$E(n) = 2.5e^{-4} m^{-1} d^{-2} n^4 , \tag{7}$$

d is the characteristic radius of curvature of the background space-time at the position of the atom. Obviously, a suitable choice of ξ may lead to a observable shift of wavelength because the variation of ξ can very large for the plasma. For example, in the plasma with the plane geometric structure if the characteristic scale length of linear density profile $L < 3.7 \mu m$ the shift of the energy levels is as large as the Lamb shift (4.4×10^{-6} eV). Thus, an atom also can be use as an instrument to detect the density or permittivity at the position of the atom.

Reference

[1]   L.Parker, L.O.Pomentel;Phys.Rev.(D),$\underline{25}$,No.12(1982),3180. and
       see Ref.[2,3,4] in it.
[2]   Shitong Zhu, Wenda Shen;J.O.S.A.(b),May,(1987),739.

Photoemission and Photoelectron Spectra from
Autoionizing Atoms in Strong Laser field

Guanhua Yao and Zhizhan Xu[*]
Shanghai Institute of Optics and Fine Mechanics, Academia Sinica, PRC

ABSTRACT: It is shown that, in strong field regime, the photoemission and photoelectron spectra from autoionizing atoms are substantially altered by the presence of higher-order ionization channels and that previous results must be revised

The interaction of intense laser fields with autoionizing states (AS) of atoms has been an active area in the last few years. Because of the poor resolution of photoelectron spectroscopy, it has been proposed that the photoemission spectrum may act as a good probe of such interaction. Many authors[1-6] have discussed the spectrum in the presence of different decay channels. However, a very important decay channel, $a-c_2$ ionization decay in Fig.1, which may substantially alter the photoemission spectrum, has not been taken seriously thus far, and discussions on the spectrum has been still too simplistic to be reliable.

We consider a model atom illustrated in Fig.1, where laser-induced transitions b-a and $b-c_1$ and configuration interaction V_{ac_1} are essential processes in the Fano model[7], and radiative damping to the third level $|f>$ has been considered by Agarwal et al.[1] We will focus here on effects of $a-c_2$ transition on the photoemission and photoelectron spectra. As this decay channel is induced by the same laser that induces resonant excitation of the AS, its decay rate γ_p is of the same order as the rate of $b-c_1$ decay γ_b, and is proportional to laser intensity. Thus as we will demonstrate, $a-c_2$ transition is particularly important in strong field regime extensively discussed in literature.

After Fano rediagonalization[8], the model is simplified to two-photon ionization via a set of intermediate structured Fano continuum states[8] $|\omega>$ which also radiatively decay to a third level $|f>$. A standard treatment as in ref.1 enables us to obtain the steady-state photoemission and photoelectron spectra. For brevity, here we consider only the total number of emitted photons N_p.

In Fig.2, we plot N_p against laser intensity I for different values

194

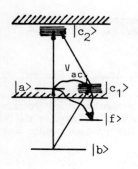

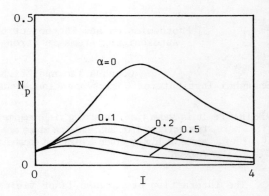

Fig.1 Model atom considered   Fig.2 Total number of emitted photons N_p vs. I

of α and for the parameters chosen in ref.1, i.e., q=1, $\delta=-1$ and $\gamma_s=0.1$
where the dimensionless intensity is defined as $I=\gamma_b/\gamma_0$, δ is detuning,
γ_0 is autoionization half width and γ_s is radiative decay rate scaled
by γ_0. Compared with the results of ref.1 (labelled $\alpha=0$ in Fig.2) when
$a-c_2$ transition is ignored, the total number of photon is drastically
reduced. Moreover, there is a strong tendancy with increasing α that
N_p decreases monotonously with I. This may be interpreted as follows:
In the presence of $a-c_2$ decay, the total decay rate is $\gamma_p+\gamma_s$. As $\gamma_p=\alpha I$
proportional to I and of order of I for the realistic case $\alpha \sim 1, \gamma_p \gg \gamma_s$,
thus more population is ionized from $|a\rangle$ to $|c_1\rangle$ with increasing I and
the number of emitted photons is reduced.

It is then natural that N_p in the strong field regime is much less
than predicted before, and the photoemission spectrum obtained in the
previous theories ignoring $a-c_2$ transition is far from reliable. The
point is that effects of this decay channel must be incorporated.

REFERENCES:

1. G.S. Agarwal et al.; Phys.Rev.Lett. **48**, 1164(1982)
2. K.Rzazewski and J.H. Eberly; Phys.Rev.A **27**,2026(1983)
3. G.S. Agarwal et al.; Phys.Rev.A **26**, 2277(1982)
4. J.H. Haus et al.; Phys.Rev.A **28**, 2269(1983)
5. G.S. Agarwal et al.; Phys.Rev.A **29**, 2565(1984)
6. Lewenstein et al.; J.Opt.Soc.Am.B **3**, 22(1986)
7. U.Fano; Phys.Rev.**124**, 1866(1961)
8. K.Rzazewski and J.H.Eberly; Phys.Rev.Lett. **47**, 408(1981)

*Also China Center of Advanced Science & Technology (World Lab.),Beijing

NATURAL RADIATIVE LIFETIME MEASUREMENTS
OF HIGH-LYING STATES OF SAMARIUM

Jiang Zhankui and Wang Dadi

Department of Physics, Jilin University,   Changchun ,

People's Republic of China

Wang Chengfei and Zhou Dafan

Changchun Institute of Applied Chemistry, Academia Sinica

People's Republic of China

Natural radiative lifetimes of 8 high-lying even parity states of Sm have been measured using stepwise excitation and fluorescence decay monitoring method in atomic beam.

natural radiative lifetimes of rare-earth elements have been studied extensively during the last few years. But most of these studies were limited to the low-lying states. Up to now in the literature there are no data on the lifetimes of high-lying states in samarium atom. The states in Sm whose lifetime has been measured   belong to odd parity states and the energy is lest than $30000cm^{-1}$. In the present paper we report lifetime measurements for several high-lying states of Sm using stepwise excitation and fluorescence decay monitoring method.

A pulsed dye laser induced 5720.19 $\overset{\circ}{A}$ transition from the stste $4f^{6}6s^{2}$ $^{7}F_{1}$ to the intermediate $4f^{6}6s6p\,^{7}F_{1}^{\circ}$ state. The second step from the $^{7}F_{1}^{\circ}$ to the investigated states was preformed with a dye laser using LD-490 dye. Both dye lasers were pumed by a Nd:YAG laser A resistively heated oven containing Sm metal produced the atomic beam in a vacuum chamber.The two laser beams were made to overlap spatially and temporally and cross the atomic beam perpendicularly. After through a suitable filter, the fluorescence decay signals released from the investigated states were detected by a fast photomultiplier, then it was fed to a time-scanning boxcar integrater.

The duration of the dye laser was less than 7 ns, which is much shorter than the lifetimes studied. Thus, the influence of laser pulse at the beginning of fluorescence decay can be neglected. In our experiment the density of the atoms in the region of interaction with the dye laser light was $5x10^{8}$ atoms per cm^{3} and the influence of collisions and radi-

ation trapping effects for the lifetime measurement can be neglected.[1]
Our measured lifetimes for the 8 high-lying states of Sm are shown in
table 1. The given lifetime values are referred to room temperature
and are not corrected for expected small effects of blackbody radiation.[2] The error limits encompass the statistical scattering obtained
from different runs as well as an additional allowance of about 2% for
possible systematical errors.

References

[1] Marek, J. and Münster, P., Astron. Astrophys. 62, 245(1978).
[2] Farley, J. and Wing, W. H., Phys. Rev. A23, 2397 (1981), Bhatia,
K.,Grafstrom, P., Levnson, C., Lundberg, H., Nilsson, L. and Svanberg,
S., Z. Phys. A303, 1(1981).

Table 1. Experimental values for the natural radiative lifetime of Sm.
The numbers in parentheses denots limits of error.

Energy level (cm^{-1})	Lifetime (ns)
37971	191(10)
38109	138(12)
38147	140(6)
38207	372(19)
38246	80(8)
38317	63(5)
38414	65(7)
38483	106(11)

TWO-STEP LASER EXCITATION OF nf RYDBERG STATES IN NEUTRAL Al AND OBSERVATION OF STARK EFFECT

Xu Lei, Zhao Youyuan, Wang guoyi and Wang Zhaoyong

Department of Physics, Fudan University

Shanghai, P.R. China

In this paper we report a spectroscopic work on nf Rydberg states (n=14-57) of group III element Al, also their electric field induced forbidden transition lines, line broadening and line shift have been observed in the spectrum.

1. Al vapour was generated in a heat pipe which was heated to $1160^{\circ}C$ and the corresponding Al pressure was about 0.1 torr. Rydberg states were reached by two-step laser excitation. Fig. 1 shows the transition scheme and the spectrum we got. One YAG was used to pump two dye lasers. The first one was frequency doubled to 309.4 nm and the second one could be tuned within 630 - 660 nm with linewidth of about 0.1 cm^{-1}. In addition, an etalon with free spectral range of 0.4993 cm^{-1} together with a Ne-He hollow cathode discharge tube was used to calibrate the wavelength absolutely. In this way, positions of Rydberg levels can be defined better than 0.1 cm^{-1} and then the first ionization limit of Al; Al^{+} ($3s^2\ ^1S_o$) was fitted to be 48279.04 cm^{-1} using the level data we got.

2. Since Rydberg states are very sensitive to electric field, when 15V bias was added to the heat pipe, we observed strong Stark effect. 3d-nd transition lines (n=22-43) which were induced by electric field, were found in the spectrum. In addition, Lu-Fano curves of nf and nd series were found to be separated in the high energy region (n large), this was also attributed to the Stark effect.

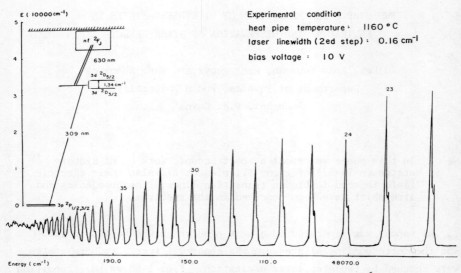

Fig. 1 Partial spectrum of Al by two-step laser excitation.
Upper left is the transition diagram.

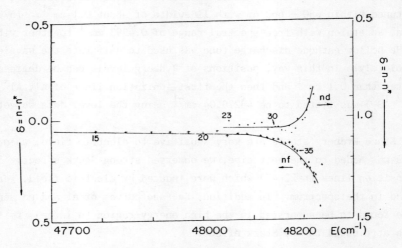

Fig. 2 Lu-Fano plot of nf and nd series of Al. The separation
of two curves in high energy region indicates the effect
of electric field (Stark effect)

MEASUREMENTS OF EXCITED SPECTRA OF THE REFRACTORY METAL ELEMENTS USING DISCHARGE SYNCHRONIZED WITH THE LASER PULSE

Zhu Lei, Du Qing, Zhao Youyuan, Wang Zhaoyong

Dept. of Physics, Fudan University,
Shanghai, P.R. China

Usually, atom beam was produced by means of heated oven or electron bombardment and laser spectra of many elements such as alkaline earth elements have been studied well in this way. But researches which are carried out into the highly refractory elements such as W, Mo and Zr were limited since it was difficult to obtain enough intense atomic or ionic beam of these elements with ordinary heated method. Here, we report a newly established system of producing intense beam for refractory metal elements by sputtering synchronized with a laser pulse and some fluorescence spectra lines of Mo atom have been got in the range of UV wavelength.

Fig.1 is a schematic diagram of the experimental set-up. When a suitable pulsed discharge is run in a hollow cathode tube, enough atoms or ions of Mo element could be produced in it in which there is a continuous flow of low pressure Ar-gas. There is a small nozzle in the centre of the cathodic plate and through which, the sputtered atomic vapor extracted into a high vacuum chamber (10^{-4} torr) by means of diffusion due to the pressure gradie-nt forms a directed beam. A dye laser (TDL-50) is pumped by a YG580 Nd: YAG laser with a repetition rate of 10Hz. After frequency doubling in a UV extension system, the wavelength of the dye laser could be tuned between 310-330nm with the linewidth of about $0.1cm^{-1}$ to complete the resonant transitions from ground state to some states of $4d^4 5s5p$ configuration and from some metastable states of $4d^4 5s^2$ configuration to some states of $4d^5 5p$ configuration. The fluorescence could then be detected using a photomultiplier (EMI9758QB) followed by a model SR235 Boxcar. Fig.1 also shows the discharge in a pulsed mode. A thyratron is triggered by a reference signal from the Nd:YAG laser. The time difference between the reference signal and the laser pulse is a few hundred μs. The delay between the discharge pulse and the laser pulse could be tuned between 0 and 100μs in order to obtain the optimizing S/N of the signal.

Fig.2 shows a part of the fluorescence spectra of Mo and its corresponding transitions. As in Fig.1, a prism is used to select the original dye laser light and the red light is diverted to a Ne lamp and a F-P interferometre. The OG signal from the Ne-Na hollow cathode and the transmitted signal from the F-P were recorded simultaneously with the fluorescence signal on a three pen chart recorder.

It shows that this new system can provide an effective method for us to study laser spectra and also lifetime measurements of the excited states for the refractory metal elements.

REFERENCE

[1] M. Baumann, H. Liening, H. Lindel, Physics Letters,
        Vol.68 A No.3, 4, 319 (1978)
[2] D.W. Duquette, S. Salih, J.E. Lawler, Phys. Lett.,
        Vol.83A No.5, 214 (1981)
[3] H. Bergstrom, H. Lundberg, A. Persson, W. Schade, Y.Y. Zhao,
        Physica Scripta, Vol.33, 513-514 (1986)
[4] P.S. Ramanujams, Physics Review Letters, Vol.39, No.19 (1977)

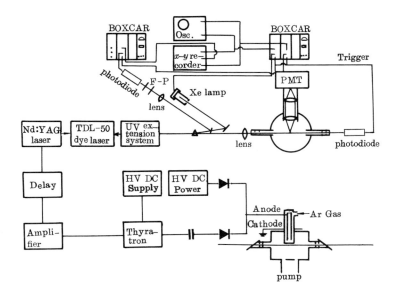

Fig 1 The Experimental Set Up

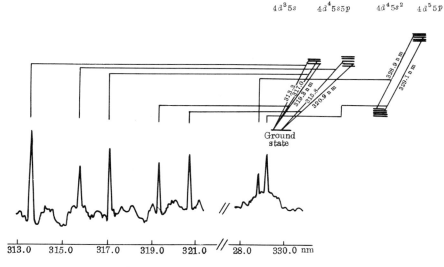

Fig 2 A part of the fluorescence spectra of neutral Mo
and its corresponding transitions

MULTIPHOTON IONIZATION OF ATOMIC LEAD AT 1.06μ

Ding Dajun, Jin Mingxing, Liu Hang, K.T. Lu

Institute of Atomic and Molecular Physics,
Jilin University, Changchun, P.R. China

Photoelectron angular distribution of non-resonant seven-photon ionization of atomic lead have been studied in our laboratory. The measurement was done in an apparatus of atomic beam intersected by a laser beam of 1.06μ. The photoelectrons emitted in the direction mutually perpendicular to both atom and laser beam were detected by a channeltron. The photoionization process is non-resonant and the intensity dependence of photoelectron signal was measured, which yielded a slope of 7 approximately, corresponding to the minimum photon number needed to ionizing neutral lead (I.P. = 7.41eV) using 1.06μ laser. The photoelectrons emitted also have an angular distribution. In the case of N-photon ionization, the angular distribution can be described by the general formula[1]

$$\frac{d\sigma^{(N)}}{d\Omega} = \sum_{i=0}^{N} a_i \cos^{2i}\theta \qquad (1)$$

where θ is the angle between the direction of ejected electrons and the polarization of laser. The angular distribution can be measured in the plane perpendicular to the direction of the laser beam by rotating a $\lambda/2$ plate for 1.06μ. The result is shown in figure 2, in the condition of 70mJ laser pulse energy. The data were fitted in the least-squares fit routine using eq.(1) firstly with N = 1, and N was increased in subsequent fits. The standard derivation was taken as a criterion and the best fit was found at N = 7. This indicates that the photoelectron angular distribution of Pb 7-photon ionization contains significant contributions from $\cos^{2i}\theta$ only for i < 7. It has been found that the angular distribution had a dependence on the laser intensity and coherence in the present experiment. The measurement of photoelectron angular distribution may give further information on the dynamics of multiphoton ionization process of atoms.

Reference:

[1]   Leuchs, G. and Walther, H., in Multiphoton Ionization of Atoms, ed. by Chin, S.L. and Lambropoulos, P., Academic Press (1984), 109.

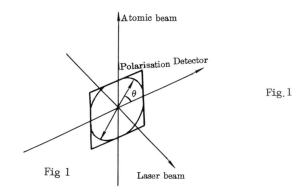

Fig 1

Fig.1

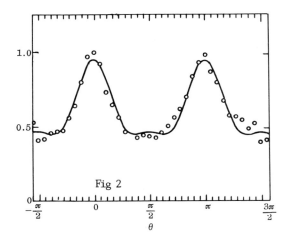

Fig 2

Fig.2

KINETIC PROCESSES
IN THE ELECTRON-BEAM PUMPED KrF LASER

H. Takuma, K. Hakuta, K. Ueda

Institute for Laser Science,
University of Electro-Communications,
Chofu-shi, Tokyo 182, JAPAN

I. INTRODUCTION

The electron-beam pumped KrF laser has been a subject of intensive study because of its high efficiency and also because of its scalability to a high power system. Although the main interest of the most of researchers has been as a next generation fusion laser, the importance of this type of laser is much more than that, becuase a high efficiency high power uv lasers are of great interests also for various applications other than fusion.

Besides developmental works which lead to the construction of kJ class systems at several laboratories in USA.[1], England[2] and Japan,[3,4] the physical processes in the electron-beam pumped KrF laser medium has also been investigated, mainly in order to establish a "kinetic model" by which the characteristics of the KrF laser medium can be calculated with sufficiently high accuracy under any given condition.[5]

The rate constants of key processes, the value of which sharply influence the characteristics of the laser medium, should be known completely with sufficient accuracy in order to develop such a comprehensive kinetic model. The kinetic processes occurring in the electron-beam pumped KrF laser medium include various reaction processes which have been of considerable interest for physical chemists, and a great deal of works have been published on experimental and /or theoretical investigations of these.

However, unfortunately most of them were given only inaccurately when we started the present series of works. It is mainly because of the complexity of such a system where several production and dissociation processes for each of many coexisting excited species are occurring simultaneously.

We have been working for accurate determination of the kinetic constants for key processes which sharply influence the laser characteristics. Such efforts have been successful only by introducing several innovative techniques with which only one can be isolated from many other overlapping processes for quantitative study.

II. EXCITATION WITH A VERY SHORT PULSE[6-8]

Even under rather high operating medium pressure of one atom to several atoms in ordinary electron beam pumped KrF lasers, the excitation by 3ns FWHM electron beam pulse generated by FEBETRON 706 can be regarded to be short enough compared with the most of kinetic processes. Moreover, in many cases we can find out certain gas mix ratio under which only one process predominates in production or destruction processes of one of the excited species.

For example, the very large cross section of dissociative attachment process

$$F_2 + e^- \rightarrow F^- + F$$

reduces the density of the secondary electrons after the turn off of the electron off of the electron-beam pulse, if F_2 density is high enough, and no formation process is expected. Thus we can determine the rate constant of the slowest quenching process with a high accuracy by observing the decay rate in the tail part of the temporal profile without any trouble. Several important quenching rates including that of the quenching of Kr_2F due to F_2 collision.

III. ABSORPTION CROSS SECTION OF Kr_2F AND Ar_2F

Becuase Kr_2F has been supposed to be the main absorber in the electron-beam excited KrF laser medium, accurate determination of its absorption cross section is extremely interesting. We have accomplished this by exciting a gas mix of Kr and F_2 with a short pulse of electron beam and observing the change of Kr_2F density due to the irradiation of a KrF laser beam[9]. The result is remarkable because the experimentally determined absorption cross section of (1.0 ±0.2)x10^{-18}cm^2 is about 1/5 of the value which has been assumed by majority of researchers[7].

Similar measurement can be done also on Ar_2F by treating a gas mix of Ar and F_2,[10] and giving an experimentally determined value of absorption cross section of (3.3±1)x10^{-18} cm^2. This is three times larger than the value which has been used by majority of researchers. Such discrepancies are not surprising by looking at the fact that there have been no experimental neither elaborate theoretical work has been done on the determination of these cross sections.

IV. STUDIES BY PHOTOASSOCIATIVE FORMATION[11]

The complexity of the electron-beam pumped excimer lasers is greatly reduced if no free electrons were involved. Such a state of medium can be prepared by forming the excimers by photoassociation without electron-beam excitation. Although the energy of a 248nm photon is insufficient to excite the thermalized Kr and F, photo-dissociation of F_2 generates energetic fluorine atoms which can climb up the potential hill of the X state so that the formation of KrF is possible by absorbing a single 248nm photon.

We have found that a KrF excimer is formed by simply irradiating a gas mix of Kr and F_2 by 248nm KrF laser beam. Moreover, we found that there are at least two types of formation process, the fast one which shows linear dependence to the laser intensity and the slower one which shows second order dependence. The fast one is considered to be a single photon association described above, and the slower one has been assigned as the two photon and/or two step excitation.

As a matter of fact, the fluorescence spectrum of the KrF excimer formed by the fast process shows that the high frequency side of the B-X transition is missing, and no D-X component is observed as shown in Fig.2(a). The temporal behaviors of the KrF density formed by this process is distinquished in the following features: 1) the density of the photo association products is proportional to the square of the laser intensity in the low power region, 2) linear to the laser intensity in the high intensity region where the F_2 absorption is perfectly saturated, and 3) formation of both of the B and C states is extremely fast, and no delay was observable with the present time resolution of 4ns.

We have found also another formation channel which is much slower in build up and shows second order relationship with the laser intensity in the high intensity region. This has been assigned as the two photon formation of the B and C states having high vibrational energy. In fact, the fluorescence spectrum of the photoaccociation products by the latter process shows the high vibrational components in the B-X line and the D-X line is clearly observed as shown in Fig.2(b).

We have determined following constants using photoassociated KrF: 1) the rate constant for B-C mixing due to Kr collision for low vibrational states as $(12\pm3) \times 10^{-11} cm^3/s$, 2) the radiative lifetime of the C state as 60 ± 10ns, 3) the quenching rate constant of the C state due to F_2 collision as $2.5 \times 10^{-9} cm^3/s$. The latter two of these constants are experimentally measured for the first time. The presently obtained B-C mixing rate is much slower than the value given by Setzer et al.[9], but the latter value has been given under a condition where higher vibrational excited states are involved, and it may be a main cause of discrepancy with the present work where only low vibrational states are involved. However, of course more detailed quantitative study is needed to completely understand the cause of this discrepancy.

The photoassociated KrF excimer involves much smaller number of excited species than the electron-beam pumping does, and there are still more to be studied using this method: 1) radiative lifetime and 2) F_2 quenching rate of the B state KrF, 3) vibrational relaxation rate in the B and C states, and so forth.

V. CONCLUSION

The rate constants of key processes in the KrF laser medium which have significant influences on the characteristics of the laser medium have been determined with high accuracy as listed in Table 1 by employing several new ideas to simplify the kinetic processes. As the result of he improved accuracy, our kinetic code for the laser medium shows excellent agreement with the experimental result obtained by an extensive measurement of the amplifier parameters.

Especially the photoassociation production of KrF excimer provides an excellent condition, and there are many to be studied in near future.

Table I. Kinetic constants determined by the present work

$KrF^*(C) \rightarrow Kr + F + h\nu$	(1.7 ± 0.3)	$\times 10^7 \quad s^{-1}$
$Ar_2F^* \rightarrow 2Ar + F + h\nu$	0.45 ± 0.05	$\times 10^7 \quad s^{-1}$
$Kr_2F^* \rightarrow 2Kr + F + h\nu$	0.48 ± 0.05	$\times 10^7 \quad s^{-1}$
$KrF^* + e_s \rightarrow Kr + F + e_s'$	3.5	$\times 10^{-7} \quad cm^3 s^{-1}$
$Ar^* + 2Ar \rightarrow Ar_2 + Ar$	(6.0 ± 0.5)	$\times 10^{-33} \quad cm^6 s^{-1}$
$Ar^* + F_2 \rightarrow ArF^* + F$	(7.4 ± 1.0)	$\times 10^{-10} \quad cm^3 s^{-1}$
$Ar^* + Kr \rightarrow Ar + Kr^*$	(2.3 ± 0.6)	$\times 10^{-12} \quad cm^3 s^{-1}$
$Kr^* + 2Kr \rightarrow Kr_2^* + Kr$	(17 ± 2)	$\times 10^{-33} \quad cm^6 s^{-1}$
$Kr^* + F_2 \rightarrow KrF^* + F$	(8.8 ± 0.5)	$\times 10^{-10} \quad cm^3 s^{-1}$
$KrF^*(C) + F_2 \rightarrow Kr + 3F$	2.5	$\times 10^{-9} \quad cm^3 s^{-1}$
$Ar_2F^* + F_2 \rightarrow products$	(1.9 ± 0.1)	$\times 10^{-10} \quad cm^3 s^{-1}$
$Ar_2F^* + Kr \rightarrow ArKrF^* + Ar$	(1.35 ± 0.05)	$\times 10^{-10} \quad cm^3 s^{-1}$
$Kr_2F^* + F_2 \rightarrow products$	(1.8 ± 0.1)	$\times 10^{-10} \quad cm^3 s^{-1}$
$KrF^*(C) + Kr \rightarrow KrF^*(B,C)$	(12 ± 3)	$\times 10^{-11} \quad cm^3 s^{-1}$

REFERENCES
1 R.J. Jensen, L.A. Rosocha, J.A. Sullivan, Proc. Soc. Photo-opt. Instru. Eng., 622, 70(1986).
2 M.J. Shaw, Private Communication.
3 K. Ueda, A. Sasaki, H. Nishioka, H. Takuma, Technical Digest, IQEC'88 p. 672-3.

4   Y. Owadano, I. Okuda, Y. Matsumoto, T. Tomie, K. Koyama, A. Yaoita,
    S. Komeiji, I. Matsushima, M. Yano, Technical Digest, IQEC'88
    p. 678-9.
5   For example, F. Kannari, M. Obara, T. Fujioka, J. Appl. Phys.,
    57, 4309 (1985).
6   K. Ueda, H. Hara, S. Kanada, H. Takuma, Jpn. J. Appl. Phys.,
    21, L500 (1982).
7   K. Udeda, S. Kanada, M. Kitagawa, H. Takuma, Review of Laser
    Engineering, 11, 576 (1983).
8   K. Ueda, S. Kanada, A. Sasaki, H. Takuma, Review of Laser
    Engineering, 12, 357 (1984).
9   K. Hakuta, H. Komori, N. Mukai, H. Takuma, J. Appl. Phys.,
    61, 2113 (1987).
10 K. Hakuta, S. Miki, H. Takuma, J. Opt. Soc. Am, B 5, 1261 (1988).
11 K. Hakuta, K. Nakayama, M. Fujino, H. Takuma, Technical Digest,
    IQEC'88 p. 226-7.

LASER-INDUCED FLUORESCENCE OF Zn_2 EXCIMER

Zhang Pei-Lin, Xie Ding-Ning and Zhao SHuo-Yan

Department of Modern Applied Physics
Tsinghua University, Beijing, China

ABSTRACT

Emission spectra from laser-induced formation bound excited Zn_2 moleculas and their time evolutions are investigated. 224-278 nm and 350-480 nm bands with five maxima at 252 nm, 359 nm, 390 nm, 424 nm and 450 nm are reported.

The zinc dimer is of interested because it exhibits shallow van der Waals ground state and deeply bound excited states, characteristic of excimer dimer. In 1931 Hamada[1] first investigated spectra of zinc dimers, using hollow cathode discharge at high temperature. Since then, only a few papers about absorption spectra of Zn_2 in matrix of the rare gases[2] [3] have been published during last decades for lack of suitable method to produce Zn_2 dimers in the gas phase. In this paper we present laser induced fluorescence emission spectra of Zn_2 dimers and also emission spectra of zinc atoms, excitation mechanism of which is related to dissociation of zinc molecules.

A frequency-doubled Nd:YAG laser was used to pump a dye laser. The dye laser output was again frequency-doubled to produce 280-330 nm radiation. The laser pulses were made incident on the zinc vapor (contained in a cross heat-pipe oven with buffer gas Ar or He) and served to pump the zinc atomic states. Zinc dimers were created by association of the metastable zinc atoms with other zinc atoms. The fluorescence emitted was resolved with a monochromator followed by a boxcar and their time dependence was registered with an optical multichannel analyzer combined with adjustable delay generator or a boxcar used in scan mode.

Fig.1 shows the fluorescence spectrum of zinc dimer between 200-500 nm with narrow linewidth excitation. There are continuum bands in the range of 224 - 278 nm and 350 - 480 nm with five maxima at 252 nm, 359 nm, 390 nm, 424 nm and 450 nm respectively. Also present are sharp discrete lines which are due to bound-bound transitions in Zn_2. From the atomic fluorescence emission lines (see what follows) we know that high atomic states can be populated through hybrid processes. Therefore we propose that corresponding high molecular states may also be populated. In order to associate the band spectra with molecular transitions we have measured the intensities of the 252 nm and 424 nm bands as a function of time following the pump laser pulse. Fig.2 shows time evolution of 252 nm band and 424 nm band respectively. They are quite different so that the two bands can not originate from the same molecular state. From the theoretical potential energy curves[4] [5] we assign 252 nm band with emission from $'\Sigma_u^+$ state (which tends to 4'P+4'S atomic limit) to ground state $X'\Sigma_g^+$, in agreement with the result of Zn_2 of Hamada[1] and [3]. Hamada also found two other bands: 307.6-376.3 nm with the maximum intensity at 368.8 nm, and 389.0-535.0 nm with λ_{max}=445.0 nm. Our experimental results are different. We observed four maxima of the continum bands in the region of 350-480 nm and some discrete sharp lines. The 359 nm, 390 nm, 424 nm and 450 nm bands are probably due to transitions between excited molecular states. The various sharp components of the band correspond to v'-v'' vibrational structure.

Spectral scans revealed fluorescence emission lines due to atomic zinc transitions, twenty-seven lines in total. The normal fluorescence emission lines observed at right angle to the laser beam have been identified to atomic transitions, $n^3D - 4^3P_J$, n=4,5,6,7,8; $m^3S_1-4^3P_J$, m=5, 6,7, J=0,1,2; $4^3P_1-4^1S_0$, $4^1D_2-4^1P_1$ and $4^1P_1-4^1S_0$. Seven of them, $4^3D-4^3P_J$, $5^3S_1-4^3P_J$ and $4^1D_2-4^1P_1$, showed stimulated amplification of spontaneous fluorescence emission, collinear with pump laser beam. Two additional fluorescence emission lines are from ionic zinc transitions, ZnII $3d^{10}4p^2P_J-3d^{10}4s^2S_{\frac{1}{2}}$. Since two pump photon have enough energy to create atomic states 4^3D and 5^3S, molecular photon absorption followed by dissociation[6] can produce the population of these states. The slowly decaying component of fluorescence emission lines suggested that at least part of these excited atoms are produced by dissociation of molecules. The mechaniam of population of atomic states higher than two pump photon energy can be explained by recombination processes after photoionization.

212

REFERENCES

[1] Hamada H., Philos. Mag., 12, 50 (1931)
[2] Duley W.W., Proc. Phys. Soc. 91, 976 (1967)
[3] Ault B.S., Andrews L., J. Mol. Spectrosc. 65, 102 (1977)
[4] Hay P.J. et al., J. Chem. Phys. 65, 2679 (1976)
[5] Tatewaki H. et al., J. Chem. Phys. 82, 5606 (1985)
[6] Komine H., Byer R.L., J. Chem. Phys., 67, 2536 (1980)

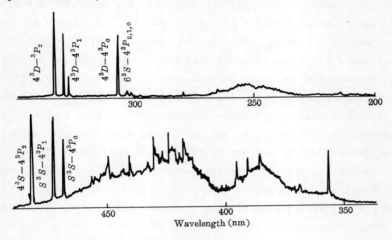

Fig.1 Fluorescence spectra of Zn_2 and ZnI.

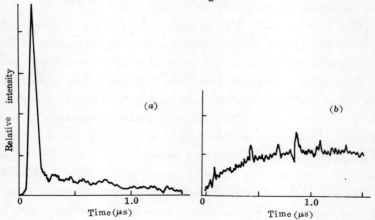

Fig.2 Time evolution of Zn_2 252nm (b) and 424nm (a) bands.

CALCULATION OF TRANSITION INTENSITY IN HETERONUCLEAR DIMER NaK : COMPARISON WITH EXPERIMENT

Zhang Limin, Wei Jilin, Gao Chengyue, Wang Hui,

Geng Yuzhen

Physics Department

University of Science & Technology of China,
Hefei, China

Xia Yuxing

Anhui Institute of Optics & Fine Mechasics,
Academia Sinica, Hefei, China

ABSTRACT

The Franck–Condon factor and the distribution of transition intensity of $D^1\Pi$ (v'=1, J'=67) to $X^1\Sigma^+$ (v''=1-17, J''=67) in NaK was computed. The comparison between computed values and experimental values of LIF spectrum induced by Ar^+ laser (5145 Å) was satisfactory. Futher analysis indicated that the transition dipole moment of heteronuclear dimer NaK was dependent on the internuclear distance r.

The intensity distribution of laser induced fluorescence (LIF) spectrum in heteronuclear dimer NaK induced by 5145 Å line of Ar^+ laser (INOVA10) was obtained. The transition was $D^1\Pi$(v'=1,J'=67) to $X^1\Sigma^+$(v'' =1-17, J''=67). The sample of NaK was produced in heat-pipe oven with temperature of 500 °C. The pressure of buffer gas He was P_{He}=23 torr. LIF spectrum was obtained by use of HRD1 double grating monochromator and lock-in amplifier. The intensity correction of spectrum was made by tunsten standard lamp.

With the Born–Oppenheimer approximation the band intensity of emission spectrum was [1]:

$$I \propto \nu^{4-2} \bar{R}_e q_{v'v''} \qquad (1)$$

where ν, $\bar{R}_e$ were the wavenumber and the average value of electron dipole moment respectively, $q_{v'v''}$ was the Franck–Condon factor of excited state and ground state with vibrational quantum numbers v' and v'' respectively:

$$q_{v'v''} = \left| \langle v' | v'' \rangle \right|^2 \qquad (2)$$

The vibrational eigenvalues and wavefunctions of excited state and ground states with Morse potentials were obtained by solving the

Schrodinger equations. The Morse potential parameters were calculated using preliminary molecular constants from ref. (1,2). The normalized Franck-Condon factor and transition intensity were computed. All calculations were completed on Altos micro-computor. By use of this method the transition intensity of $D^1\Pi(v'=1, J'=67)$ to $X^1\Sigma^+(v''=1-17, J''=67)$ was computed.

Both transition intensity $(I_{cal})_{v'v''}$ computed and $(I_{exp})_{v'v''}$ of LIF spectrum were shown in Fig.1. The agreement between computed values and the experimental values was statisfactory. In more detail, (I_{exp}/I_{cal}) showed a increase with internuclear distance r. This was due to that in the calculations of $(I_{cal})_{v'v''}$ the r-dependence of the transition moment $R_e(r)$ was omitted. By use of r-centroid approxamation (3) we deduced that:

$$\sqrt{\frac{(I_{exp})_{v'v''}}{(I_{cal})_{v'v''}}} \propto \frac{R_e(\bar{r}_{v'v''})}{R_0} = 1 + \frac{\alpha}{R_0}\bar{r}_{v'v''} \qquad (3)$$

where $\bar{r}_{v'v''}$ = $<v'$ r $v''>$ / $<v'$ $v''>$,α, R_0 were constants. By least-squares fitting according to eq.(3) the constant α/R_0 was obtained. The result was $|\alpha/R_0|$=0.38 D/Å (3.91 Å $<$ r $<$ 4.44 Å). This value was consistent with the value in ref.(4). So that the dipole transition moment for heteronuclear NaK ($D^1\Pi-X^1\Sigma^+$) had a larger increasing with internuclear distance r than for the homonuclear $Na_2(B^1\Pi, A^1\Sigma_u^+-X^1\Sigma_g^+)$. In conclusion, the calculations of Franck-Condon factor and the transition intensity of $D^1\Pi-X^1\Sigma^+$ of NaK with Morse potentials were convenient and had a satisfactory agreement woth experiment. Moreover the changing of the dipole transition moment of NaK can be obtained from the comparison of computed and experimental values by use of r-centroid approxamation.

We are grateful to Dr. Gao Yuming (AIOFM) and Zhang Yunsheng (USTC) for helping us in the calculation and in the experiment, respectively. The work was supported by Center of Structure and Element Analysis of USTC.

APPENDIX

The potentials used in Schrödinger equations were U'(r)=0.59794(1-exp(

215

$-0.54825(r-4.1813)))^2$ eV and $U''(r)=0.97392(1-\exp(-0.64834(r-3.4967)))^2$ eV.

REFERENCES

(1) G.Herzberg, "Molecular Spectra and Molecular Structure I. Spectra of Diatomic Molecules", D. Van Nostrand, (1953)

(2) A.J.Ross, C.Effantin, J.D'incan, R.F.Barrow, Mol. Phys. 56(1985), 903

(3) C.Noda, R.N.Zare, J.Mol.Spectrosc. 95(1982), 254

(4) L.B.Ratcliff, D.D.Konowalow, W.J.Stevens, J.Mol.Spectrosc. 110(1985), 242

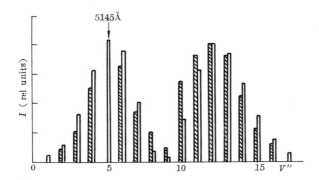

Fig.1 Comparison of intensities between the values computed (⸺) and the values of LIF spectrum (▨)

LASER-INDUCED FLUORESCENCE OF CCl$_2$ CARBENE

ZHOU Shi-Kang, ZHAN Ming-Sheng, SHI Ji-Liang,* Qiu Yue-Wu,**
WANG Chen-Xi

Laboratory of Laser Spectroscopy, Anhui Institute of
Optics and Fine Mechanics, Academia Sinica, P.O. Box 25, Hefei, P.R.C.

This paper reports the gas phase spectra of CCl$_2$ studied by laser ind-
uced fluorescence (LIF) in a crossed molecular beam apparatus. This is
the first observation of CCl$_2$ carbene produced by the reaction of F(^{2}p)
atom with CH$_2$Cl$_2$ and the reaction of O(^{3}p) atom with CF$_2$=CCl$_2$.
The apparatus is essentially the same as that described in ref.[1].
Briefly, the F (or O) atoms were produced by a microwave discharge in
CF$_4$ (or O$_2$) diluted in Ar. The CH$_2$Cl$_2$ beam crossed with F beam (or CF$_2$=
CCl$_2$ with O). The CCl$_2$ molecules were synthesized by the two step reac-
tions:

$$F + CH_2Cl_2 \longrightarrow CHCl_2 + HF \qquad H = -168KJ/mole$$
$$F + CHCl_2 \longrightarrow CCl_2 + HF \qquad H = -204KJ/mole$$

and the one step reaction:

$$O + CF_2=CCl_2 \longrightarrow CCl_2 + CF_2O.$$

The light source was a ring dye laser. The LIF signal was collected by
a photomultiplier, a photon counter, a multichannel analyzer and a computer.

Figure 1 is the LIF spectrum of CCl$_2$ observed by the reaction F + CH$_2$Cl$_2$
in the 540-630 nm region. Six peaks near 540, 550, 560, 570, 580, 590nm
were observed. The average space between two bands was 310 ± 10 cm^{-1}.
They can be assigned to be $\tilde{A}^1B_1(0v_2'0) \longleftarrow \tilde{X}^1A_1(000)$ transition of CCl$_2$
carbene. Comparing with the spectrum in the literatures obtained by the
matrix-isolation method, it suggests that the spectrum in solid phase
has a blue shift relative to the spectrum in gas phase: $\Delta V_{00} = V_{00}(M) -$
$V_{00}(G) = 190$ cm^{-1}. The vibrational frequency of bending mode of CCl$_2$ is
$V_2 = 310\pm10$cm^{-1}, which is consistent with the result of matrix-isolation.

The spectra of CCl_2 have a long V_2 progression. That is the feature of all halocarbenes. But comparing the gas phase spectra of carbenes with that of matrix, it is clear that the wave number of electronic transition in matrix has a shift. This indicates that there is a difference in the interaction potential between the matrix molecules with the ground electronic state of halocarbenes and with excited electronic state. But there is no obvious matrix shift for vibrational fundamental frequency for CCl_2 carbene. This is also consistent with the results of other halocarbenes.[1]

The spectra at 578 nm and 540–550 nm for reaction $F + CH_2Cl_2$ are the $\tilde{A}^1B_1(000) \leftarrow \tilde{X}^1A_1(000)$ and $(010) \leftarrow (000)$ transitions of CHF. This indicates that there is another channel of secondary reaction for F atom reaction with CH_2Cl_2:

$$F + CHCl_2 \longrightarrow Cl_2 + CHF \qquad\qquad H = -44kJ/mole$$

No spectrum of CHF was found for experimental of $O + CF_2CCl_2$.

This project was supported by the Science Fund of the Chinese Academy of Sciences.

REFERENCE

[1] Qiu,Y.W., Zhou,S.K. and Shi,J.L., Chem. Phys. Lett., 136,93(1987).

[2] Jacox,M.E., J. Phys. Chem. Ref. Data, 13,945(1984).

----

* Shanghai Institute of Organic Chemistry, Academia Sinica, Shanghai.

** Physics Department of Tongji University, Shanghai.

STUDY OF MULTIPHOTON IONIZATION SPECTRUM OF
BENZENE AND TWO-PHOTON ABSORPTION CROSS SECTION*

Zheng Bo, Lin Mei-rong, Zhang Bao-zheng

and Chen Wen-ju

Institute of Modern Optics, Nankai Univ.
Tianjin, P.R.China

Multiphoton ionization (MPI) spectroscopy developed in seventies is a most powerful technique to study excited states of molecules in gas phase (including triplet and high lying Rydberg states). The two-photon absorption (TPA) cross section is one of significant parameters in MPT technique. The method of multiphoton absorption[1] and fluorescene[2] to measure TPA cross section which have been reported are only applicable to valence states, studies about TPA cross section fot Rydberg two-photon transition have not been reported before. We present here a new method for attaining TPA cross section of molecular excited states by fitting experimental ion decay curve.

In this paper, the two-photon resonance MPI signal as a function of time has been got with density thepry. It is

$$\frac{dN_3(t)}{dt} = \frac{N_o' \, \sigma^{(1)} \sigma^{(2)} I^3}{\sigma^{(1)} I + \gamma - \sigma^{(2)} I^2} \{ e^{-\sigma^{(2)} I^2 t} - e^{-\left(\sigma^{(1)} I + \gamma\right) t} \} \tag{1}$$

where $dN_3(t)/dt$, $\sigma^{(\omega)}$, $\sigma^{(1)}$, γ, I, N_o^1 denote ion current, TPA section, OPI cross section, non-radiative relaxation rate, excited intensity, population of ground state at t=0 respectively.

A Quantal TDL_{50} Nd^{3+}:YAG laser pumping dye laser is used. The laser beam is focused into an ion cell by a quartz lens. The ion signal is obtained from a load resistance, then is fed to PAR 162/165 boxcar, and is recorded by X-Y recorder to measure the time dependence of ion signal. Firstly, we have measured two-photon resonance four photon ionization decay curve of valence state $^1B_{2u} 14_o^1$ of benzene, and then found best fitting according to equation (1). Fig.1 shows the measured ion decay curve and theoretical fitting curve of state $^1B_{2u} 14_o^1$. The TPA cross section of state $^1B_{2u} 14_o^1$ is about 1.59×10^{-52} ($cm^4 s \, ph^{-1} \, mol^{-1}$), which is averaged over measurements for several times. This is in agreement with the theoretical calculation of Honing et al[3] who predicted 0.4 <

$\sigma^{(2)} < 100 \times 10^{-51} (cm^4 s \ ph^{-1} \ mol^{-1})$. We have also for the first time obtained TPA cross section of Rydberg state $4R_g$ 0-0 of benzene. The result is $4.77 \times 10^{-54} (cm^4 s \ ph^{-1} \ mol^{-1})$. It is consistent with Parker et al[4] theoretical prediction: TPA cross section of Rydberg states $\sigma^{(2)} \leq 10^{-54} (cm^4 s \ ph^{-1} \ mol^{-1})$.

The experiment shows that the method is applicable to valence states, especially effective to measure the MPA cross section of Rydberg states. Since the ion signal is measured instead of light signal with this method, the detection accuracy and signal-noise ratio is greatly improved.

References:

1. P.R.Monson, W.M.Mcclain, J.Chem. Phys. 53(1979),29

2. L.Wunsch, H.J.Neusser, and E.W.Schlag, Chem. Phys. Lett. 31(1975), 433

3. B.Honig, J.Jortner and A.Szöke, J.Chem. Phys. 46(1967),2714

4. A.H.Zewail, Advances in Laser Chenistry, (Spring-Verlag New York, 1978), P.327

* This project is supported by National Natural Science Fundation of China

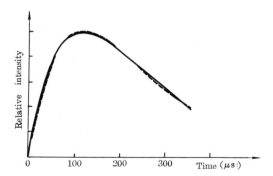

Fig.1. The measured ion decay curve and theoretical fitting curve of states $^1B_{2u} 14_0^1$

DICKE NARROWING OF N_2O LINEWIDTH PERTURBED BY N_2 AT 10μm BAND

Cai Peipei   and   Shen Shanxiong

Department of Physics, East China Normal University

Shanghai,200062, China

ABSTRACT

Dicke narrowing of spectral linewidth is observed in high J vibration rotation transitions of $10\,\mu m$ band of N_2O with N_2 as the collision partner. The narrowing occurs at the reasonably low pressure of about 300 Torr and has been obtained through measurement of resonance absorption spectroscopy.

Introduction

In the research of molecular absorption spectroscopy, the broadened linewidth and its mechanism are widely discussed. Usually, the more the gas pressure, the more broadening the spectral linewidth is, since the increase of collision chance enables the lifetime of transition energy level to decrease. When collisions do not affect the quantum state of radiating or absorbing system and mean free path is smaller than the wavelength of interacting radiation, linewidth will be narrowed, which was discussed by Dicke in 1953[1].The Dicke narrowing of N_2O gas pertubed by N_2 with the help of resonance absorption spectroscopy is researched in this paper

Experiment

The experiment apparatus employed is the same as Shen[2]. A grating tuning N_2O laser is used as N_2O 10 R(16) radiation resource and the resonance absorption spectroscopy of N_2O gas is observed. The gas sample cell is a White cell with changeable pathlength from 40 m to 1000 m. After the cell is vacuumized, N_2O gas of 20 Torr pressure is injected, then pure N_2 gas is added up to the total pressure of 343Torr.

Results and analysis

The least square method is used to fit with Lorentzian line shape formula for the absorption coefficients measured at pressure lower than 250 Torr. Line strength, self-and N_2-broadening coefficients are obtained.

Using these data, the linewidth observed can be calculated from absorption coefficients obtained at the pressure higher than 250Torr. The results are shown in Fig.1 where (a) shows the observed linewidth vs pressure, (b) shows calculated collision broadened linewidth,and (c) shows Dicke narrowing since the linewidth observed should be contributed by both of collision broadening and Dicke narrowing. Linewidth narrowing can be observed effectively when the masses of collision partners are same. In our experiment the J-value of laser beam is higher so that the narrowing can be still observed though there is a great difference between two masses.

References
[1]   Rao,D.R. and Oka,T., J.Mol.Spectrosc. 122, 16 (1987)
[2]   Shen,S.X.,International Journal of Infrared and Millimeter Waves, 6,423 (1985).

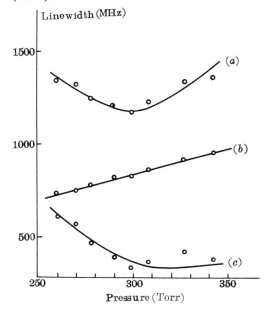

Fig. 1

POLYATOMIC MOLECULAR IONS STUDIED BY LASER PHOTODISSOCIATION SPECTROSCOPY

Chen Lingbing

Fudan University, Shanghai, P. R. China

H. Figger, W. Quint, H. Walther

Max-Planck-Institut für Quantenoptik, Garching, F. R. Germany

Compared to the enormous amount of spectroscopic data accumulated for neutral molecules, relatively little information is available for ionic molecules especially for polyatomic ions. In particular it is supposed that the unidentified diffuse interstellar absorption lines might be caused by predissociating states of molecular ions present in the interstellar space. We report here a laser photodissociation spectroscopic study on polyatomic ions $HCOOH^+$, $DCOOD^+$, CH_3I^+ and CS_2^+.

A photodissociation spectrometer including three tandem quadrupole mass spectrometers was set up along with an UV argon laser pumped tunable stilbene dye laser for the measurement. Molecular beam were injected into the ion source through a nozzle of 50–300 μm in diameter. Ion samples were prepared by low energy electron impact and directed into the first quadrupole mass filter, where the parent ions to be studied were selected out and transferred to the second stage by the ion optics.

The second rf-free quadrupole stage focus the ions, which interacted with the collinearly running CW dye laser beam and partially dissociated afterwards. Parent and fragment ions then entered the third mass selective quadrupole system, where the fragment ions were filtered out and transferred to the secondary electron multiplier for detection. The fragment ion current and the laser power were recorded while scanning the wavelength of the laser from 410 to 460 nm.

Photodissociation spectra for $HCOOH^+$ and $DCOOD^+$ were obtained for the first time, fragmenting to HCO^+ + OH and DCO^+ + OD respectively. The signal to noise ratio of the measurement reached 7:1 and 8:1 respectively.

Dissociation cross sections were estimated to be less than 10^{-20} cm^2. CH_3I^+ and CD_3I^+ fragmenting to CH_3^+ + I and CD_3^+ + I provided the most informative high resolution photodissociation spectra for polyatomic ions we studied.

Absorption transition occurred from the lower spin-orbit component of the ground state $\tilde{X}(^2E_{3/2})$ to the first excited state $\tilde{A}(^2E_{1/2})$. Three vibrational progressions of C-I stretching mode for both molecular ions were assigned. The vibronic bands were rotationally resolved. Vibrational constants for the A state of CH_3I^+ were derived by a least square fit: $\nu_2=1157$ cm^{-1}, $\nu_3=302$ cm^{-1} and $\chi_{33}=2.56$ cm^{-1}. The symmetric top rotational constants for the (0,1,10) band were determined by computer simulation: $A=4.785$ cm^{-1}, $B=0.176$ cm^{-1} and $\zeta_e=0.51$. Two new dissociation channels in the visible were detected by laser photodissociation for CS_2^+ with the signal to noise retio higher than 100:1. The two channels are CS_2^+ fragmenting to $CS^+ + S$ and $CS + S^+$. $\tilde{B}(^2\Sigma_u^+)$ was assigned the lower state of the photoabsorption transition and $\tilde{D}(^2\Sigma_u^+)$ the upper state.

TRANSVERSE-OPTICALLY PUMPED ULTRAVIOLET S_2 LASER

Yu Junhua,   Sun Shangwen,   Cheng Yongkang,
Tang Chen,   Ma Zuguang

Institute of Opto-Electronics
Harbin Institute of Technology, Harbin, P.R. China

ABSTRACT

A series of six laser lines at 330-390nm on the B-X transition
in sulfur dimer vapor transverse-pumped by a XeCl laser at
308.1nm was first observed and the corresponding absorption
coefficient and small signal gain coefficient were measured.
The mean gain is about 0.30 cm^{-1}.

S_2 laser was attractive as a blue-green laser for ocean
optics, but it is also a promising candidate for an efficient ultraviolet laser system. The shortest wavelength of
UV lasing on the B-X transition in S_2 at 365nm was achieved.

We realized the anticipated extending of the S_2 UV lasing region towards the shorter wavelength for about 35nm in
a T-shaped quartz tube with a plane-plane cavity and two-
temperature heating-zones by transverse-optically pumping
from an EMG 201MSC XeCl excimer laser. The detail structure of this laser was described in reference 3. To verify
the possibility to lase in UV region, we measured the absorption coefficient and the small signal gain coefficient.
It was found that the gain coefficient is always larger
than the absorption in the spectral region from 300nm to
450nm (Fig.1).   There are two absorption peaks near 311.6nm
and 391.9nm, but the latter one results from the trimer S_3,
so the most efficient pumping wavelength for S_2 should be
selected near the former absorption peak. In our case the
XeCl laser at 308nm is very suitable pumping source for S_2.
The average value of the gain coefficient in the whole UV
region is about 0.30 cm^{-1}.

We have achieved UV laser oscillation on the B-X transition of sulfur dimers, which has obvious threshold effect
with laser intensity stronger than the fluorescence one by

a factor of 1000. Obtained six UV laser bands peak at
3804.2Å, 3531.6Å, 3448.1Å, 3401.8Å, 3377.3Å and 3309.3Å,
respectively (Fig.2).

REFERENCES

[1] Leone S.R., Kosnik K.G., "A Tunable Visible and Ultra-violet
    laser on S_2", Appl. Phys. Lett., 30, 346 (1977)
[2] Girardeau-Montaut J.P., Moreau G., "Optically Pumped Super-
    fluorescence S_2 Molecular Laser", Appl. Phys. Lett.,
    36, 509 (1980)
[3] Yu Junhua, Sun Shangwen, Cheng Yongkang, Zhou Li and Ma Zuguang,
    "An Optically Pumped S_2 Blue-Green Laser", Chinese J.
    Lasers, 15, 112 (1988).

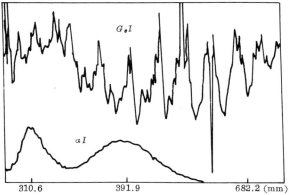

Fig 1 Dependence of single pass absorption
and gain of S_2 on the wavelength

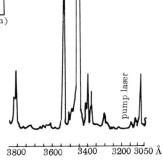

Fig 2 S_2 UV laser spectrum
on the B X transition

MULTIPHOTON IONIZATION OF PROPANAL
BY HIGH POWER LASER

Liu Houxiang, Li Zhaolin, Qin Yile, Li Shutao
and Wu Cunkai

Laboratory of Laser Spectroscopy, Anhui Institute
of Optics & Fine Mechanics, Academia Sinica, Hefei,
China.

The process of unimolecular fragmentation of C_3H_6O species have been studied in a few works. Despite of this, little detail is known about the fragmentation mechanism of propanal ion with the oxygen atom on the first carbon atom in it. We have studied the multiphoton ionization dynamics of acetaldehyde and formaldehyde with a diffuse molecular beam. We report here, for the first time, the experimental results about the multiphoton ionization of propanal molecule by using a pulsed molecular beam and a high power laser.

The propanal vapor at room temperature seeded in 3.0 atm. helium gas is introduced into ionization region through a pulsed nozzle of 0.125 mm aperature. The time averaged pressure is about 10^{-6} Torr in the time-of-flight mass spectrometer chamber when the pulsed nozzle is operating. A high power of XeCl excimer laser beam was focused on the ionization region and crossed with the molecular beam perpendicularly. Photoions were detected by the time-of-flight mass analyzer and the resulting signals were processed and recorded by a transient recorder and a multichannel analyzer system.

Two dimensional laser-intensity-dependent mass spectra of propanal were measured. A much extensive fragmentation pattern including ion peaks corresponding to C^+, CH_3^+, C_2^+, C_2H^+, $C_2H_2^+$, $C_2H_3^+$, $C_2H_5^+$ and/or CHO^+, CH_3O^+ and $C_3H_5O^+$ was observed experimentally at high power of laser radiation. However, only $C_2H_3^+$, $C_2H_5^+$, and/or CHO^+ and $C_3H_5O^+$ ions were observed at low power of laser radiation.

The laser intensity dependences of some ion signals, m/e = $27(C_2H_3^+)$, m/e = $29(C_2H_5^+$ and/or $CHO^+)$ and m/e = $57(C_3H_5O^+)$, were measured, that gives the laser intensity indeces as 4.2, 3.3 and 2.6, respectively.

Based on our experimental results, the multiphoton ionization mechanism of propanal molecule was analyzed and the calculation results by rate equation approach is consistent with the experimental observation.

UV MPI MASS SPECTROSCOPY AND DYNAMICS OF PHOTODISSOCIATION OF SO_2

Li Zhaolin Liu Houxiang Li Shutao Qian Yile and Wu Cunkai

Laboratory of laser Spectroscopy

Anhui Institute of Optics and Fine Mechanics, Academia Sinica

Sulfur dioxide (SO_2) is one of the most interesting molecules in the atomspheric physics and chemistry. In this paper the multiphoton ionization time of flight mass spectroscopy of SO_2 is experimental investigated under super sonic conditions using XeCl excimer laser. The mechanism of multiphoton processes of SO_2 is proposed.

The supersonic nozzle beam crosser is at a right angle with the focused 308 nm excimer laser. The photo-generated ions are detected and amplified by a time of flight mass spectromter. The output is fed to a Biomation fast transient recorder which is linked to a multichannel analysis to average the signals and to reapear the figure.

The mass peaks of SO_2 appearing m/e=32, m/e=48 and m/e=64 are assigned to S^+, SO^+ and SO_2^+ respectively. The dependence of intensity of ion signals produced by MPI of SO_2 to the laser power is studied. The index of laser power are 2.5, 2.7, and 1.2 respectively. Then the dependence of the ratio of each ions intensity on the laser power is obtained.

According to the energy levels of SO_2 and the experimental results, the MPI mechanism of SO_2 could be proposed as 3, (3+1) and (3+1+1) photon processes. First the SO_2 molecules are excited to the high Rydberg State H and ionized to ground state parent SO_2 ($X\ ^1A_1$) following absorb three 308 nm photons. Then the immediately photodissociation into SO ($X\ ^2\Pi_2$) + $O(^3P_2)$ and S^+ ($^4S_{3/2}$) + $O_2(X\ ^3\Sigma)$ could take place because the absorbed energy is much more than the ionization/ dissociation threshold. Both of SO^+ and SO_2^+ could absorb the fifth photon to become S at the strong laser intensity So the branch ratio of S^+ is increased following the enhancement of laser power while the SO is decreased in the mean time

According to the experimental results and compuation we have obtained the dependence of the mean energy <E> obsorbed by every parent molecule ion on the laser power. By comparing the different mass patterns of SO_2 produced by one-photon ionzation and MPI at 220 nm and 308 nm, we could consider that the MPI processes of SO_2 are dominated by laser wavelength rather than the pure statistical.

The results are important to both of study on MPI processes of small molecules and atomospheric photochemistry.

MULTIPHOTON IONIZATION-FRAGMENTATION PATTERNS OF
ETHYLAMINE AND DIMETHYLAMINE ISOMERS

S.T.Li, H.X.Liu, Z.L.Li, W.X.Ke,* J.C.Han and C.K.Wu

Laboratory of Laser Spectroscopy, Anhui Institute of

Optics & Fine Mechanics, Academia Sinica, Hefei, China

With a supersonic molecular beam and mass spectrometrically detecting system, we found very different multiphoton ionization fragmentation (MPI-F) patterns for ethylamine (EA) and dimethyl-amine (DMA). 308 nm ionization produced intense signals at m/e=45, 39 and 30 and weak ones at m/e=43 and 28 for EA, and intense signals at m/e=44, 43, 39, 28, 18 and 15 and weak onse at m/e=45, 41, 30 and 27 for DMA. 442 nm ionization produced a more extensive fragmentation patterns for both EA and DMA, and some new additional ion signals appear at m/e=41, 26 and 15 for EA, and m/e=29 and 26 for DMA.

The linear dependences of various ion intensities on sample pressure measured show that MPI-F of EA and DMA is unimolecular collisionless process. The laser intensity dependences of various ion signals were measured, which gives laser intensity indices as n=1.63 (m/e=45), 1.34 (m/e=44), 1.63 (m/e=43), 1.17 (m/e=39), 1.83 (m/e=30), 1.34 (m/e=28), 1.17 (m/e=18) and 1.64 (m/e=15) for DMA at 308 nm; n=2.3 (m/e=45), 2.5 (m/e=43), 3.5 (m/e=30) and 4.2 (m/e=28) for EA at 442 nm.

Based on our experimental results, it is suggested that MPI-F mechanisms of both EA and DMA are parent ion fragmentation model. DMA ion is formed by (2+1) MPI via intermediate state (n,σ*) at 308 nm, and EA ion is formed by (2+2) and (2+1+1) MPI processes via intermediate state (n,3s) and (n,σ*), respectively, at 442 nm. The dynamics analysis indicates that the (2+1+1) MPI is a favorable channel for EA ion formation

at 442 nm. The extensive daughter ion in DMA at 308 nm is induced only by the excess energy that total available energy from initial (2+1) MPI event exceeds the appearance potential of parent ion. But, for EA at 442 nm this further fragmentation is masked by additional photon absorption of parent ion or daughter ions.

It is explained that the primary fragmentation processes in DMA ion and EA ion are dissociation reactions lossing H atom and methyl, respectively. Postulated fragmentation processes are presented, as shown in Fig.1 (a) and (b), and discussed.

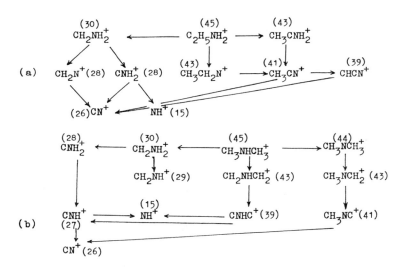

Fig.1 Postulated reaction mechanisms of unimplecular decompositions of (a) ethylamine (EA) and dimethylamine cations.

CARS MESUREMENTS OF SF_6 PUMPED BY A CO_2 LASER PULSE

Iwao KITAZIMA

Department of Electronics, Fukui University

Bunkyo 3-9-1, Fukui City, Japan

The SF_6 molecule, a typical heavy polyatomic one, has 15 vibrational modes (degenerated into $\nu_1 \sim \nu_6$) which result in formation of the quasicontinuum from a comparatively low energy level (3000 or 5000 cm^{-1}) to the dissociation limit due to stochastization of vibrational energy by anharmonic shift, splitting and interactions among the modes. Although the quasicontinuum is very important for the multiphoton dissociation, the dynamic properties are not yet completely understood. Today it is supposed that the vibrational distribution in the state is nonequilibrium and that the energy relaxation depends nonlinearly on the rate of excitation [1,2]. The molecular collision effects on excitation and relaxation must be also considered because collisions are inevitable in most experiments. We found in the measurements cn the ν_3 mode nonlinear absorption of SF_6 (5 Torr) by a TEA CO_2 laser that the absorption spectra was red-shifted with increasing the laser intensity and the significant difference between the additive effects of He and N_2 suggested the stochastization limit effectively lowered down to about 2000 cm^{-1} [3].

Now we will report on the ν_1-CARS measurements of SF_6 pumped by a tunable CO_2 pulse laser [4,5] for observations of the energy transfer between ν_1 and ν_3 modes, the relaxations from higher levels and quasicontinuum, and collisional effects on them. The SF_6 gas (with additive gases) contained in a cross-shaped cell at the room temperature pumped by a TEA CO_2 laser pulse (10.4 μm P(20), 0.8J/cm^2, 250 ns duration) focused into the beam diameter of 2 mm^{ϕ} , in the cell centre. The gas is probed in the transverse direction by collinear focused beams (0.1 mm^{ϕ}) at the second harmonic from a YAG : Nd laser (532 nm, 10 ns) and a dye laser ($\simeq$555 nm, 10 ns).

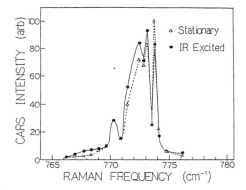

Fig.1. CARS spectra of SF_6
600 Torr probed at time
delay 20 μs after pumped
by a CO_2 P(20) laser pulse
($1J/cm^2$, 250ns), and at
stationary condition
without pumping.

Fig.1 shows the experimental result on the CARS spectra of SF_6 600 Torr probed at the time delay $\tau_d = 20$ μs after pumped by a CO_2 laser, and at the stationary condition (without pumping). The hot bands coupled with ν_4, ν_5 and ν_6 at 771~773 cm^{-1} were enhanced by the IR excitition, and also the low intensity broad band at about 766~770 cm^{-1} attributed to the quasicontinuum was enhanced, due to the energy transfer from the ν_3 mode by the V-V exchange. On the other hand, the fundamental band of the ν_1 mode at 774 cm^{-1} was quenched by the IR irradiation, so that the Raman transition from the ground state to the $\nu_1 = 1$ was supressed by the decrease of the ground state population due to the $\nu_3 = 0 \rightarrow 1 \rightarrow 2$ excitation.

The delay time dependence of the hot band enhancement in SF_6 50 Torr was also observed with/without additive gases (He, Ar, N_2, CO_2), as well as that of the ν_1 band quenching. The former decay time is about 10~25 μs and the latter, more strongly affected by the V-T relaxation of the ν_6 mode, is about 35~60 μs, depending on the gas collisional conditions.

References
[1] V.S.Letokhov; "nonlinear laser chemistry" ed.by F.P.Schäfer
     (Springer, 1983) p.181.
[2] C. dÁbrosio, R.Bruzzese, A.Ferrigno and S.Solimeno; Optics Comm.55
     (1985) 159.
[3] I.Kitazima; Optics Comm.53 (1985) 27.
[4] R.V.Ambartsumyan, S.A.Akhmanov, A.M.Brodnikivskii, S.M.Gladkov,
     A.V.Evseev, V.N.Zadkov, M.G.Karimov, N.I.Koroteev, and A.A.Puretskii
     ; JETP Lett.35 (1982) 170.
[5] S.S.Alimpiev, S.I.Valyanskii, S.M.Nikiforov, V.V.Smirnov, B.G.Sarta-
     kov, V.I.Fabelinskii, and A.L.Shtarkov; JETP Lett.35 (1982) 291.

ANGULAR DEPENDENCE OF PHASE CONJUGATION OF CO_2 LASER ON SF_6 GAS

C.L.Cesar, J.W.R.Tabosa, M.Ducloy*, and J.R.Rios Leite

Departmento de Fisica, Universidade Federal de Pernambuco

50739 Recife-PE, Brazil

Abstract

Sub-Doppler degenerate four-wave-mixing of continuous CO_2 laser on SF_6, giving excelent signal of conjugation for angles up to 15 degrees on a 2cm cell, permitted checking theory for linewidth and reflectivity angular dependences.

Resonant degenerate four wave mixing was used to produce phase conjugation of the 10.6 μm P(18) CO_2 line on a low pressure SF_6 cell,as represented in figure 1. The CW laser was tunable over 60 MHz covering the A_2^1 P(33) line of SF_6 and thus giving sub-Doppler lineshape for the reflected beam, as shown in figure 2. The signal was detected with an HgCdTe photodiode** followed by a lock-in amplifier referenced to a 600 Hz mechanical chopper on the probe (object) beam. A vibrating mirror (240 Hz) on the probe path was critical to eliminate the interference between parasitic reflexions of the pumps and the conjugate fields reaching the detector. backward pump intensity of 250 mW in 7 mm diameter, the lineshapes had the dispersion dominated doublet form(figure 2a) due to the saturation effect. To compare our experiments with theory the backward laser beams was attenuated to give single peaked lines as in figure 2b. A residual power broadening of the linewidth still persisted as discussed bellow. The polarization of the conjugate field was verified to be in accordance to predictions[3] for different configurations of pump and probe polarizations.

Theoretical predictions for the phase conjugation in a Doppler broadened gas medium has been done using density matrix calculation for two-level system[2](see other reference in 2). The relevant constants for the SF_6 line are known[1]: Doppler width 29 MHz, collision broadening 17±3 MHz/Torr and saturation intensity 8.0 W/cm^2Torr^2. Assuming a two level model for the SF_6 transition and using equations from ref.[2] and adjusting the signal at $\beta = 0°$ to normalize the detector calibration one obtains the curve of figure 3. The corresponding linewidths of the signal are given in figure 4. The zero angle linewidth expected at 200 mTorr is 3.4±0.6 MHz and we measured 4.5 MHz. Such excess is due to residual power broadening of the line.

These results are the first study, without optical pumping[4] and using CW low power laser[5], to verify the angular dependence of phase conjugation in a two level atomic gas.

Thus we have demonstrated that with low power CW CO_2 laser it is possible to obtain phase conjugation by Doppler-free degenerate four-wave-mixing with good enough signal to noise to confront theoretical predictions for angular dependence of reflectivity and linewidth. Further work is under way, directed to higher angle (90 degress), saturation effects and ultranarrow frequency conjugation in SF_6.

_____

*Permanent adress:Laboratoire de Physique des Lasers,(CNRS LA 282)
Universite Paris-Nord,Villetanese, France.
**The HgCdTe detector was lend by SAT-France.

REFERENCES

!   J.L.Boulnois,P.Aubourg   and   A.Van Lerberghe;Appl.Phys.Lett.,42 (1982)225.

2   M.Ducloy   and   D.Bloch;J.Physique42(1981),711.

3   M.Ducloy   and   Bloch;Phys.Rev.(A),A30(1984),3107.

4   L.M.Humphrey,J.P.Gordon   and   P.F.Liao;Opt.Lett.,5(1980)56.

5   D.A.Goryachkin,U.P.Kalinin,I.M.Kozlovskaya,F.A.Komin and N.A. Romanov;Opt.Spectrosco. USSR,60(1986),198.

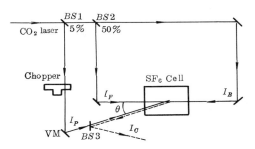

Fig.1 Experimental diagram. BS are beam, splitters. I_F forward beam, I_B backward beam I_P probe beam and I_C conjugate beam.

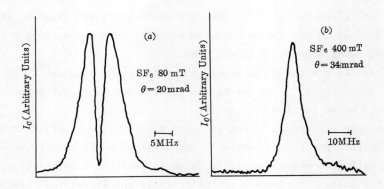

Fig 2 Conjugate signal on SF_6: (a) I_F=55mW I_B=250mW and the line is
splitted by saturation; (b) I_F=55mW and I_B=10mW I_P=10mW.

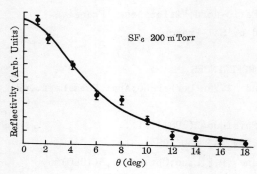

Fig.3 Angular dependence of
reflectivity with low pump
power. The continous line
is the theoretical results
from reference 2, with the
SF_6 parameter. Maximum refle
ctivity was typically 10^{-3}.

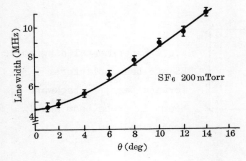

Fig.4 Linewidth angular depend
ence for same conditions of fig.
3. Continous line is theoretical
prediction from reference 2, with
1.1MHz added to the zero degree
linewidth (atributed to power
broadening) and no adjustable
parameter.

RESOLUTION OF STRETCHING-VIBRATIONAL AND TRANSLATIONAL
RAMAN BANDS OF LIQUID WATER BY MEANS OF POLARIZATION FOUR-
PHOTON SPECTROSCOPY

A.F.Bunkin, A.S.Galumian, D.V.Maltsev, K.O.Sursky
General Physics Institute,Ac.Sci.USSR,117942,Moscow,USSR

1. Polarization CARS and RIKES spectra of the O-H stretching-vibrational Raman band of liquid water ($2000 \div 4100$ cm^{-1}) have been obtained when the pump laser waves were within the water transmission region[1]. The structure of the band was resolved (with 5 lines detected) and deformations of the CARS and RIKES spectra were investigated as a function of temperature over a range of 5 to 60 °C and LiI concentration. The CARS and RIKES spectra were compared with a number of mixed models of liquid water but neither of these models, however, could yield a quantitative description of our experimental results. We calculated spectroscopic parameters of overlapping lines. Fig.1 shows the temperature-induced deformations of the polarization CARS spectra of the water stretching band.

2. For the first time, RIKES spectra of liquid water have been obtained in the region $0 \div 200$ cm^{-1} at 7 °C and 22 °C. Four Raman lines with frequencies $71 cm^{-1}$, 82 cm^{-1}, 12 cm^{-1}, and 16 cm^{-1} have been registered. These lines were attributed to the symmetrical and antisymmetrical hydrogen-bond bending vibrations of $(H_2O)_2$ (71 cm^{-1}, 82 cm^{-1}) and $(H_2O)_4 \div$ $(H_2O)_6$ (12 cm^{-1}, 16 cm^{-1}). The RIKES spectra of the intermolecular vibrations of liquid water at a temperature of 7 °C are demonstrated in Fig.2.

R e f e r e n c e

1. Bunkin A.F., Vlasov D.V., Galumyan A.S., and Sursky K.O.
   In: Oceanic Remote Sensing, Eds.F.V.Bunkin and K.I.Voliak, Nova Science Publishers, N.Y.,74-88 (1987).

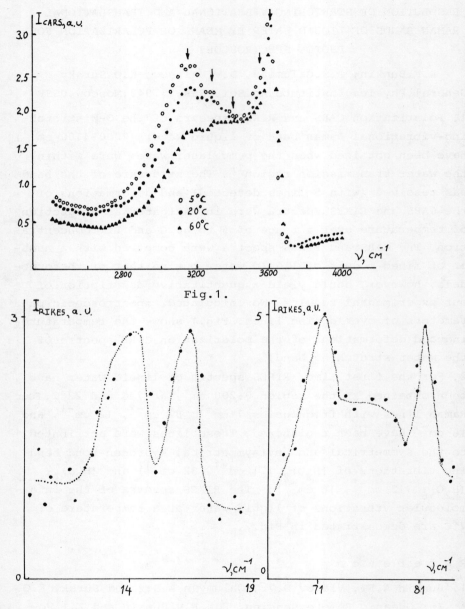

Fig.1.

Fig.2.

LASER-PRODUCED PLASMA AS AN EFFECTIVE SOURCE
FOR X-RAY SPECTROSCOPY

I.I.Sobel'man

P.N.Lebedev Physical Institute

Academy of Sciences of the USSR

Leninsky prospect 53, Moscow, USSR

Laser-produced plasma (LPP) - the plasma created by sharp focusing a high-power laser beam on a solid target very effectively radiates in extreme ultraviolet and soft X-ray spectral regions. For neodimium glass laser flux density $q \sim 10^{13}-10^{15} W/cm^2$ and high Z_A targets the X-ray conversion efficiency could be as high as 0.1-0.5. The effective radiation temperature T is of the order of 100-200 eV.

The possible applications of the LPP as an effective source of soft X-rays for spectroscopy and same other experiments were discussed in many papers during the last years but very serious limitations are imposed on such applications due to the large divergency of LPP radiation $\sim 2\pi$ and the lack of transmiting, reflecting and focusing "optics" for soft X-rays. Due to recent progress in technology of multilayer coatings and in technology of high quality substrates it seems possible now to fabricate multilayer spherical mirrors with reflectivity about tens of percent at least for λ 10-60 nm.

In our experiments the Mo-Si multilayer mirrors with period 2d=18 nm, radius of curvature r=150 mm, reflection coefficient for normal incidence 0.2, spectral selectivity $(\lambda/\Delta\lambda)=14$ are used[1]. The beam of the second harmonic neodmium glass laser with pulse energy up to 20 J and pulse duration 3 nsec is focused in a spot with diameter ~ 30 microns on Re target $(Z_A=75)$. The soft X-ray radiation of the LPP is collimated by spherical multilayer mirror. The energy output $\sim 10^{-5} J$ in spectral region 17-19 nm and the divergency of the collimated beam better than 10^{-3} rad are obtained. When radiation of the LPP is focused by multilayer mirror the flux density $10^7 W/cm^2$ is measured. For λ 18 nm such flux density is extremely high.

In experiments with carbon plasma the strong spectral line λ 18.2 nm

of CVI coincides with the maximum reflectivity of the multilayer mirror.
.As a result the picture of LPP in spectral line λ 18.2 nm is registered giving the distribution of the hydrogen like carbon ions.
The possibility to collimate and focus soft X-ray radiation of the LPP
using spherical multilayer mirrors is very important for different
applications. One of the most interesting application seems EXAFS spectroscopy. The very soft X-ray spectral region is especially of interest
to EXAFS spectroscopy of light atoms. in particular oxygen atoms in
high-temperature superconductors.

Reference
[1]   V.Artsimovitch, S.Gaponov, U.Kas'yanov, B.Luskin, N.Salaschenko,
A.Shevelko, I.Sobel'man; JETP Letters, Vol.46(1987)391-394.

ROTATIONAL STRUCTURE OF THE LOW LYING
ELECTRONIC STATES OF SAMARIUM MONOXIDE

Guo Bujin and C. Linton

Physics Dept., University of New Brunswick

Fredericton, N.B. Canada E3B 5A3

Our investigation into the spectroscopy of Samarium Monoxide is part of a continuing program that has successfully explained the electronic structure of several rare earth oxides (1) in terms of a Ligand Field Theoretical model (2). We have recently (3) identified, assigned and linked eleven low lying states and shown that the results are consistent with Ligand Field Theory predictions. In this paper, we present the detailed rotational analysis of several transitions involving three of the lowest states, $X0^-$, $(1)1$ and $(2)0^+$ (0^-, 1, 0^+ are the Ω-values of the states).

High resolution spectra were obtained using Wavelength Selected Excitation Spectroscopy. The transitions studied are shown in Figure 1. There was significant Ω doubling in both $\Omega=1$ states and in the $\Omega=2$ state and this added to the complexity of the spectra. However, for the $[16.9]1\leftarrow(1)1$ transition it was possible to obtain separate spectra for the e and f parity components by careful selection of the monochromator wavelength. For example, if the monochromator were set to detect the Q branch of the $[16.9]1-X0^-$ transition, only transitions to the e-levels of the upper state would be detected. This is illustrated in Figure 2 which shows a portion of the P branch obtained when (a) the f-levels and (b) the e-levels of the $[16.9]1$ state were detected. Transitions for each of the 7 stable isotopes are clearly resolved.

All the rotational lines were assigned in all the transitions. For each isotope, all the lines were fitted simultaneously using a least-squares technique and the rotational and Ω doubling constants determined for each state. The results for ^{154}SmO are listed in Table I.

There are numerous perturbations. This is to be expected as there are predicted to be > 24000 electronic states below 100000 cm^{-1}.

242

We have already reported (3) a perturbation at J=5 in the f-levels of
the [16.9]1 state. Fig. 2 clearly shows a perturbation at J~19 in the
e-levels. It is very likely that the same state is responsible for
both perturbations and that the investigation in progress will enable
us to assign and calculate constants for the perturbing state.

References

(1) C. Linton, M. Dulick, R.W. Field, P. Carette, P.C. Leyland and
     R.F. Barrow. J. Mol. Spectrosc. 102, 441-497 (1983)

(2) M. Dulick. Ph.D. Thesis, MIT. 1982

(3) C. Linton, Guo Bujin, R.S. Rana and J.A. Gray. J. Mol.
     Spectrosc. 126, 370-392 (1987)

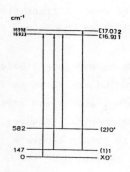

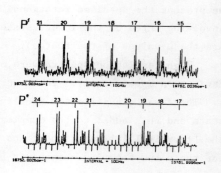

TABLE I. Spectroscopic Constants (all in cm⁻¹) for ¹⁵⁴SmO

State	T_0	B_0	$10^7 D_0$	$10^3 q$	$10^6 p$
X0	0	0.352552(24)			
(1)1	146.9755(49)	0.356557(38)	2.47(50)	7.20(13)	0.314(27)
(2)0⁺	582.2251(58)	0.355968(71)	10.2(18)		
[16.9]1	16923.4325(51)	0.340035(59)	-104.1(19)	1.85(34)	1.06(13)
[17.0]2	16998.3906(52)	0.346029(50)	-6.3(14)	0.023(18)	-1.50(37)

Numbers in parentheses are one standard deviation uncertainties in the
last two digits.

EFFECTS OF POLING AND STRETCHING ON SECOND-HARMONIC GENERATION IN AMORPHOUS VINYLIDENE CYANIDE/VINYL ACETATE COPOLYMER

HEIHACHI SATO, KAZUO OKA, IWAO SEO[*] AND HIDEYA GAMO[**]
DEPARTMENT OF ELECTRICAL ENGINEERING, NATIONAL DEFENSE
ACADEMY, YOKOSUKA 239, JAPAN.
[*] MITSUBISHI PETROCHEMICAL CO. LTD., IBARAKI 300-03, JAPAN.
[**] DEPARTMENT OF ELECTRICAL ENGINEERING, UNIVERSITY OF
CALIFORNIA AT IRVINE, IRVINE, CALIFORNIA 92717, U. S. A.

ABSTRACT

Dependence of second-harmonic generation(SHG) power on
poling- and stretching processings is experimentally investigated for amorphous vinylidene cyanide/vinyl acetate
(VDCN/VAc) copolymer using a pulsed Nd:YAG 1.06 μm
laser.

1. INTRODUCTION

Organic copolymer such as vinylidene fluoride/trifluoroethylene
(VDF/TrFE) and vinylidene cyanide/vinyl acetate(VDCN/VAc) have
been recently attracted as a new materials towards nonlinear optics,[1,2]
in addition to conventional uses of their ferro-, piezo- and pyroelectric properties. In particular, the VDCN/VAc copolymer exhibits
amorphous or paracrystalline phase, though, it has significantly
strong piezo- and ferroelectricities. We have newly observed the
second-harmonic generation(SHG) in this copolymer by using Nd:YAG
laser. Its nonlinear dipole-moment has been expected to be much
enhanced by poling and stretching. In the present paper we shall
thus experimentally study how the SHG power is increased with both
processings above.

2. EXPERIMENT

The experimental setup used is the same as in our previous paper.
A pulsed Nd:YAG 1.06 μm laser was used as an optical source, having
the peak power 1 kW and the pulse width 100 μs. Systematically
conditioned fifteen specimens were prepared, of which the thickness
is 30 - 50 μm, having the typical dimension of 30 mm length along
the x direction perpendicular to the beam-traveling direction(the y

direction) and tapering from 3 mm to 0 mm along the beam direction. The poling field from 45 to 75 kV/mm was applied along the film normal, i.e. the z direction. The direction of stretching was chosen so as to accord to the x-, and the x- and y directions, respectively, for uni- and biaxially stretched specimens.

In Fig. 1 the obtained SHG power was depicted as a function of the piezoelectric coefficient D_{33} measured with the conventional RF method at 30 MHz, showing its quadratic dependence. The similar characteristics have also been obtained between the SHG power $P_{2\omega}$ and the input fundamental beam power P_ω. The SHG power is depicted as a function of the poling field E, together with the relation between D_{33} and E in Fig. 2, where $\bar{a} \times \bar{b}$ denotes a stretching condition, in which $\bar{a}$ corresponds to the uni- or biaxial stretching, while $\bar{b}$ implies the mechanical amount of the stretching σ. We can roughly say that the SHG power tends to increase in proportion to E, but D_{33} increases proportionally with $E^{1/2}$. In Fig. 3 the SHG power was measured as a function of σ for the uniaxially stretched specimens, together with the relationship between D_{33} and $P_{2\omega}$.

3. DISCUSSIONS AND CONCLUSION

Taking the poling and stretching processings, it is expected to be the so-called planar-zigzag conformation belonging to the point symmetry mm2, so that its nonlinear polarization along the film normal is proportional to the corresponding optical nonlinear suscepti-bility χ_{33} for the input beam having its electric field along the z direction. Thus, the SHG power $P_{2\omega}(y)$ can be expressed for the path length of the specimen y as[3]

$$P_{2\omega}(y) = K\chi_{33}^2 P_\omega^2(0) y^2 \; \text{sinc}^2(\Delta k y/2)/S, \qquad (1)$$

where K is the proportional constant, $P_\omega(0)$ the fundamental input power, Δk the phase mismatching of the wave number for both input- and SHG beams and S the beam cross section. From the quadratic dependence of the SHG power on D_{33} and of Eq. (1) it is found that the optical nonlinear susceptibility χ_{33} must linearly correspond to D_{33}. From the experimental results in Figs. 2 and 3 we can deduce the following relation ;

$$\chi_{33} \; \propto \; (E\sigma)^{1/2}. \qquad (2)$$

Therefore, the linear dependence of the SHG power upon E and σ is well consistent with the obtained results. Under the most appropriate situation , about one order improvement of the conversion efficiency will be , so far, expected through both poling and stretching.

REFERENCES

1. H. Sato and H. Gamo, Jpn. J. Appl. phys. 25, L990(1986).

2. H. Sato, T. Yamamoto, I. Seo and H. Gamo, Opt. Lett. 12, 579 (1987).

3. Y. R. Shen,"The Principles of Nonlinear Optics", Wiley, New York, 1984, p. 123.

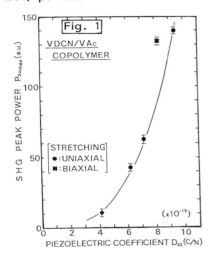

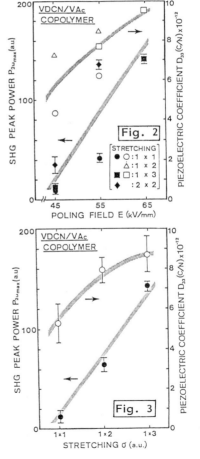

Fig. 1   Relation between SHG power and D_{33}.

Fig. 2   Relation between SHG power or D_{33} and poling field E.

Fig. 3   Relation between SHG power or D_{33} and stretching σ.

Laser Induced Spectroscopy of Cardiovascular Tissues

G.H. Pettit, R. Sauerbrey, and F.K. Tittel
Department of Electrical and Computer Engineering
Rice University, Houston, Texas, 77251-1892
M.P. Sartori, and P.D. Henry
Baylor College of Medicine, Houston, Texas

Laser angioplasty is a new medical application of lasers that is being studied in an increasing number of laboratories worldwide. Laser radiation, delivered via flexible optical fiber, can remove fatty and fibrous deposits that build up inside the arteries and obstruct the normal flow of blood. Laser recanalization of vessels, which involves threading a small catheter percutaneously through the bloodstream to the occluded site, is an attractive alternative to invasive surgical procedures. A serious problem with this technique, however, is uncontrolled damage to the vessel wall, including perforation. Appropriate targeting of diseased plaque and monitoring of the ablation process can help avoid this type of catastrophic injury. Hence real-time diagnostic information about the diseased arterial site is virtually essential.

This work reports on laser induced tissue autofluorescence analysis which allows discrimination between diseased and healthy vessel wall, and which could aid in the safe application of ablative laser energy (1). The experiments so far have been conducted with cadaver tissues to establish the technique. Such spectral analysis will be particularly useful in laser angioplasty if similar results can be obtained under *in vivo* conditions.

Optical fiber guided argon ion laser radiation at 458 nm and XeF excimer laser radiation at 351 nm have been used to elicit autofluorescence spectra of fresh arterial wall samples in air and under blood. The autofluorescence profiles were detected by a spectrometer coupled to an optical multichannel analyzer. The experimental set-up is shown in Figure 1. Tissue sites were classified as either healthy arterial wall, lipid-rich tissue (early stage of atherosclerosis), or calcified plaque.

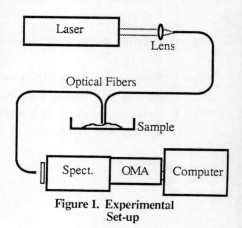

Figure 1. Experimental Set-up

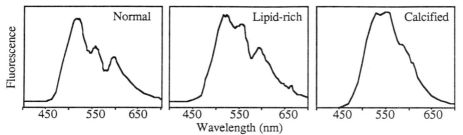

Figure 2. Argon Ion Laser Induced Arterial Wall Fluorescence

Typical 458 nm argon ion laser fluorescence profiles from all three tissue types are shown in Figure 2. Spectra excited at this wavelength from healthy wall exhibited three distinct peaks, at 520 nm, 555 nm, and 595 nm. This pattern of three discrete peaks was not seen at the diseased sites. In addition to this difference in spectral shape, the absolute fluorescence intensity was found to be much higher with atherosclerotic tissue. Changes in spectral shape and intensity corresponded to the degree of atherosclerosis, and made it possible to differentiate all three tissue types by fluorescence analysis.

Fluorescence profiles obtained by 351 nm excimer laser excitation are shown in Figure 3. The distinction between healthy and diseased tissue spectra is more subtle with this excitation wavelength. The disparity between the two profiles is more apparent in the difference curve in the lower portion of Figure 3. While healthy sites exhibit a consistent autofluorescence response, fatty or calcific sites produce a variety of spectral shapes. Careful subtraction analysis of spectra normalized to the same absolute intensity scale results in accurate histologic determination, as with the 458 nm fluorescence data. This is important because the excimer laser is an

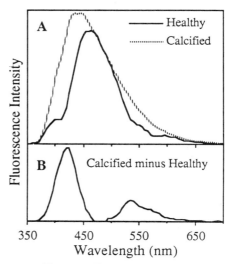

Figure 3. Excimer Laser Arterial Autofluorescence

attractive candidate for angioplasty, due to its photoablative properties. Successful excimer-based diagnostics would allow for a clinical system requiring a single laser.

The cause of the fluorescence response is open to debate. Most likely, many molecular agents within the arterial wall contribute in a complex process of photon emission and reabsorption which leads to the bulk tissue reaction observed. Flavins are a possible fluorescence source (2). Oxyhemoglobin, present in the wall, may play a reabsorptive roll in forming the "three peak" spectral shape seen in 458 nm induced profiles (3).

Using this fluorescence information, detailed images of the luminal surface of diseased arteries were made which clearly delineated regions of plaque (4). Sample were fixed flat with the endothelial surface exposed. Argon ion laser autofluorescence spectra were obtained at sites spaced every 2.5 mm along a two dimensional square grid. The maximum intensity in each spectrum at 520 nm was entered into a digital image processor and color-coded peak intensity maps of the luminal vessel surface were produced. These images clearly showed fine details in the distribution of the calcified tissue as well as grossly diseased areas. Presently, the obtained spatial resolution is approximately 1 mm. Using smaller diameter optical fiber it should be possible to reach 100 μm to 10 μm discrimination. This type of information could greatly aid in targeting laser ablation.

Spectral analysis can also be used to determine the appropriate endpoint of an ablation procedure, that is, when the diseased tissue layer has been completely removed and the healthy wall reached. Argon ion laser induced fluorescence spectra of a region of calcified plaque before and after excimer laser ablation are shown in Figure 4. An abrupt decrease in absolute intensity as well as change in profile shape are observed in samples when the calcified tissue has been completely penetrated (as confirmed by

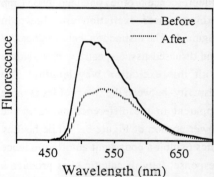

Figure 4. Autofluorescence at a Diseased Arterial Site before and after Ablative Plaque Removal

histologic analysis). Similar effects are seen with excimer laser analysis. These results agree with those obtained by mechanical removal of the calcified plaque. The 3 tissue types have been tested in this procedure, and peak intensity results are shown in Table 1.

As presented in Table 1, the fluorescence intensity at normal sites rises with ablation into the tissue. It is possible that the different histologic layers of the arterial wall produce variations in the spectral response. This could prove advantageous for excimer laser-based analysis. Because of the shallow penetration depth of ultraviolet light, a UV source could better discriminate thin layers of tissue by spectral investigation.

Spectroscopy can also be used to directly monitor ablation. Excimer laser ablation of plaque in air produces a localized plasma plume containing ionic forms of calcium. Fluorescence profiles

Table 1. Argon Ion Laser Autofluorescence Peak Intensity at Ablated Arterial Wall Sites

Arterial Tissue Type	Before Ablation	After Ablation
Healthy	2.9	5.2
Lipid-Rich	2.6	3.5
Calcified	23.5	6.5

taken under these conditions will have sharp lines superimposed on the tissue response curve. These lines are only seen at calcified sites and only during ablative plume formation. This phenomenon could furnish very useful feedback information during the laser angioplasty procedure if it can be reproduced in a blood environment.

In summary, laser spectroscopy has been shown *in vitro* to provide information useful in directing photoablative recanalization. Autofluorescence analysis can be used to diagnose wall tissue as atherosclerotic and to assess the progress of plaque removal. This information should be important in implementing a safe laser angioplasty system.

References:

1) C. Kittrell, R.L. Willet, C. de los Santos Pacleo, N.B. Ratliff, J.R. Kramer, E.G. Malk, and M.S. Feld, "Diagnosis of fibrous arterial atherosclerosis using fluorescence," Applied Optics 24:2280 (1985)

2) R.C. Benson, R.A. Meyer, M.E. Zaruba, and G.M. McKhann, "Cellular autofluorescence - is it due to flavins ?," J. Histochem. Cytochem. 27:44-48 (1978)

3) Y. Hoshihara, S. Fukuchi, K. Hayakawa, N. Yamada, Y. Yoshida, M. Hashimoto, T. Ogawa, and T. Nishisaka, "The influence of tissue hemoglobin on argon laser induced fluorescence at gastrointestinal mucosa," Nippon Shokaki Gakkai Zasshi 82:1853-1857 (1985)

4) M. Sartori, R. Sauerbrey, S. Kubodera, F.K. Tittel, R. Roberts, and P.D. Henry, "Autofluorescence maps of atherosclerotic human arteries - a new technique in medical imaging," IEEE Journal of Quantum Electronics QE-23:1794-1797 (1987)

5) R.H. Clarke, J.M. Isner, T. Gauthier, K. Nakagawa, F. Cerio, E. Hanlon, E. Gaffney, E. Rouse, and S. DeJesus, "Spectroscopic characterization of cardiovascular tissue," Lasers in Surgery and Medicine 8:45-59 (1988)

LASER-EXCITED MALIGNANCY AUTOFLUORESCENCE FOR TUMOUR
MALIGNANCY INVESTIGATION AND ITS ORIGIN

Yang Yuanlong, Ye Yanming, Li Liming, Xia Jingfang

Physics Department, Fudan University, Shanghai, China

In this paper we will report on the characteristic autofluorescence of malignants tissues. The term 'autofluorescence' is used here to denote the fluorescence emitted by the tissue when it is excited by laser beam without HpD or other fluorescent drug administration. With the 365 nm line of a pulsed Xe ion laser as the excitation source we have measured the autofluorescence spectra of normal and cancerous tissue specimens of stomach, esophagus, mandible, glimas, tongue etc. and have detected clinically the malignancy for patients with oral or gastric tumours. The main results are presented in the following.

A) Characteristic peaks of autofluorescence can be taken as a criterion for cancer diagnosis:

We have measured the autofluorescence spectra from the patients with oral malignant tumour. It is obvious that similar to the case of the specimen the 630 nm and 690 nm characteristic peaks only appear in the spectra of cancerous tissues but not in the spectra of the corresponding normal region or of the tissues with benign lesion. By means of laser-OMA-endoscope system we detected the autofluorescence spectra of gastric cancers. The characteristic peaks near 630 nm and/or 690 nm still exist but are on a strong background fluorescence from normal tissue, as shown in Fig.1. There are no characteristic peaks in the autofluorescence spectra of superficial gastritis or superficial-atrophic gastritis.

B) The autofluorescence spectra may be considered as an indication of premalignancy:

In buccal malignancy investigation, we found that characteristic peaks appear in the spectrum of tissue with serious atypical hyper plasia, but not in that of the mild case. We also found that among patients with chromic atrophic gastritis some gave characteristic peaks in the autofluorescence spectra, as shown in Fig.2. Further biopsy examination confirms that serious atypical hyperplasia had appeared in these patients.

C) The endogeneous porphyrin may be responsible for the characteristic autofluorescence:

The investigation conducted by us has led us to the preliminary conclusion that the characteristic autofluorescence of malignant tumour may originate from some prophyrin compounds which are formed in human body and localized in the cancer region.

D) As blood-brain barrier exists in human brain we are interested to know whether the brain cancer gives the characteristic autofluorescence or not. Preliminary experiments have been done on the rats which have been implanted with sarcoma in the brain. After a week of implantation we detected the autofluorescence spectra of the brain tumour of rats after craniotomy. The characteristic peaks are evident and similar to those of H_pD as shown in Fig 3 According to our previous work we believed that in this case of sarcoma in the brain is also due to the endogeneous porphyrin It means that the endogeneous porphyrin can still be localized and retained in the saroom formed in the brain despite of the bloodbrain barrier.

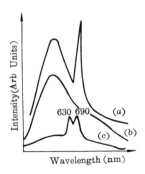

Fig.1 Typical illustration from patient.

 (a) gastric cancer region,

 (b) normal region,

 (c) spectrum of HpD as wavelength reference.

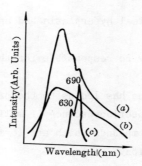

Fig.2 Characteristic auto-fluorescence taken from patient.

    (a) with chromic atrophic gastritis.

    (b) normal region,

    (c) spectrum of HpD as wavelength regerence.

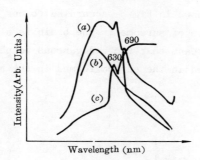

Fig.3 Autofluorescence spectra of sarcoma implanted in the
    brain of the rats.

    (a) sarcoma in the brain,

    (b) normal region,

    (c) spectrum of HpD as wavelength reference.

A STUDY ON SEVERAL HEMATOPORPHYRIN DERIVATIVES
BY TIME-RESOLVED SPECTROSCOPY

W.L. Sha,   J.T. Chen,   Z.X. Yu,   S.H. Liu

Ultrafast Laser Spectroscopy Laboratory,
Zhongshan University, Guangzhou, P.R. China

ABSTRACT
The wavelength dependence time-resolved fluorescence studies of several hematoporphyrin derivatives (HpD) in solution condition with 10ps time resolution are presented. A fast decay process with time constant less than 45ps is observed. As far as our knowledge, this is the first report. Simultaneously, a subnanosecond and a slow (about 10ns) decay processes are observed. Further analysis shows that two fast decay processes which time constants are shorter than nanosecond are from head-to-tail aggregates.

Photochemotherapy with hematoporphyrin derivative (HpD) is increasingly being used for the local treatment of malignant tumors[1]. For clear understanding and further developing of the photodynamic therapy of malignant tumors, it is of essential importance to study the micromechanics of its effect upon the tumors. This paper is about the wavelength dependence time-resolved fluorescence studies of several home-made HpD offered by the Yangzhou biochemical factory and Beijing medicine institute.

The experimental apparatus is shown in Fig.1. The exciting source is synchronously pumped tunable Rh-6G dye laser with cavity dumper (tuning rang 570-620nm, pulsewidth 4ps, repetition 400Hz - 4MHz). A polychromator was attached between a sample optics and a synchroscan streak camera with 10ps time resolution. The excitation pulses used repetition 800 KHz at 570 nm, to avoid photodegeradation due to irradiation around 400 nm, is focused by a lens into the sample cell. The line of focus is directed just inside the output face of the cell to minimize any reabsorption of the fluorescence. The fluorescence at 90° with respect to the input direction is focussed on the input slit of the polychromator.

Table 1 Fluorescence decay time constants several
3.7 x 10^{-4} M HpD in 10 mM PBS at pH 7.4 excited 570 nm

HpD samples	τ_1 (ps)	τ_2 (ns)	τ_3 (ns)
No.1	< 45	0.55+0.01	11.35+0.07
No.2	< 35	0.35+0.01	8.09+0.02
No.3	< 36	0.48+0.01	11.10+0.13
No.4	< 45	0.45+0.03	11.60+0.39

By analysing the wavelength dependence time-resolved fluorescence spectra I (λ,t) for HpD at high concentration in 10 mM PBS, we can decompose the fluorescence spectra into two sets of fluorescence spectra with similar band shape and quite different fluorescence decay time constants. And the fast decay one has a red-shift in relative to the slow one.

This was also confirmed by measuring wavelength-resolved fluorescence decays for the different wave band shown in Fig.1.

It is the only slow fluorescence decay component (about 10 ns) observed in the methanol solvent and the micellar system where HpD is known to exist exclusively in the monomeric form. The fluorescence decay time contants are the same at both emission peaks and not to change with the concentration of HpD. We contribute the two fast decaying processes to aggreations. And according to Kasha's theory, the aggregates are head-to-tail, not face to face.

REFERENCES

[1] Yamashita, M., Nomura, M., Kobayashi, S., Sato, T. and
    Aizawa, K., IEEE J. Quan. Elec., 20, 1363 (1984).
[2] Kasha, M., Radia. Res., 20, 55 (1963).

INTENSITY(a u)

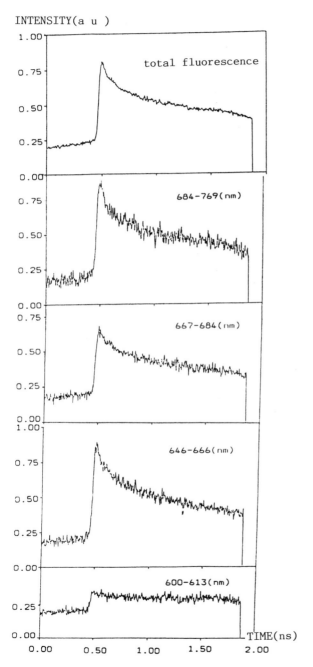

total fluorescence

684-769(nm)

667-684(nm)

646-666(nm)

600-613(nm)

TIME(ns)

Fig.1 Fluorescence decays at different wave bands for 3.7x 10^{-4}M HpD in 10mM PBS at pH7.4

RESEARCH ON STRONG FIELD PROCESSES WITH
A SUBPICOSECOND 400 GW ULTRAVIOLET SOURCE

T. S. Luk, K. Boyer, G. Gibson, H. Jara, I. A. McIntyre
A. McPherson, X. J. Pan, R. Rosman, X. M. Shi, and C. K. Rhodes
Laboratory for Atomic, Molecular, and Radiation Physics
Department of Physics, University of Illinois at Chicago
P. O. Box 4348, Chicago, Illinois   60680, U. S. A.
and
J. C. Solem
Physics Division, MS–B210, Los Alamos National Laboratory
Los Alamos, New Mexico   87545, U. S. A.

ABSTRACT

Developments in short pulse high power ultraviolet sources are enabling the production of electric field strengths far above an atomic unit, a regime in which atoms and molecules are expected to exhibit complex behavior. The general amplitude under study is $NY + X \rightarrow X^{q+} + qe^- + Y'$, a class of processes which involves ionization with the collateral production of energetic electrons and radiation. Among the phenomena that are expected to be observed in the high field studies are hard x–ray generation, electron–positron pair production, the induction of nuclear reactions, and new modes of electromagnetic propagation in plasmas.

The trend of current developments[1–4] in ultraviolet sources is moving rapidly toward the attainment of focal intensities[5] on the order of $\sim 10^{21}$ W/cm^2, an extremely high value corresponding to an electric field amplitude of ~ 170 e/a_o^2. In a force field of this extraordinary strength, the induced electronic motions will become strongly relativistic even at ultraviolet wavelengths. This is a completely unexplored regime of interaction experimentally and one in which current theoretical descriptions provide little guidance.

The basic features of such a laser system are shown in Fig. (1). It shows a 745 nm dispersion compensated hybridly mode–locked dye laser, which produces transform–limited pulses of ~ 200 fs duration, the dye amplifier, the crystals used for up conversion to 248 nm, and the two–stage amplifier chain. The final amplifier (Prometheus) has a cross sectional area of ~ 100 cm^2 and is driven typically with a pulse whose energy is in the 0.1 − 1.0 mJ range.

The power amplifier has two x–ray preionized discharge channels of 2.5 m in length, each operating at a discharge voltage of 180 keV with a total gas pressure of 1216 torr. The measured gain is 1.5×10^{-2} cm^{-1}, a value in accord with expected performance. Under normal operating conditions an input pulse of ~ 300 μJ produces an output of ~ 350 mJ with an associated production of amplified spontaneous emission at about ten percent of the short pulse energy. With a pulse duration of ~ 800 fs, a peak power of ~ 400 GW is obtained.

The generation of high focal intensities requires that the spatial quality of the output radiation be undistorted. Measurements at 633 nm show that the distortion introduced by the Prometheus system is less than $\lambda/13$ over the full aperture. In addition, the intensity on the output CaF_2 window is not greater than 10 GW/cm^2, a value significantly below the measured damage limit of ~ 150 GW/cm^2. Experiments have shown, however, that propagation through air for a distance of ~ 4 m strongly affects the spatial profile of the beam. After propagating for a total distance of ~ 10 m, the principal manifestation is the appearance of filaments with diameters in the 50 – 100 μm range. These filaments form a surprisingly stable pattern and appear to exist for lengths on the order of several tens of centimeters or, perhaps, longer. At the current power level, the replacement of the air path with helium completely eliminates the channel formation and restores the beam to good spatial character. As the power of this system is developed to higher levels, however, even the relatively low nonlinear index of helium may not be sufficiently low and it may be necessary to propagate the beam through an evacuated path.

Experiments have begun with a large aperture system and subpicosecond pulses have been used in preliminary spectroscopic studies of argon, krypton, and xenon. The spectrum of argon shown in Fig. (2) clearly indicates the excitation

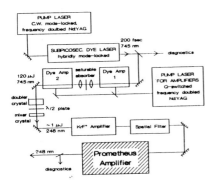

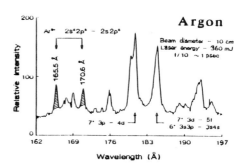

Fig. (1): Basic features of a subpicosecond KrF* 248 nm source showing the 745 nm hybridly mode–locked source, the crystals for up conversion, and the ultraviolet amplifiers.

Fig. (2): Spectrum of Ar observed in a gaseous target with subpicosecond 248 nm pulse produced by the 10^2 cm² aperture system.

of the argon L–shell in the Ar^{9+} system by the identification of a well known $2s^2 2p^5 - 2s2p^6$ doublet. It is noted that the production[6] of ground state Ar^{9+} from Ar^{8+}, by the removal of a 2p electron, requires a minimum of 422 eV and that the total energy investment needed to produce Ar^{9+} from a neutral atom exceeds a kilovolt. Plainly, very high energy deposition rates are achievable by this means and estimates[7] indicate that they are sufficient for the generation of significant amplification at x–ray wavelengths.

A new technology of optical pulsed power is presently being developed that will enable the production of focal intensities on the order of 10^{21} W/cm^2 with laboratory scale apparatus. In recent times, the field of strong–field multiquantum processes seems to have taken a turn in a rather unexpected direction. New modes of thinking about this general class of phenomena are emerging and experiments are reaching far into unexplored territory. It is expected that many high energy processes will occur with very substantial rates under the conditions which these experiments will involve. Among these are (a) hard x–ray generation, (b) electron–positron pair production, (c) the induction of nuclear reactions,[8] and (d) new modes of channeled electromagnetic propagation in plasmas.[9] All of these processes will provide important information on the fundamental behavior of atomic matter in the strong–field regime.

The authors acknowledge fruitful discussions with G. Wendin. This work was supported by the U. S. Office of Naval Research, the U. S. Air Force Office of Scientific Research, the Lawrence Livermore National Laboratory, the Directed Energy Office and the Innovative Science and Technology Office of the Strategic Defense Initiative Organization, and the National Science Foundation.

1. Schwarzenbach, A. P., Luk, T. S., McIntyre, I. A., Johann, U., McPherson, A., Boyer, K. and Rhodes, C. K., Opt. Lett. 11, 499 (1986).
2. Glownia, J. H., Misewich, J. and Sorokin, P. P., Opt. Lett. 4, 1061 (1987).
3. Szátmari, S., Schäfer, F. P., Müller–Horsche, E. and Müchenheim, W., Opt. Comm. 63, 305 (1987).
4. Watanabe, S., Endoh, A., Watanabe, M. and Surakura, N., Opt. Lett. 13, 580 (1988).
5. Rhodes, C. K., Science 229, 1345 (1985).
6. McPherson, A., Gibson, G., Jara, H., Johann, U., Luk, T. S., McIntyre, I. A., Boyer, K. and Rhodes, C. K., J. Opt. Soc. Am. B 4, 595 (1987).
7. Boyer, K., Jara, H., Luk, T. S., McIntyre, I. A., McPherson, A., Rosman, R. and Rhodes, C. K., Revue Phys. Appl. 22, 1793 (1987).
8. Boyer, K., Luk, T. S. and Rhodes, C. K., Phys. Rev. Lett. 60, 557 (1988).
9. Solem, J. C., Boyer, K. and Rhodes, C. K., "Generation of Coherent X–Radiation with Charge–Displacement Self–Channeling," submitted to Phys. Rev. Lett.

GROWTH, DECAY, AND QUENCHING OF STIMULATED RAMAN SCATTERING IN TRANSPARENT LIQUID DROPLETS

Jia-biao Zheng,* Wen-Feng Hsieh,† Shu-chi Chen,‡ and Richard K. Chang

Yale University

Section of Applied Physics and Center for Laser Diagnostics

New Haven, Connecticut 06520-2157, USA

ABSTRACT

The time behavior of the various multiorders of stimulated Raman scattering (SRS) and the input laser pulse was determined with a streak camera. The observed time delay between the 1st-order Stokes and the input pulse is the buildup time of the SRS originating from spontaneous Raman noise. The short time delay of the 1st order and higher-order Stokes is indicative that the higher order Stokes originates from signals resulting from the four-wave-mixing process. The SRS photons are trapped within the droplet and decay according to the droplet cavity leakage rate and the intensity-dependent depletion rate. Quenching of the SRS was observed before the onset of atomic emission, which is indicative of laser-induced breakdown within the droplet. The optical absorption introduced by the internal plasma is effective in quenching SRS.

INTRODUCTION

The micrometer-sized droplet acts as a lens to concentrate the input radiation just within the droplet shadow face and serves as a spherical resonator to provide optical feedback at specific wavelengths within the linewidth of the Raman gain profile. Stimulated Raman scattering (SRS) from single droplets has been observed with surprisingly low input intensity.[1] The laser generated 1st-order Stokes SRS intensity, which circumvents the droplet rim, is also intense enough to generate 1st-order Stokes with a Raman shift cor-

---

*On leave from the Physics Department, Fudan University, Shanghai, People's Republic of China.

†Present address: Electronic Physics Department, Chao-Tung University, Hsin-Chu, Taiwan, Republic of China.

‡On leave from the Physics Department, Hong Kong Baptist College, Hong Kong.

responding to twice that of the 1st-order Stokes (commonly referred to as 2nd-order Stokes SRS). Multiorder Stokes, up to the 14th order, has been reported for CCl_4 droplets.[2]

In this paper, we integrate the results of several investigations on 1st- and multiorder SRS from single droplets and review the following characteristics: (1) the time delay between the growth of the SRS and the input pulse;[3-5] (2) the decay time of the SRS trapped within the droplet;[6] and (3) the quenching of SRS by the plasma created during the initial phase of laser-induced breakdown (LIB) within the droplet.[3,7]

SRS BUILDUP TIME

Figure 1 shows the experimental configuration used in simultaneously measuring the temporal behavior of the input laser pulse along with the 1st-, 2nd-, and 3rd-order Stokes SRS associated with the symmetric stretching mode of the NO_3^- ions dissolved in water droplets (with ≈ 40 μm radius). A portion of the incident laser beam is channeled by an optical fiber to the top portion of the entrance slit of the streak camera and the intense 90° elastic scattering from the droplet is blocked by a long pass filter. The multiorder Stokes SRS from the droplet is collected at 90° by lenses and dispersed by a spectrograph. The dispersed spectrum is channeled by a fiber ribbon with one end aligned horizontally at the spectrograph exit plane and the other vertically at the streak camera entrance slit. The streak camera sweeps the entire slit image at 5 ns/mm, and the output phosphor of the streak camera is read by a vidicon camera.

Figure 2 shows the single pulse output from the streak camera with time along the horizontal axis, wavelength along the vertical axis, and the relative intensity plotted in the third dimension. Note that the 1st-order Stokes is significantly delayed with respect to the input laser pulse while the time delays between the 2nd- and 1st-order Stokes and between the 3rd- and 2nd-order Stokes are much shorter (within the 1 ns resolution of the streak camera).[3] Similar results have been observed with a photomultiper which is able to measure only the time delay between the laser pulse and one of the multiorder Stokes.[4,5]

The long time delay between the 1st-order SRS and the input laser pulse is the buildup time of the amplified Raman radiation originating from spontaneous Raman scattering. With increased input intensity, we have observed that this time delay decreases.[3] The relatively short time delays between the 2nd- and 1st-order SRS and between the 3rd- and 2nd-order SRS suggest that the starting signal of the higher-order SRS is different from that of the 1st-order SRS. For example, the starting signal at the 2nd-Stokes wave-

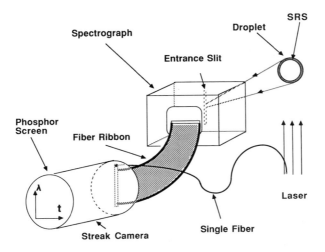

Fig. 1.   Experimental configuration used to detect simultaneously the time behavior of the input laser pulse and the multiorder SRS (1st-, 2nd-, and 3rd-order). A spectrograph disperses the SRS and a ribbon fiber channels the dispersed SRS spectra to the streak camera. A single fiber channels some of the laser pulse to the top of the streak camera slit.

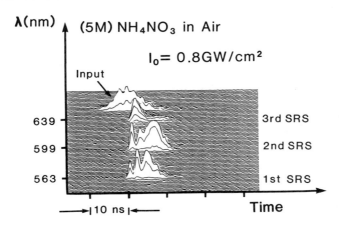

Fig. 2.   Single laser pulse time-resolved data from a water droplet containing 4M NH_4NO_3. A long time delay is observed between the 1st-order Stokes of the SRS and the input laser pulse. Only a small time delay is observed in the multi-orders of SRS with respect to 1st-order Stokes.

length has two contributions: the spontaneous Raman scattering signal resulting from the 1st-order SRS serving as the pump radiation, analogous to that from the input laser radiation; the parametric scattering signal resulting from the four-wave-mixing of the two 1st-Stokes electric fields (E_S) and the input electric fields (E_o), i.e., $\chi^{(3)}E_S E_S E_o^*$, where $\chi^{(3)}$ is the third-order nonlinear susceptibility. The short time delay between the 2nd- and 1st-order SRS implies that the parametric signal is initially larger than the spontaneous Raman signal and that the amplification of the 2nd-order SRS starts from the parametric signal.[3]

SRS DECAY TIME

The amount of leakage from the droplet is commensurate with the morphology-dependent resonances. The detected SRS signal is proportional to the internal intensity and the leakage rate. Using picosecond laser pulses and a streak camera with 2 ps resolution, we have observed that the time decay of the SRS within the droplet is dependent on the input laser intensity, i.e., the decay is faster at higher input intensities.[6] Such intensity dependence suggests that the decay time of the internal Raman radiation is decreased by leakage from the droplet as well as by nonlinear conversion to the next higher-order Stokes.

The decay times of the 1st-, 2nd-, and 3rd-order SRS are shown in Fig. 2 with 1 ns resolution. The SRS at various multiorder Stokes can continue to circumvent the droplet even when the pump radiation is off. The decay time at each order of SRS is dictated by the morphology dependent resonances coincident with the wavelength of the SRS and the amount of intensity dependent depletion needed to amplify the next higher-order SRS.

SRS QUENCHING BY PLASMA

Radiation within a transparent droplet is lost via leakage and nonlinear conversion to another wavelength. However, the plasma generated by LIB can introduce enough optical absorption to totally quench the SRS. Because of the focusing effect of the spherical illuminated face, LIB is initiated just within the droplet shadow face and during the rising portion of the high intensity input pulse. The resultant plasma absorbs the remaining portion of the input pulse, and a plasma wave is generated which travels from the shadow face toward the illuminated face.[7] The LIB generated plasma also absorbs the internal SRS each time this radiation traverses the localized region in which LIB is initiated.

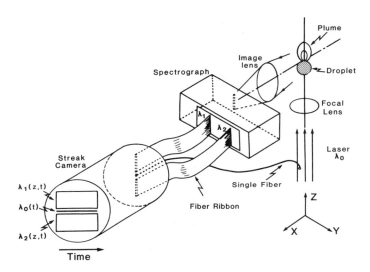

Fig. 3.   Experimental configuration used to measure the time behavior of the spatially resolved (along one dimension) signal at two selected wavelengths (the first at the resonance Na and the second at the Balmer H_α and the SRS of water). The two fiber ribbons preserve the image along the spectrograph entrance slit, and the time behavior of the laser pulse is detected by channeling part of the laser pulse through the single fiber.

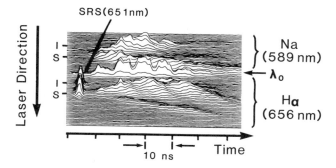

Fig. 4.   Single laser shot results indicating that SRS buildup occurs during the initial portion of the laser pulse and that SRS quenching occurs soon afterwards, before detection of the plasma-related Na and H_α.

Figure 3 shows the experimental configuration used to monitor the time behavior of the laser pulse and the time and space behavior of the 1st-order SRS and of the fluorescence emission of atomic species within the plasma (resonance Na and Balmer H_α). An optical fiber channels part of the input radiation, as is shown in Fig. 1. The two fiber ribbons channel the spatially preserved signals at two selectable wavelengths, i.e., one ribbon is set at the resonance Na wavelength and the other at the H_α wavelength, which also includes some of the SRS associated with the O-H stretching mode of water.

Figure 4 shows the time resolved and one-dimensional space-resolved data. SRS is concentrated along the droplet rim, i.e., at the illuminated face (I) and shadow face (S). At the high input intensity, SRS occurs at the beginning of the input laser pulse, i.e., the time delay is very short in contrast to the long time delay shown in Fig. 2 for much lower input intensity. Figure 4 shows that the SRS is rapidly quenched and that soon afterwards the fluorescence of two atomic species (Na and H) within the plasma becomes detectable. It is evident from Fig. 4 that SRS quenching by the plasma is more sensitive than our detection system in observing the presence of resonance Na and H_α emission. At higher input intensity, the plasma absorption becomes significant and prevents SRS buildup.

We thank David H. Leach and Jian-Zhi Zhang for helpful discussions and gratefully acknowledge the partial support of this work by the U.S. Air Force Office of Scientific Research (Grant No. 88-0100) and the U.S. Army Research Office (Contract No. DAAL03-87-0076).

REFERENCES

1) Snow, J.B., Qian, S.-X., and Chang, R.K., Opt. Lett. 10, 37 (1985).

2) Qian, S.-X. and Chang, R.K., Phys. Rev. Lett. 56, 926 (1986).

3) Hsieh, W.-F., Zheng, J.-B., and Chang, R.K., Opt. Lett. 13, 497 (1988).

4) Chýlek, P., Biswas, A., Jarzembski, M.A., Srivastava, V., and Pinnick, R.G., Appl. Phys. Lett. 52, 1642 (1988).

5) Pinnick, R.G., Biswas, A., Chýlek, P., Armstrong, R.L., Latifi, H., Creegan, E., Srivastava, V., Jarembski, M., and Fernández, G., Opt. Lett. 13, 494 (1988).

6) Zhang, J.-Z., Leach, D.H., and Chang, R.K., Opt. Lett. 13, 270 (1988).

7) Chang, R.K., Eickmans, J.H., Hsieh, W.-F., Wood, C.F., Zhang, J.-Z., and Zheng, J.-B., Appl. Opt., in press.

8) Zheng, J.-B., Hsieh, W.-F., Chen, S.-C., and Chang, R. K., Opt. Lett., in press.

LAYER CONDENSED AMMONIA STUDIED BY PHOTOACOUSTIC SPECTROSCOPY

Yu Liming, Xu Ying, Hu Changwu, Zhu Angru, Wang Zhaoyong

Institute of Modern Physics, Fudan University,
Shanghai, P.R. China

1. INTRODUCTION

Photoacoustic spectroscopy (PAS) is used to revisit ammonia adsorbed on Ag surface. Both higher sensitivity and resolution, stemming in part from radiation characters of the laser shows that PAS is very appropriate to surface study.

2. EXPERIMENTAL

The apparatus which we have used to carry out the PAS measurement is totally home-made and the details can be found in somewhere else[4]. The experiments were done under ultra high vacuum with base pressure less than 5×10^{-8} Pa. A tunable cw CO_2 laser (with power of 1 W for most strong lines) was employed as an excitation source. The PA signal was detected by a piezoelectric transducer. A polycrystalling Ag in high purity was used as the substrate and was cooled down to 115 K for NH_3 adsorption. The dosing gas was guided by a thin tube to the front of the substrate surface closely, which was crucial for physisorption.

3. RESULTS AND DISCUSSION

The Adsorption Kinetics   The physisorption isotherms of NH_3 at 115K with different pressure have been studied. The adsorption and desorption isotherms of NH_3 showed that there were many steps in equilibrium curves, it means that the phase transitions in whole layer was the first order one. Besides, we have also observed the fluctuation between two successive layers, which were in good agreement with those results obtained from Monte Carlo simulations[5].

Layer Discriminated Vibrational Spectra   Due to discontineous of lasing branch lines, a part of the ν_2 spectrum would fall into the blank interval between branch lines. At higher coverage rather complete feature appeared. In account of the general feature of the spectrum and by using the curve fitting technique, the spectrum of each individual layer could clearly been discriminated. In contrast to those results obtained from HREELS [1-3], the peak position, FWHM and the intensity (spectrum area) corresponded to each layer are shown in Fig.1.

The frequency of the first layer is at 1063 cm^{-1} and that of the second one is 1072 cm^{-1}. In comparison to that of only one layer condensation is at 1066 cm^{-1}. These variations reflect the interactions between the molecule dipole and its mirror image induced in the metal surface and those of molecule dipoles among different layers. It can be suggested that the frequency will be higher when height of the NH$_3$ pyramid is lowered and the latter depends on the interaction of dipoles. The change can also be seen on FWHM's.

Inversion Splitting of the Resonance Excitation   When the NH$_3$ absorbed on preoxidized Ag surface, the ν_2 spectrum of first layer does not show any significant difference from that on clean Ag surface. Nevertheless, the second layer spectrum displayed a substantially different feature shown in Fig.2. It was splitting into two peaks on either side of the first layer peak. The frequency is 1065 cm^{-1} for the first layer and they are 1048 and 1076 cm^{-1} for the second layer respectively. The sum area of the latter two peaks equals to that of the former one. Due to the frequency difference of two peaks is 28 cm^{-1}, one can attribute that these two peaks belong to the inversion doubling. It means that once the dipole interaction is reduced to certain degree, the NH$_3$ pyramid can stand on the surface in either orientation. The condition on which the inversion doubling need to meet is crucial, relatively higher temperature of the substrate and rather high translation energy of the dosing gas the easier to establish the inversion state. Moreover, the inversion doubling always occurs in higher layers on the clean Ag surface. It is worthwhile to note that the unique peak appeared in the first one layer or two could be attributed to the antisymmetric mode, so that its resonant state is at higher energy level, furthermore, it possesses a longer lifetime than the lower one.

References
[1] B.A. Sexton, G.E. Mitchell, Surf. Sci., 99, 523 (1980)
[2] J.L. Gland, B.A. Sexton, G.E. Mitchell, Surf. Sci., 115, 623 (1982)
[3] S.T. Ceyer, J.T. Yates, Jr., Surf. Sci., 156, 584 (1985)
[4] Wu Mingchen, Yuan Duping, Zhu Angru, Lu Huizong, Yu Liming,
        Wang Zhaoyong, China Phys., 7, 965 (1987)
[5] I.M. Klin, D.P. Landan, Surf. Sci., 110, 415 (1981)

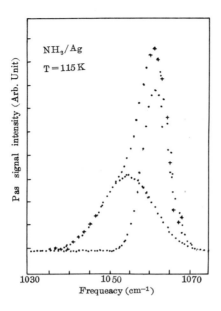

Fig. 1

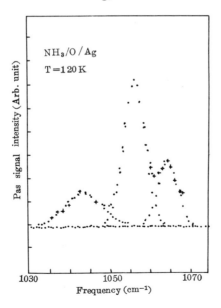

Fig. 2

HIGH EFFICIENCY RAMAN CONVERSION OF XeCl LASER RADIATION IN LEAD VAPOR

Qihong Lou   Hongping Guo

(Shanghai Institute of Optics and Fine Mechanics,
Academia Sinica, P.O. Box 8211, Shanghai, China)

ABSTRACT

Raman energy conversion efficiency in lead vapour
as high as 52% was obtained when the pumping XeCl
excimer laser adopts the unstable resonator
configuration. The characteristics of SERS in Pb
vapour pumped by XeCl laser with unstable and
stable resonators were compared in detail.

INTRODUCTION

Stimulated electronic Raman scattering of XeCl excimer laser in Pb vapour have fund to be an efficient way to obtain high power blue-green laser from UV radiation. [1] In ref.(2) one joule of blue radiation with 50% Raman energy conversion has been obtained by using injection-locked excimer laser system. In this paper, we report our recent results about XeCl/Pb system. When unstable resonator configuration was adoped to improve the pumping beam quality and some special technology was taken to improve the uniformity of the temperature distribution along the Raman cell, high Raman energy conversion efficiency as high as 52% was obtained with lower pumping energy. The comparision between the pumping laser with stable cavity and unstable one was also given in detail.

EXPERIMENTAL SETUP

The pump laser is an X-ray preionized transverse gas flow XeCl laser with output energy from 350-500mJ and FWHM of 60ns. An unstable resonator was used to reduce the beam spread angle to about 0.2-0.3 mrad. The Raman cell, a 1.6m long stainless tube, can be heated up to 1200°C by a electric oven. The water cold apparatus placed at the two ends of the Raman cell and to allow the confinement of the lead vapor to the hot zone without contamination of the windows.

RESULTS AND DISCUSSION

Fig. 1 shows the dependence of the Raman shifted radiation energy on Pb vapor pressure in the Raman cell. The pumping energy is fixed to 190 mJ/pulse, within the cell temperature from 1040 to 1200°C, the Stokes output energy is linearly proportional to the Pb vapor pressure. By comparision, the experimental results with stable laser cavity was also given in Fig. 1. It is shown that, with unstable resonator, the conversion efficiency will increase linearly as the lead vapor pressure increases, while with stable resonator, the conversion efficiency is becoming flat at high vapour pressure and the increasement of SERS output mainly depends on the pump light intensity rather than the number density of the lead vapour.

In order to investigate the effect of the buffer gas pressure on the Stokes output, the dependence of the Stokes output on the He gas pressure was measured. The results are given in Fig.2. When the He pressure is above 80 torr, an evident decreasement of Stokes energy was measured. This is due to the collision between lead atom and the buffer gas atoms. The existence of the buffer gas widened the Raman linewidth, therefore the conversion efficiency was reduced. The calculated relative output energy of the Raman shifted radiation as a function of the collision-induced linewidth was given in Fig.2. It is in good agreement with the experimental data.

REFERENCE

[1] Burnham, R. and Djeu, N., Optics Letters, 3, 215(1978).

[2] Brosnan, S.J. Komine, H., Stappaerts, M.J., Plummer, M.J. and West, J.B., Optics Letters, 7, 154(1982).

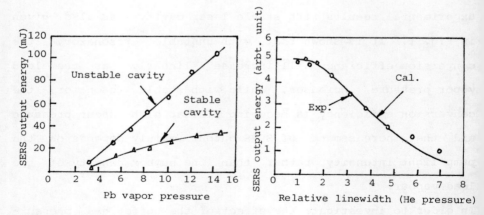

Fig.1. Comparision of the Raman shifted radiation energy as a function of Pb vapor pressure between the pumping laser adopting unstable and stable cavity.

Fig.2. Calculated relative output energy of the Raman shifted radiation energy as a function of the collision induced linewidth and He gas pressure

COMBINED EFFECT OF STIMULATED SCATTERING AND PHASE MODULATION ON GENERATION OF SUPERCONTINUM

R. Zhu, Q.X. Li, W.H. Qin, Z.X. Yu

Ultrafast Laser Spectroscopy Laboratory,
Zhongshan University, Guangzhou, P.R. China

Here we report the cross phase modulation phenomenum in the propagation of ultrashort laser pulse in water and discuss the combined effects of stimulated Raman scattering and phase modulation in the generation of supercontinum in water.

The experimental arrangement is shown in our former paper. A single 6ps laser pulse at 530nm of a few milijoules is generated from the second harmonic of mode-locked Nd: glass laser system. This pulse is taken as input pulse, focused by a lens with focal length of 30cm into 15cm long a sample cell containing pure water. Then the light signal produced in the cell is sent to a OMA II system through a color filter used to cut out input pulse and a neutral filter to be detected and analyzed. The energy of input pulse is measured by a energy meter.

Three spectral distribution curves of Stokes field produced from stimulated Raman scattering by input ultrashort pulse in water are shown in Fig.1, corresponding to input pulse energies of 0.2mj, 1mj, and 4.8mj respectively. In these three curves, there is a sharp peak for each lying at about 6405A. By identifying carefully, we understand that this peak field is generated from the excitation of stretch Raman vibration mode in liquid water molecules. We can also find easily from figure that the Stokes peak is broadened widely under the higher input energy, with the spectral distribution shape like a bell and the largest spectral width observed to be about 1000A. It is caused by the intense input pulse, that is to say the cross phase modulation is the main mechanism. In the stimulated amplification propagation, for the nonlinear refractive index changed by the intense input laser field, the phase of Stokes field is modulated and the energy of Stokes field is transferred to other frequencies. In our experiment, we do not find any other mode being excited. So, the cross phase modulation is the interaction between input laser field and single mode Stokes field. P.A. Cornelius et al.[1] studied theoretically this interaction. The spectral distributions of Stokes field observed in our experiment agree well with their calculation results.

More precise theoretical discription of cross phase modulation was presented by Lu Jianhua et al.[2] The result shows that the spectral width of Stokes field increases linearly with the input laser power. We observed this phenomenum experimentally and Fig.2 shows the varitation of spectral width of Stokes field with the energy of input pulse. For lower energy, the spectral width increases linearly with the energy, which confirms very well the theoretical result. But for higher energy, the spectral width tends to be saturated. It is caused probably by the excitation of higher order Raman modes. All theories didn't take this process into consideration. Therefore, this saturation could not be predicted theoretically.

We also analyzed the relation between intensity of broadened Stokes field and input energy and find the former changes linearly with the later. It reflects the same characteristic as that in the case of usual transient stimulated Raman scattering[3] and shows that the conversion efficiency to Stokes scattering is not affected by the phase modulation.

This research is supported by NSFC.

References
[1] P.A. Cornelius, C.B. Harris, Opt. Lett., Vol 6, p 128.
[2] Lu Jianhua, Li Yulin, Jiang Jialin, Opt. Quantum Electronics, Vol. 17, p 187, 1985.
[3] W. Kaiser, M. Maier, in 'Laser Handbook', edited by F.T. Arecchi and E.O. Schulz-Dubois (North-Holland, 1972), p 1077.

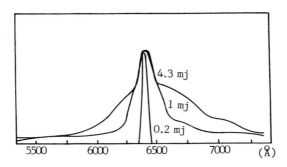

Fig 1 Spectral distribution curves at sveral pump energies

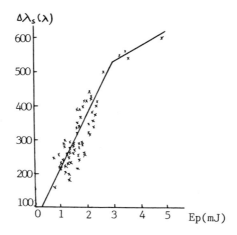

Fig 2 Observed relation between spectral width $\Delta\lambda_s$ and pump energy Ep

RESONANT MULTIWAVE MIXING IN SODIUM VAPOR

Wang Shumin, Qian Shixiong, Li Yufen

Physics Department, Fudan University,
Shanghai, P.R. China

Forward emission spectrum near D lines in Na vapor has been investigated recently by several groups. Shevy et al [1,2] ascribed these new lines to the excited state Raman scattering (ESRS) and resonantly enhanced three - photon scattering (RETPS) and observed ESRS to the second order. T.J. Chen et al[3] reported even high-order scattering from the Na vapor. Here we report our observation from Na vapor under near resonant pumping and a theory of three field-three level scheme is developed to explain the main experimental results.

We used a tunable pulsed dye laser with linewidth 0.15 cm^{-1} to pump Na vapor in a 70cm long heat-pipe oven filled with Ne buffer gas. Two methods, photograph and optoelectronic recording, were employed simultaneously to detect emission signals. A typical emission spectrum with pumping near D_2 line is shown in Fig.1 in which three lines at the blue side of the resonant pump beam and four lines at the red side can be seen clearly. We found that the frequency intervals between adjacent lines were near the same (equal to that of $3P_{3/2}$ and $3P_{1/2}$ states) and independent upon the the detuning. The spatial divergence angle was about several mrad in the forward direction and there was a strong dependence of the intensities of multiline on the pumping. We have observed the 4th Stokes and the 3rd anti-Stokes lines. From this we conclude that this is a stimulated ESRS (ESSRS) process.

We suggest that the mechanism for ESSRS, schematically shown in Fig.2, is as following. The intense pump field near D_2 line at first pumps the Na atoms to the upper level $3P_{3/2}$, due to small saturation intensity of that transition, strong saturation takes place. This causes the large population difference between $3P_{3/2}$ and $3P_{1/2}$ states and thus induces the 1st anti-Stokes Raman scattering (see Fig.2a). In this process, the $3S_{1/2}$ level acts as the near resonant intermediate state to greatly enhance the Raman gain. With enough Na atom density and intense pump field, the Raman gain for this anti-Stokes line is large enough to produce the stimulated process.

For the 1st Stokes line, however, it is generated through four-wave mixing (FWM) process shown in Fig.2b. The pump field and the ESSRS 1st anti-Stokes field can mix together to generate the 1st Stokes emission. With nearly double resonance and intense pump field, efficient intense 1st-order Stokes emission can be produced. But because this line is very near the D_1 line, due to large absorption at high temperature, the emitted 1st Stokes radiation has been greatly reduced.

For high-order Stokes emissions, they are produced by FWM process similar to the 1st Stokes line, while for high-order anti-Stokes emissions, they are produced by both FWM and direct ESSRS processes. The similar multi-line structure was observed for pump field near D_1 line, we have detected 3 orders Stokes and 2 orders anti-Stokes ESSRS emission.

With model of interaction between three-field and a V-type three-level atomic system, we solve the density matrix equations as follows

$$\partial P_{jj} / \partial t = - \Gamma_{jj} (P - P(0))_{jj} + i[P,H]_{jj} / \hbar$$
$$\partial P_{jk} / \partial t = = \Gamma_{jk} P_{jk} + i[P,H]_{jk} / \hbar \qquad j,k = 1,2,3$$

where Γ is relaxation rate. Making near resonant excitation, steady-state and rotating wave approximation and using perturbation theory, we have got the gain expression for 1st anti-Stokes ESSRS emission. This gain is proportion to the population difference of two excited states and reaches large value when the laser is tuned to near D lines.

276

So it well explains our main experimental results.

REFERENCE
[1] Y. Shevy, M. Rosenbluh, H. Friedmann, Opt. Lett., 11, 85(1986).
[2] Y. Shevy, M. Rosenbluh, Opt. Lett., 12, 257(1987).
[3] T.J. Chen et al, Digest of International Conference on Lasers, Xiamen, P.R. China, p.353(1987).

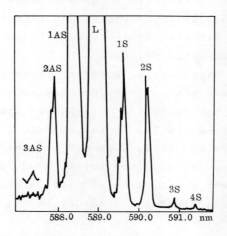

Fig 1 Emission spectrum ofESSRS near D_2 line(pump wavelength 5890 4Å)

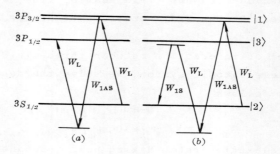

Fig 2 Schematical diagram for (a) the 1stanti Stokes
(b) the 1st Stokes ESSRS emissions

HIGH PRESSURE BRILLOUIN SCATTERING IN LIQUID TOLUENE

A. Asenbaum, P. Soufi-Siavoch, R. Aschauer,

Institute for Experimental Physics, University of Vienna,
Boltzmanngasse 5, A-1090 Vienna, Austria

and Emmerich Wilhelm

Institute of Physical Chemistry, University of Vienna,
Währinger Straße 42, A-1090, Vienna, Austria

ABSTRACT

Brillouin spectra of liquid toluene have been measured at 303.15 K and at pressures from 1 to 1625 bar and 90° scattering geometry. From the experimental spectra the hypersound speed, the relaxing heat capacity, the relaxing and the nonrelaxing bulk viscosities, and the energy relaxation time were determined as a function of pressure.

I. INTRODUCTION

From recent Brillouin scattering experiments on liquid carbon tetrachloride[1] and Benzene[2] it has been shown that the energy relaxation time τ_E decreases with increasing pressure. Toluene is a simple derivative of benzene with a somewhat greater shape anisotropy, a very small permanent electric dipole moment and a sound absorption α/f^2, which is smaller by an order of magnitude than for Benzene. Therefore it seemed interesting to study the behavior of τ_E of Toluene at high pressures by means of Brillouin scattering.

II. EXPERIMENTAL

The experimental set-up is shown in figure 1. A 150 mW argon ion laser in single mode operation (λ = 514.5 nm) was used as light source for the experiment. Toluene was contained in a high-pressure cell equipped with three windows in a 90 ° ± 10' scattering geometry. The high-pressure system consisted of a hand-pump for P ≤ 700 bar, and a manually operated screw-press for the pressure range up to 2000 bar. The pressure was measured with a Bourdon gauge and with a

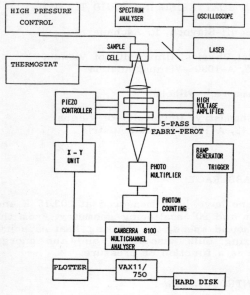

Figure 1

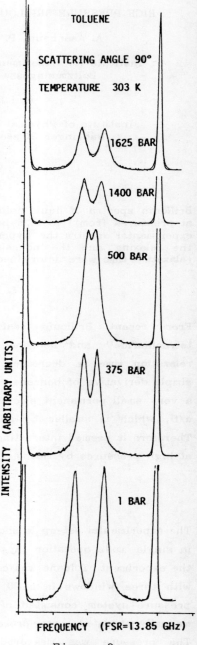

Figure 2

temperature compensated manganin cell, both with a maximum error of ± 12 bar. The scattered light was analyzed with an electronically stabilized five-pass Fabry-Perot interferometer with a free spectral range of 13.85 GHz and a finesse of about 40. The scattered light was detected by a cooled photomultiplier and stored in a multichannel analyzer (MCA). Dark current and background count rates were of the order of 3 counts/s. The MCA was triggered by the Rayleigh line to prevent broadening of the spectral components of the Brillouin spectra due to the drift of the laser frequency relative to the interferometer pass frequency.

III. RESULTS AND DISCUSSION

From the experimental spectra, shown in figure 2, the hypersound speed v_H was determined as a function of pressure, as were the relaxing heat capacity C_i, the relaxing and the nonrelaxing bulk viscosities $\eta_{v,r}$ and $\eta_{v,nr}$, the corresponding relaxation time τ_v, and the energy relaxation time τ_e, as shown in figure 3. The three quantities v_H, $\eta_{v,nr}$ and τ_v were obtained through use of Mountain's theory MOUETA[3] of polarized light scattering in dense fluids with internal degrees of freedom. The remaining quantities were determined by incorporating low-frequency sound absorption results. All necessary

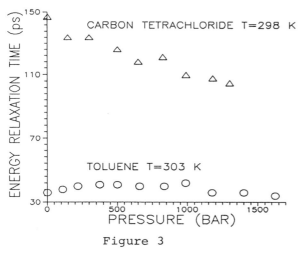

Figure 3

thermodynamic data have been critically evaluated from recent high precision measurements. We find a weak increase of C_i with increasing pressure, which is accompanied by a decrease of $\eta_{v,nr}$ as well as a nearly pressure independent behavior of τ_e (figure 3) and τ_v. Finally, the ratio $\tau_e(P)/\tau_e(P=1$ bar$)$ is compared with results obtained from the isolated binary collision theory for vibrational relaxation in conjunction with a simplified cell model of dense fluids for estimating the collision frequencies.

REFERENCES

1. A.Asenbaum and H.D.Hochheimer, J.Chem.Phys. 74, 1 (1981)
2. A.Asenbaum and H.D.Hochheimer, Z..Naturforsch. 38a, 980 (1983)
3. R.D.Mountain, J.Res.Natl.Bur.Stand. 70A, 207 (1966)

OPTICAL NONLINEARITIES AND BISTABILITY IN GOLD COLLOID

Li Chunfei, Wang Liming and Liu Yudong

Department of Physics, Harbin Institute of Technology

Harbin, China

ABSTRACT

We have measured the third-order nonlinear susceptibility in gold colloid both by degenerate four wave mixing with YAG pulse laser and by self-focusing with Ar^+ laser. The values of the third order nonlinear susceptibility are 5.5×10^{-8} esu and 1.2×10^{-8} esu respectively. We have also demonstrated transverse optical bistability in gold colloid

Introduction

There has been a growing interest in developing low dimensional optical nonlinear materials in recent years. Materials made by suspending dielectric, semiconductor and metal clusters in solution as colloidal particles have been shown to possess large nonlinearities[1)-3)]. We present here the study of optical nonlinearities and bistability in the gold colloid. We developed two different method to measure the third-order susceptibility. Using the YAG pulse laser, we obtained the value as 5.5×10^{-8} esu by degenerate four wave mixing. The response time is 10 ns. On the other hand, we also measured the third-order susceptibility as 1.2×10^{-8} esu in a slow process by self-focusing and self-trapping. the response time is 0.1 ms. With the same apparatus of the slow process. We demonstated transverse optical bistability in the gold colloid

Experimental

Made by chemical reaction the gold colloid density is $1.8 \times 10^{11}/cm^3$ with the colloidal particle diameter of 35 nm. The Ar^+ laser beam with line 514.5 nm was directly focused on the sample box. When the incident laser power P_{in} reached 300 mW, the self-focusing was observed. When the power was over 300 mW the beam inside the gold colloid became trapping, shown in Figure 1. The nonlinear third-order susceptibility, calculated according to the measured threshold value of self-trapping, was 1.2×10^{-8} esu. The nonlinear mechanism is mainly light pressure forces[4)] and thermal effects. Detecting the transmitted light from a 10 mm long medium through a small aperture, we obtained optical bistability for $P_{in} > 300$ mW, as shown in the Figure 2. the switch on and switch off are both 0.1 ms.

In the fast nonlinear process in gold colloid we found the nonlinear mechanism is the surface plasma resonance[3] and quantum size regime[2]. The thermal effect can be neglected for a 15 ns (FWHM) YAG laser pulse. By measuring the signal and probe light intensities in the standard degenerate four wave mixing circuit, we calculated the third-order susceptibility as 5.5×10^{-8} esu.

Conclusion

In conclusion, we have measured the third-order susceptibilities in gold colloid both by degenerate four wave mixing and self-focusing for the two different nonlinear optical process. We have also demonstrated transverse optical bistability in the gold colloid. Our research indicates that since there are more factors influence the optical nonlinerity in colloidal materials than that of bulk ones, the colloidal and other low dimensional materials will be found more useful in the nonlinear optics and optical computing.

Reference

1 Smith, P.W., Ashkin, A., Bjorkholm, J.E. and Eilenberger, D.J.,Opt. Lett. 10, 131 (1984).

2. Sarid, D., Rhee, B.K., McGinnis, B.P. and Sandroff, C.J., Appl. Phys. Lett. 49, 1196 (1986).

3. Richard, D., Roussignol, Ph. and Flytzanis, Chr., Opt. Lett. 10, 511 (1985).

4. Ashkin, A., Dziedzic, J.M. and Smith, P.W., Opt. Lett. 7, 276 (1982).

282

a   b

Fig.1. Experimental results of self focusing and self trapping; a. P_{in}=300mW, b. P_{in}=800mW.

Fig. 2. Optical bistable loop for P_{in}=300mW.

SUM-FREQUENCY GENERATION FOR SURFACE VIBRATIONAL SPECTROSCOPY

P. Guyot-Sionnest, R. Superfine, J. H. Hunt, and Y. R. Shen

Department of Physics, University of California
Center for Advanced Materials, Lawrence Berkeley Laboratory
Berkeley, California 94720 USA

ABSTRACT

Infrared-visible sum-frequency generation can be used for
surface vibrational spectroscopy of molecules adsorbed at
various interfaces. The highly surface-specific nature of the
technique is demonstrated.

In recent years, optical second harmonic generation (SHG) has been
proven to be a useful tool for surface and interface studies.[1] The
technique is based on the principle that a second-order optical process
is highly surface-specific since it is forbidden in a medium with
inversion symmetry but is necessarily allowed on a surface. Compared
with conventional surface probes, it has quite a number of unique
advantages. In particular, the technique can be employed for in-situ,
remote sensing of a surface in a hostile environment, and can be
applied to all types of interfaces accessible by light. With the help
of ultrashort laser pulses, it can even be used to probe ultrafast
surface dynamics and reactions. By scanning the pump laser frequency,
spectroscopic studies of surfaces and interfaces by SHG also become
possible. Unfortunately, the SH signal from a surface is often weak,
so that a photomultiplier is usually required for its detection. This
then limits the applicability of the technique to SHG with an output
wavelength $\lambda_{2\omega} \lesssim 1$ μm. Therefore, only surface electronic transitions
can be probed. The large bandwidths of electronic transitions make
selective detection of adsorbed molecular species rather difficult.
For the latter purpose, surface vibrational spectroscopy is needed.

We have recently succeeded in developing the IR-visible sum-
frequency generation (SFG) technique for surface vibrational
spectroscopy.[2] It is a simple extension of SHG; instead of the two
input frequencies ω_1 and ω_2 being the same, we now have $\omega_1 \neq \omega_2$ (Fig.
1). One is tunable in the infrared and the other is in the visible,
resulting in an output also in the visible. Scanning of the infrared

frequency then allows us to probe vibrational resonances by SFG. Like SHG, SFG is surface-specific and sensitive.

As seen in Fig. 1, the resonant SFG can be considered as a combined process of infrared and Raman excitations between the ground and excited states. Thus the resonances to be probed must be both infrared and Raman active. The SFG output is given by

$$S \propto \left| \chi_S^{(2)}(\omega_1 + \omega_2) \right|^2 \tag{1}$$

where the surface nonlinear susceptibility $\chi_S^{(2)}$ can be explicitly written as

$$\overleftrightarrow{\chi}_S^{(2)} = \overleftrightarrow{\chi}_{S,nr}^{(2)} + \sum_i \frac{\overleftrightarrow{A}_i}{(\omega_1 - \omega_i + i\Gamma_i)} . \tag{2}$$

Here, $\overleftrightarrow{\chi}_{S,nr}^{(2)}$ is the nonresonant part of $\overleftrightarrow{\chi}_S^{(2)}$, and $\overleftrightarrow{A}_i$, ω_i, and Γ_i are the strength, frequency, and damping coefficients of the ith resonance, respectively. The expression in Eq. (2) resembles that for coherent antiStokes Raman scattering (CARS).[3] One should therefore expect that SFG has the same general spectral shapes as CARS. For molecules adsorbed on a substrate, $\overleftrightarrow{\chi}_{S,nr}^{(2)}$ has, in general, contributions from both the molecules and the substrate. In cases where the substrate contribution dominates, the relative sign between $\overleftrightarrow{\chi}_{S,nr}^{(2)}$ and $\overleftrightarrow{\chi}_{S,r}^{(2)}$ (from molecules) is a direct indication of the polarity of the molecules with respect to the substrate.

Demonstrative experiments of SFG surface spectroscopy have been carried out in the 10 μm and 3 μm regions.[2] In the latter case, picosecond tunable infrared pulses can be generated by optical parametric amplification in $LiNbO_3$ crystals pumped by picosecond Nd:YAG laser pulses at 1.06 μm. The same laser can provide picosecond green pulses at 0.53 μm from a frequency doubler. The infrared and green pulses are then overlapped on the surface to be investigated and the resultant SF output is generated in the directions specified by the wavevector relation $\vec{k}_{1,\parallel} + \vec{k}_{2,\parallel} = \vec{k}_{sf,\parallel}$, where the subscript $\parallel$ denotes wavevector components parallel to the surface. After proper filtering, the SF output is detected by a photomultiplier connected to a gated integrator. With such a setup, the spectrum of allowed CH stretch vibrations of a polar-oriented molecular monolayer in the neighborhood of 3 μm can be readily detected.

To show the potential of this new spectroscopic technique,[3-6] we consider here, as an example, the application of the technique to monolayers of surfactant molecules adsorbed at liquid/solid interfaces.[4] Figure 2 describes the SFG spectra of three systems: (1) a hexadecane/fused silica interface, (2) a monolayer of octadecyltrichlorosilane (OTS) molecules adsorbed at a hexadecane/fused silica interface, and (3) a monolayer of OTS adsorbed at a CCl_4/fused silica interface. In the first case, no identifiable spectral feature is present even though in linear IR spectroscopy, one would expect strong peaks in this region arising from the many CH stretch vibrations on hexadecane. Here, the overall contribution of the CH stretches in the SFG spectrum vanishes because the hexadecane molecules are randomly oriented in the liquid phase. The spectra of cases (2) and (3) are nearly identical, both being the spectra of an ordered monolayer of OTS. The results clearly demonstrate the surface specificity of the technique.

By comparison with the Raman spectrum of OTS in solution, the three peaks in the spectra of Fig. 2 can be assigned to the CH_3 symmetric stretch (2875 cm^{-1}), the asymmetric stretch (2964 cm^{-1}), and the Fermi resonance between the s-stretch and the overtone of the CH bending mode (2942 cm^{-1}). These spectra resemble those of other long-chain molecules, such as pentadecanoic acid, on water in a full monolayer form.[5] They indicate that the OTS molecules have their long alkane chains straightened and aligned normal to the interfaces.

Different input-output polarization combinations can lead to SFG spectra with very different relative peak strengths. This suggests that the spectra can be used to deduce information about the orientation of the adsorbed molecules.[5] An example is shown in Fig. 3 for OTS monolayers at CCl_4/silica and CH_3OH/silica interfaces.[4] The strengths of the three peaks of OTS are clearly different for different polarization combinations. From the ratio of the s-stretch intensities of the (s_{vis}, p_{IR}) and (p_{vis}, s_{IR}) polarization combinations in Fig. 3(2) and 3(3), we can calculate the average tilt angle θ between the symmetry axis of the CH_3 terminal group and the surface normal. We find $50° < \theta < 40°$, indicating that the alkane chain is inclined at no more than $15°$ from the surface normal. This is consistent with the usual picture that the chain is along the surface normal; the apparent larger tilt obtained here is presumably due to roughness of the silica surface.

Figure 3b also exhibits an additional peak at 2834 cm^{-1}, which can be identified as the CH$_3$ s-stretch mode of methanol. Since no SF signal should be expected from the bulk methanol as suggested by the experiment on hexadecane mentioned earlier, we must attribute this peak to the methanol adsorbates at the interface. It suggests that some methanol molecules must have adsorbed at the interface with some degree of polar ordering. The relative strength of the CH$_3$ s-stretch mode of OTS in Fig. 3(2b) and 3(3b) however indicates that the OTS orientation at the interface is not affected by the presence of methanol.

From the above examples, it is clear that SFG surface spectroscopy has an enormous potential for surface studies. The technique is also capable of subpicosecond time resolution. It can therefore be used for research on ultrafast surface dynamics and reactions with selective detection of molecules. This opens the door to many interesting research opportunities in many branches of surface science. The future is only limited by one's imagination.

This work was supported by the Director, Office of Energy Research, Office of Basic Energy Scineces, Materials Sciences Division of the U.S. Department of Energy under Contract No. DE-AC03-76SF00098.

REFERENCES

1. See, for example, Shen, Y. R., J. Vac. Sci. Technol. B3, 1464 (1985); Ann. Rev. Mat. Sci. 16, 69 (1986).
2. Zhu, X. D., Shur, H., and Shen, Y. R., Phys. Rev. B 35, 3047 (1987); Hunt, J. H., Guyot-Sionnest, P., and Shen, Y. R., Chem. Phys. Lett. 133, 189 (1987).
3. See, for example, Shen, Y. R., The Principles of Nonlinear Optics (J. Wiley, New York, 1984), p. 267.
4. Guyot-Sionnest, Superfine, R., Hunt, J. H., and Shen, Y. R., Chem. Phys. Lett. 144, 1 (1988).
5. Guyot-Sionnest, P., Hunt, J. H., and Shen, Y. R., Phys. Rev. Lett. 59, 1587 (1987).
6. Harris, A. L., Chidsey, C. E. D., Levinos, N. J., and Loiacono, D. N., Chem. Phys. Lett. 141, 350 (1987); Superfine, R., Guyot-

Sionnest, P., Hunt, J. H., Kao, C. T., and Shen, Y. R., Surf. Sci. Lett. 200, L445 (1988).

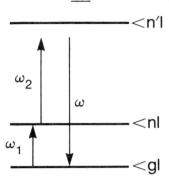

Fig. 1 Schematic of a sum-frequency genera-tion process in which ω_1 is a tunable IR frequence near resonance.

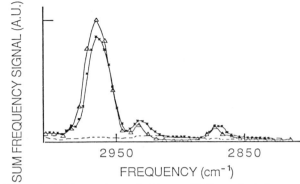

Fig. 2 SFG spectra at three different inter-faces in the (p_{vis}, p_{IR}) polarization combina-tion. Dashes: hexadecane/silica interface. Solid squares: OTS at CCl_4/silica interface. Triangles: OTS at hexadecane/silica interface.

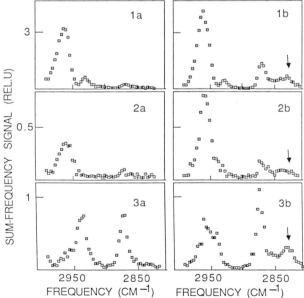

Fig. 3 SFG spectra at (a) the CCl_4/OTS/silica interface and (b) the CH_3OH/OTS/silica interface. The different panels refer to different polarization combinations: (1) (p_{vis}, p_{IR}); (2) (p_{vis}, s_{IR}); (3) (s_{vis}, p_{IR}). The arrows in (b) indicate the methyl s-stretch mode of the adsorbed methanol molecules.

Optical Studies of Molecule/Surface Interactions

M. M. T. Loy, T. F. Heinz, J. A. Prybyla and J. A. Misewich

IBM Research Division
Thomas J. Watson Research Center
Yorktown Heights, New York, 10598

ABSTRACT Nonlinear optical techniques such as second-harmonic generation and multiphoton ionization offer unique opportunities for the study of molecule-surface interactions in ultrahigh vacuum. We will discuss our recent application of these techniques to the study of surface diffusion of atoms and desorption of molecules from surfaces.

Recently, there has been much interest focused on the dynamics of molecules interacting with surfaces. This follows from the rapid advances in surface sciences, particularly in the structures and electronic spectroscopies of clean, well-characterized surfaces in ultra-high vacuum. With this knowledge, one can then attempt to understand what happens when a molecule is brought near the surface. A sample of the questions include: the sticking probability of the molecule, its dependence on the energy of the molecule, the sticking site on the surface, etc. These challenging questions are most interesting scientifically; they are also crucial to a better understanding and control of materials surface processing methods, such as deposition and etching, which are of great importances technologically. In this paper, we describe two experiments where we applied optical techniques to study some aspects of this general problem of molecule/surface interaction. In the first experiment, we obtained a complete characterization of a thermal desorption of NO molecules from the Pd(111) surface, using the resonantly-enhanced multiphoton ionization technique. The results of this experiment, together with the principle of detail balancing, allow us to answer the question how the sticking probability depends on energy distributions (i.e., rotational, vibrational and translational) of the incoming NO molecule.[1] In the second experiment, we used the surface second-harmonic technique to study surface diffusion of Si atoms on a Si(111) surface. This allows us to determine the activation energy needed for the Si atom to diffuse to the epitaxial site.

The measurements of NO desorption from a Pd(111) surface was performed under ultrahigh vacuum conditions with the apparatus which has been described previously.[2] The cleanliness and order of the Pd(111) surface was verified by LEED (low-energy electron diffraction), thermal desorption spectroscopy, and Auger spectroscopy. NO was adsorbed on the Pd(111) surface at room temperature; thermal desorption occurred upon heating the substrate with a 5 nsec pulse of radiation at 532 nm. Peak substrate temperatures up to 1,200 K could be obtained with a uniform laser spot produced by Fourier filtering. Care was taken to restrict the amount of NO desorbed during each heating pulse to < 0.01 monolayer, thereby ensuring our detection of the nascent distribution of the desorbed molecules, without perturbation from post-desorption gas-phase collisions. The NO molecules in each rovibrational level of the $X^2\Pi_{1/2, 3/2}$ ground electronic states were detected by means of resonantly-enhaced multiphoton ionization through the $A\Sigma$ state. The tunable UV radiation (220 to 226 nm) was generated by doubling of blue dye laser radiation using a KPB crystal, and more recently, the far superior BBO crystal.

Among the principal results of this study is a clear demonstration of the phenomenon of rotational cooling in thermal desorption, i.e., of molecules emerging with rotational energy below the surface temperature. We found that the rotational population of the desorbed NO followed a Boltzmann distribution, but at a temperature ($T_r = 660$K) significantly below the surface temperature ($T_s = 1150$K). The qualitative difference between our result and the previous measurement of direct thermal desorption of NO from Pt(111),[3] which showed no rotational cooling, is striking. One possible explanation is the different regime under which the desoption occurred: in our case, the desorption took place at high surface temperature on the timescale of nanosecond while in the earlier

NO/Pt(111) work, the NO was desorbed on a timescale of seconds at a much lower surface temperature. It has been predicted in molecular dynamics simulations[4] that the degree of rotational cooling will increase with increasing surface temperature. It would be most interesting to see whether even more rotational cooling can be observed with even shorter desorption timescale, i.e., using a mode-locked laser pulse. Such an experiment is currently underway in our laboratory.

Since the desorption timescale was short compared to the transit time of the desorbed molecule, the translational energy distribution can be obtained simply by varying the delay time between the ionizing and desorbing laser pulses. This distribution (independent of internal energy) was also strongly cooled with respect to the surface temperature. The origin of this effect lies in the short (0.1 psec) duration of the interaction of the molecule in the final stage of escaping from the surface potential and the associated inefficient energy transfer. Equivalently, detailed balancing allows us to associate the translational cooling with a probability of sticking probability that decreases with increasing translational energy. At higher incident energy, it is more difficult to dissipate the excess energy required to trap the molecules. Our findings for the internal energy distributions of the desorbed molecules correspond to a sticking probability that decreases with increasing rotational energy, but is unaffected by vibrational excitation, since no vibrational cooling was observed. This behavior points to a picture of a strong coupling of the rotational and translational motion of a molecule in the vicinity of the surface, together with a relatively weakly coupled vibrational motion.

In the above experiment, we studied effects determined by the molecule-surface energy potential along a direction normal to the surface. In the next experiment, where we study surface diffusion of Si atoms using surface SHG, we turn our attention to energy barriers the atom or molecule must overcome to move about on the surface. In particular, when Si atoms are deposited on a clean Si(111) surface at room temperature, they lie randomly on the surface since the thermal kinetic energies of the atoms are less than the activation barrier for diffusion. Only at higher substrate temperature can the

Si atoms diffuse on the surface and get to the sites which register with the Si crystal leading to epitaxial growth. This activation barrier is then a limiting factor in Si film growth by MBE (Molecular Beam Epitaxy) and its determination is the subject of the following study.

The high sensitivity of the SHG process to the atomic structure and symmetry of the surface layer has been demonstrated in several studies of modification and reactions on semiconductor surfaces. One class of applications, including the present experiment, involves assessing the degree of surface ordering by means of the strength of the SH response. (See Ref. 5 for a more detailed review on this subject.) In the present work, the Si(111)-7x7 surfaces were prepared and studied in an ultrahigh vacuum chamber where the surface structure and cleanliness of the sample can be verified. The pump radiation for the SH measurements from a laser was directed at normal incidence onto the sample surface. SH was detected showing the characteristic anisotropic angular pattern determined by the surface structure. SH signal was then monitored as Si atoms were deposited onto the surface. With the Si substrate at room temperature, the SH intensity decreased indicating the formation of an amorphous adlayer of Si atoms covering the underlying crystalline Si surface. The sample thus prepared was then annealed at above 500 C. The recovery of the SH signal indicated the reestablishment of the surface ordering with the adatoms diffusing to epitaxial sites. By performing such measurements over a range of substrate annealing temperatures, we determined the activation energy for crystallization in such a two-dimensional layer. The activation energy was found to be significantly below that for solid phase epitaxy, as anticipated, but significantly higher than the 0.2 ev value suggested by the early study of Abbink et al.[6]

In conclusion, we have shown that optical techniques can be most useful in the study of aspects of molecule/surface interactions. In complement to electron spectroscopy techniques, optical methods such as SHG are particularly attractive for real-time, in-situ studies, and can be used under non-ultrahigh vacuum conditions. Other optical techniques such as resonantly-enhanced ionization detection offer high sensitivity and energy resolution enabling the complete characterization of the various degrees of freedom of molecules at or near surfaces.

This work was supported in part by the Office of Naval Research.

REFERENCES

1. Prybyla, J.A., Heinz, T.F., Misewich, J.A., and Loy, M.M.T., "Dynamics of molecular Thermal Desorption: NO/Pd(111)," to be published.

2. Misewich, J., Roland, P.A., and Loy, M.M.T., Surface Sci. 171, 483 (1986).

3. Mantell, D.A., Cavanagh, R.R., and King, D.S., J. Chem. Phys. 84, 5131 (1986).

4. Muhlhausen, C.W., Williams, L.R., and Tully, J.C., J. Chem. Phys. 83, 2594 (1985).

5. Heinz, T.F., and Loy, M.M.T., "Optical Second-Harmonic Generation from Semiconductor Surfaces", Proceedings of the Third International Laser Science Conference, to be published.

6. Abbink, H.C., Broudy, R.M., and McCarthy, G.P., J. Appl. Phys. 39, 4673 (1968).

OPTICAL SECOND HARMONIC GENERATION WITH
COUPLED SURFACE PLASMONS FROM A MULTI-LAYER SILVER/QUARTZ GRATING

Zhan Chen* and H. J. Simon
Department of Physics and Astronomy
The University of Toledo
Toledo, Ohio 43606

*On leave from Fudan University, Shanghai, PRC

ABSTRACT

We report a theoretical and experimental study of enhanced optical second harmonic generation (SHG) due to excitation of coupled surface plasmons from a thin silver grating bounded by two quartz crystals. With incident radiation from a Nd:YAG laser, diffracted SHG in the n = 1, 0, and -1 modes is observed for the coupled surface plasmon fundamental modes. Nonlinear reflectance ratios calculated by using the reduced Rayleigh equations yield good agreement with the experimental results. Enhancement of the SHG is limited by the scattering of the long-range surface plasmon mode in the thinnest silver films studied.

INTRODUCTION

The properties of surface electromagnetic modes associated with multi-layer thin films continues to be a subject of interest. Particular attention has been given to the resonances associated with coupled surface plasmons which propagate on opposite sides of a thick metal film bounded by dielectrics with the same index of refraction. Of the two modes, corresponding to symmetric and antisymmetric magnetic field profiles, the former has a surprisingly low attenuation for a sufficiently thin metal film and has been named the long-range surface plasmon (LRSP)[1]. Recent attention has focused on the use of grating structures for coupling to the LRSP mode. The present authors have studied attenuated total reflectance (ATR) from a thin silver grating multi-layer structure.[2]

The enormous enhancement of optical second-harmonic generation (SHG) due to prism-coupling of the fundamental LRSP mode in a quartz/silver/liquid dielectric structure was first demonstrated five years ago.[3] Subsequently the use of metallic gratings to excite surface plasmons for enhancing SHG has been studied. Surface SHG from air/Ag grating interfaces has received the greatest attention,[4] while bulk SHG with surface plasmons in the infrared from GaAs/Al grating interfaces has also been reported.[5] Recently the present authors have also conducted a study of bulk SHG with surface plasmons from a single-boundary silvered quartz grating[6] in which excellent quantitative agreement was found between the experimental results and a theoretical calculation based on the reduced Rayleigh equations. Here we present results of an experimental and theoretical study of enhanced SHG due to grating excitation of coupled surface plasmons from a quartz/Ag/quartz grating structure.

EXPERIMENT

The source of fundamental radiation was a continuous mode-locked Nd-YAG laser with an acousto-optic Q-switch which yielded a pulse train of approximately twenty-five 100 psec pulses at a 500 Hz rate. The beam passed through a polarizer and was incident on the grating which could be rotated by means of a stepping motor in increments of 0.01°. SHG data were taken with a boxcar averager the output of which was proportional to the ratio of SHG in the signal to reference channels. The diffraction grating used in this experiment was prepared by standard holographic technique to obtain a line spacing of 1200 lines/mm. The grating pattern in the photoresist was then transferred to a quartz crystal substrate by Argon ion etching thus producing a groove depth of 270 Å. The quartz crystal was x-cut with the grating grooves parallel to the crystal y-axis and the grating wavevector parallel to the crystal z-axis. Silver films were thermally evaporated onto the quartz grating substrate thus replicating the grating profile on both silver surfaces. A second quartz crystal of y-cut quartz with a flat face was contacted using index matching liquid on top of the silvered grating to complete the multilayer structure. The angles of incidence at which the fundamental coupled surface plasmon modes are excited by the n=1 diffracted mode were determined from the linear ATR curves[2]. Diffracted SHG was observed in the harmonic n=1,0 and -1 modes. The inset to Fig. 1b illustrates these three diffraction geometries. Note that the diffracted radiating harmonic mode with wavevector closest to twice the fundamental coupled surface plasmon wavevector is the n=1 mode.

RESULTS

Our results are displayed by plotting the nonlinear reflectance R_n, defined as the ratio of the harmonic reflected power in the diffracted order labelled by n to the square of the incident fundamental power, versus the angle of incidence in air. The observed SHG is calibrated to the theoretical value by comparing to the transmitted SHG in a wedged z-cut quartz crystal near normal incidence. We display in Fig. 1a the reflected SHG diffracted into the n=1 mode for a metal film thickness of 520 Å. The splitting of the SHG resonance into the two coupled mode resonances is clearly observed. The solid curve and solid dots are the results of the theoretical calculation and the experimental data respectively for the LRSP mode. These numerical values are given on the left scale. The dashed theoretical curve with the solid squares displays the results for the SRSP mode. These numerical values are given on the right scale which has been expanded by a factor of approximately three relative to that of the left scale for display purposes only. The level of the background SHG observed off resonance is created by the small misalignment of the grating grooves from parallelism with the crystal y-axis.

In Fig. 1b we show on a logarithmic scale the reflected SHG diffracted into the n=1,0 and -1 orders in the case of the LRSP mode for a metal film thickness of 230 Å. Again the agreement between theory and experiment at the peak of the resonance is excellent for all three diffracted orders. As reported in earlier experiments with this same grating a 4% admixture of the third spatial harmonic of the grating sinusoidal profile was included to account for the anomalously high SHG level observed in the n = -1 order. Part of the angular broadening may be attributed to increased scattering of the LRSP from the grating grooves as the LRSP propagation length increases. Comparison of Figures 1a and 1b shows that coupling to the LRSP mode in thinner films increases the enhancement of the SHG as expected; however, as the metal film thickness was further decreased to 130 Å the enhancement actually decreased. The scattering of the LRSP mode in these thin films degrades the amplitude of the mode and consequently also the SHG enhancement.

CONCLUSION

In conclusion we have observed enhanced diffracted SHG due to excitation of the fundamental coupled surface plasmon modes in a grating multi-layer structure. The reduced Rayleigh method may be used to accurately predict the SHG diffracted into all orders. Enhancement of the SHG due to coupling into the LRSP mode for thin silver films is limited by the scattering of the LRSP mode from the grating grooves.

296

REFERENCES

1. J. C. Quail, J. G. Rako and H. J. Simon, Opt. Lett. **8**, 377–379 (1983).
2. Zhan Chen and H. J. Simon, J. Opt. Soc. Am. B **5**, (1988).
3. J. C. Quail, J. G. Rako, H. J. Simon, and R. T. Deck, Phys. Rev. Lett. **50**, 1987 (1983).
4. J. C. Quail, and H. J. Simon, J. Opt. Soc. Am. B **5**, 325 (1988).
5. Chen Zhenghao, Cui Dafu, Lu Huibin, and Zhou Yueliang, Opt. Lett. **8**, 563 (1983).
6. H. J. Simon, C. Huang, J. C. Quail, and Z. Chen, submitted to Phys. Rev. B.

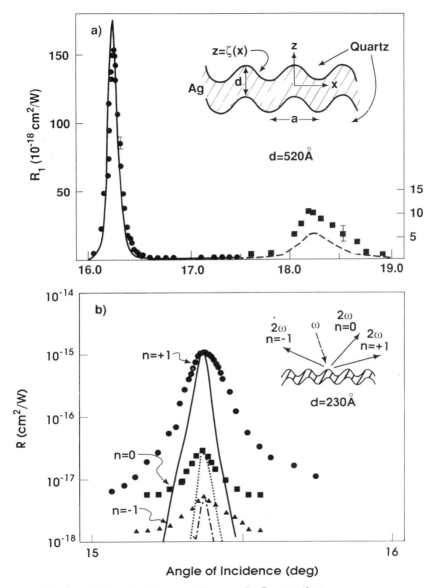

Fig. 1. Diffracted Second Harmonic Generation
from Quartz / ag / Quartz Grating; a) d=520Å, b) d=230Å

EVIDENCE OF SILVER CLUSTER AND ITS ROLE IN
SURFACE ENHANCED RAMAN SCATTERING (SERS)

Dong Shuyan, Wang Wencheng and Wang Gongming

Lab of Laser Physics & Optics

Fudan University, Shanghai, P.R. China

The microstructure in cluster form on a rough silver surface offers an enhancement factor of 10^2-10^3 for the surface enhanced Raman scattering (SERS) through the modification of the adsorbed molecules.

1. It is generally recognized that there exists short range enhancement in SERS and the scale of this short range is estimated to be around 4-40Å in distance. This enhancement depends greatly on the topography of the rough surface, that is, metal atoms (especially for Ag, Au, etc.) could be composed to become clusters on these rough surfaces and the molecules adsorbed on these clusters will serve as "Raman active sites". Though there are a few experimental results which confirmed that some low frequency shifted Raman bands should be assigned only from the contributions of rough microstructures themselves, however there still lack of experimental data to reveal the close relationships between the clusters fromed on rough surfaces whth the enhancement of SERS.

2. In our experiment, we employed different methods to fabricate three kinds of rough silver surfaces. The first one is electrochemically roughened silver electrode, the electrolyte is 0.1M KCl. The second is to evaporate a small quantity of Ag on a glass substrate to form an island film (about 60Å in mass thickness) under the vacuum environment of 10^{-5} - 10^{-6}torr. The last sample is chemical reactionary deposited Ag film with average roughness about 100Å, and with mass thickness of about 500Å. The CW argon ion laser beam of 100mw at 5145Å or 4880Å is focused on the samples at an angle of about 70°. The scattering light is collected by a lens into a double monochromator and detected by a photocounter. We have found three vibration bands of 116cm^{-1}, 104cm^{-1} and a weak one of 150cm^{-1} which sould be assigned no other than the eigen

bands of Raman scattering from the silver rough microstructures (see Fig. 1). The correlation between these bands with SERS can be revealed in the following. Fig. 2 shows the intensities of pyridine bands of 1008cm^{-1} and 1036cm^{-1}, after dipping all these three rough samples in to the solution of 0.05M of pyridine for four hours. The fact that the intensities of the bands 1008cm^{-1}, 1036cm^{-1} from the three different kinds of rough surfaces are quantitatively the same(see Fig.2 b, c, d) and which implies that the SERS enhancement mechanism of these three samples are the same. Curve e of Fig. 2 gives the SERS intensity of standard case: pyridine involved in electrochemical reaction, that is believed to be being enhanced $10^{-5} - 10^{-6}$ relative to bulk scattering but only 10 times more intense than the cases of b, c, and d of Fig.2. We suppose this factor of 10 is of chemical enhancing and the another two enhancement channels: that is the local plasmon excitation and the cluster-molecules formed Raman active sites will incooperate to offer a factor of $10^4 - 10^{-5}$. The idea that these clusters will serve as Raman active sites and can be firmly identified by the experimental results showed in Fig.3.The similar behavior of the intensity variations for the bands 116cm^{-1} and 1008cm^{-1} vs the applied voltage V_{sce} demonstrated that the amount of clusters on the rough surface will dominate the intensity of SERS.

3. Conclusion can be made now that there exist stable clusters of Ag_n or Ag_n^+ ($3 < n < 6$) on the rough silver surface, which will greatly modify the Raman tensors of the adsorbed molecules through electronic affections. This will give an increased scattering cross-section and offer an enhancement of a factor of $10^2 - 10^3$.

4. We also calculated the eigen Raman bands for the cluster Ag_4^+ in pyramidal form as an example and found the values of bands to be 168.3cm^{-1}, 119.0cm^{-1}, 107.6cm^{-1} and 76cm^{-1}, which show that two of them fit quite well with our experimental data for the bands of 116cm^{-1} and 104cm^{-1}.

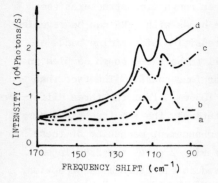

Fig.1 Raman signals from samples of:
(a) Mechanically polished Ag plate
(b) Island film of massthickness 60Å
(c) Chemical reactionary deposited Ag film
(d) Ag electrode roughened by one ORC*in O.1M KCL

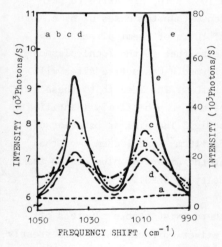

Fig.2 Raman signals of Pyridine's bands from samples of
(a) Mechanically polished Ag plate
(b) Island film of massthickness 60Å
(c) Chemical reactionary deposited Ag film
(d) Ag electrode roughened by one ORC in O.1M KCL
(e) Ag electrode roughened by one ORC in O.1M KCL/O.05M Py.

*ORC—Oxidation Reduction Circle

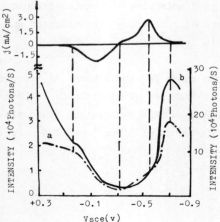

Fig.3 Raman signals of bands
at a) 116cm⁻¹ vs. V̄sce
b) 1008cm⁻¹

STUDY ON COLD-EVAPORATED SILVER SURFACES WITH
SECOND-HARMONIC-GENERATION

Li Le, Yu Gongda*, Liu Yanghua, Wang Wencheng and Zhang Zhiming

Lab of Laser Physics & Optics, Fudan University

Shanghai, P. R. China

Cold-evaporated silver surfaces are studied with Second-Har-monic-Generation during the annealing process from 120°K to 400°K under ultrahigh vacuum environments. Both local and non-local surface plasmon polaritons are excited in evidence experimentally.

1. Study on annealing properties of cold-evaporated metal surfaces has attracted wide spread interests since they were recognized to exhibit huge enhancement of the Raman scattering cross-section. As to the electromagnetic effects responsible for the surface-enhanced Raman scattering, it is believed that the long-range electromagnetic effects can lead to the significant enhancement. However, recently Ushioda and his co-workers[1] reported their experimental results by using the Attenuated Total Reflection (ATR) technique on the study of cold-evaporated silver films and believed that the localized surface plasmon polaritons (SPP) are in response for such an enhancement. We will report on the study of cold-evaporate silver surface by Second-Harmonic-Generation (SHG) which is a non-linear optical process as compared to the ATR technique and will give directly the evidence of the excitation of SPP. We have found both local and non-local SPP could be excited during the annealing process of a cold-evaporated Ag surfaces as predicted from our experimental results.

2. The experiments were conducted in an ultrahigh vacuum system of 10^{-10} torr in pressure. The details of the experiments will be described eleswhere[2]. A p-polarized fundamental wave of 1.06 m from a Q-switched YAG laser was impinged onto the surface of the samples and the reflected SHG signals were measured by a photomultiplier. The sample

was a polished Ag block of purity 4N for easy to cool down to the required temperature, this sample block was taken as the substrate for supporting the silver thin film which was cold-evaporated onto this substrate at a rate of 2.5Å - 5Å per second when this block was cooled to the tempreature of 120^OK. The purity of this evaporated Ag film was also 4N. SHG signals were then measured during annealing process as a function of the temperature which was raised slowly from 120^OK to over 400^OK. The measured SHG intensities decreased from the very beginning in an exponential-like way as the surface temperature raised. A typical spectrum is shown in Fig.1 which was taken under the incident angle of 41°. Also, some peaks of the SHG signals were observed overlapping upon this decaying signal curve. When we lowered the temperature of the Ag surface in the reverse direction from 400^OK to 120^OK, all the two kinds of signals disappeared and the SHG signals kept at a constant level corresponding to the signal intensity at 400^OK, thus it meant that they were irreversible. As Ushioda had pointed out in their results that the decaying SHG signals against the temperature could be recognized from the excitations of localized SPP, but those peaked SHG signals, as we have observed, were believed to be contributed from the long-range non-localized SPP. To confirm this assignment, a simple model was proposed that such cold-evaporated Ag surfaces were porous and might have micro-sturctures which could be regarded as consisiting of various grating structures with different groove spacings. The SHG peak signal was expected to be originated from one of the grating structures which matched with the incident angle of the fundamental wave as given by the well known wave vector relation and a non-local SPP then could be excited. As the spacing of the grooves would be changed during the annealing process, the coupling of the laser field to SPP or the positions of the peak signals located upon the decaying curve should be in dependence with the incident angle, which is the important charater for a non-localized SPP. Besides, we could estimate that for higher temperature, the grooves would be more coarse and the peak position of the SHG signals would be then located at some larger angles of incidence. Such an incident angle correlation characterists were observed by us in our experiments as shown in Fig.2, in which the peak shifts were evidently demonstrated by the corresponding arrows respectively for six different incident angles. In conclusion, a local SPP would cause the induced EM field near the top of the micro-structures and be incident angle independent, but a non-local SPP will be in angular dependence with the micro-structures on the surface.

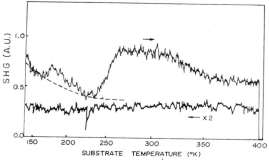

FIG.I SHG SPECTRA OF ANNEALING A COLD–EVAPORATED SILVER FILM (UPPER ARROR) AS WELL AS THE REVERSED PROCESS (LOWER ARROR)

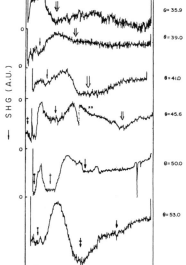

FIG.2 THE ANNEALING SHG SPECTRA OF COLD EVAPORATED SILVER FILMS FOR SIX INCIDENT ANGLES SHOWING THE PEAK SHIFTS, AS INDICATED BY THE ARROWS, RESPECTIVELY

References:

1. C.E. Reed, et al, Phys. Rev. B31, 1873 (1985)

2. L. Le, et al, Phys. Rev.B (to be published)

* On leave from Suzhou University, China

STUDY OF OPTICAL SECOND-HARMONIC-GENERATION AT METAL SURFACE WITH POLARIZATION STATES

Zheng Wanquan, Li Le, Dong Shuyan, Wang Wencheng and Zhang Zhiming

Lab of Laser Physics & Optics

Fudan University

Shanghai, P. R. China

We have measured and analysed the optical Second-Harmonic-Generation (SHG) at metal surface as function of polarization states, the value of phenomenological parameter "a" characterizing the normal surface current is determined to be -4 for Ag under ultrahigh vacuum condition at 1.06 μm incident radiation.

1. The sources for the reflected optical SHG at a metal surface consist of three parts[1]: the bulk current which extents about an optical skin depth into the metal, and two surface currents, one is normal and the other parallel to the surface, which are confined to the surface region of only a few angstrons. The phenomenological parameters "a" and "b" are introduced to estimate the value of these surface currents respectively and it is of great interest to determine the value of "a" for it is the most sensitive to the details of the surface charge distribution. A recent quantum-mechanical calculation[2] predicts a =-28 for Al and a =-12 for Ag at low frequency, but which is an order of magnitude larger than the earlier calculation with free-electron model[1] or the hydrodynamic model[3]. Experimental measurements were also carried out recently[4] and showed that a=1.5 for Al and 0.9 for Ag which seems to be consistent with the classical model but not the quantum one. We report here our experimental study on measuring the SHG intensity changes with the azimuth angle of ploarization Ø which defines as the angle between the electric vector of the incident wave and the plane of incidence. By fitting with the theoretical calculations, we can get the parameter "a". Our results show explicitly that the value of "a" for Ag and Al should be between -2 and -7 under several surface conditions for incident 1.06 μm radiation. We have also

analysed the variation of the SHg signals as function of the incident angle θ and the value of "a" so determined confirms our above measurements.

2. Use a mode-locked YAG laser output as the fundamental beams, the polarization state of the incident beam can be varied by a λ/2 rotator and energy is restricted to 1 mj/pulse. Measurement is done on an ultra-clean Ag surface which is prepared by evaporating Ag on a substrate in the vacuum of 10^{-10} torr. The results are shown in Fig.1, each experimental point here represents an average over about one thousand shots. The reflected SHG wave from vacuum-metal interface is given by:

$$|E_{2\omega}^{R}| = |\frac{e\,\omega_{p}^{2}\sin\theta\cos^{2}\theta E_{o}^{2}}{mc\omega^{3}\epsilon(\omega)\epsilon(2\omega)^{\frac{1}{2}}F(\omega)F(2\omega)}\{\epsilon(\omega)\,[\cos^{2}\emptyset + \frac{F^{2}(\omega)\sin^{2}\emptyset}{g^{2}(\omega)}\,]$$

$$+ 2a\epsilon(2\omega)\cos^{2}\emptyset - 4b\cos^{2}\emptyset\,[\epsilon(\omega) - \sin^{2}\emptyset]^{\frac{1}{2}}[\epsilon(2\omega) - \sin^{2}\emptyset]^{\frac{1}{2}}\}|$$

where E_{o} is the incident field amplitude and

$$F(\omega) = [\epsilon(\omega)]^{\frac{1}{2}}\cos\theta + [1 - \frac{\sin^{2}\theta}{\epsilon(\omega)}]^{\frac{1}{2}}\,,$$

$$g(\omega) = \cos\theta + [\epsilon(\omega) - \sin^{2}\theta]^{\frac{1}{2}}\,, \qquad \omega_{p} = [\frac{4\pi n_{o}e^{2}}{m}]^{\frac{1}{2}}$$

is the plasma frequency. We reckon $b = -1$ [3] and the dielectric constants of Ag are taken from ref. (4). It is found when a =-4.0 ± 0.8, the experimental data can fit quite well with the theoretical curve. The theoretical results for a=0.9, -5, and -10 are also drawn for comparison which are clearly incompatible with our experimental results. We have also measured the SHG signals from the glass-Ag,vacuum-Al and glass-Al interfaces, and we get 2≤ -a ≤7 under all these surface conditions. Our experimental results also demonstrate that the value of "a" increases slightly with the angle of θ for the glass-Ag interface, as an example, we show the variation in the Table 1. The dependence of the SHG signals on the angle of θ at glass-Ag interface is given in Fig. 2. The value of "a" is determined to be -6.4 at ∅ = 0, which is compatible with our previous measurements.

3. In conclusion, we have shown experimentally that the parameter "a" has the value between -2 and -7 for Ag and Al under various surface conditions and especially a= -4.0 for an ultra-clean Ag surface. As our experimental measurements does not gear with the present theoretical and experimental resutls, we feel that more detailed work in this field will be required.

Table 1. Dependence of "a" on the incident angle θ for glass-Ag interface

θ	48	51	54	57	60	66	68	71
-a	3.2	3.8	4.2	4.0	4.6	4.4	5.4	5.6

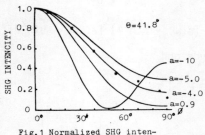

Fig.1 Normalized SHG intensities from ultra-clean Ag surface change with the angle ∅ at 1.06 µm incidence

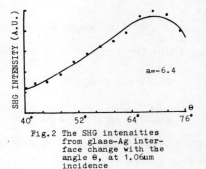

Fig.2 The SHG intensities from glass-Ag interface change with the angle θ, at 1.06µm incidence

REFERENCES:

(1) J. Rudnick and E.A Stern, Phys. Rev. B4, 5274, (1971)
(2) M. Weber and A. Leibsch, Phys. Rev., B35, 7411, (1987)
(3) J.E. Sipe, et al, Phys. Rev., B21, 4389, (1980); J.E. Sipe and G.I. Stegeman, in "Surface Polaritons", edited by V.M. Agranvich and D.L. Mills, North-Holland, (1982)
(4) J.C. Quail and H.J. Simon, Phys. Rev., B31, 4900, (1985)

SPECTROSCOPIC STUDIES OF J-AGGREGATES OF
PSEUDOISOCYANINE IN MOLECULAR MONOLAYERS
IN THE RANGE 300 TO 20 K

Alexander Müller, Liang Peihui*, Zhang Weiqing*

Max-Planck-Institut fur Biophysikalische Chemie,
Abt. Laserphysik, Postfach 2841, D-34 Gottingen
Federal Republic of Germany

*Shanghai Institute of Optics and Fine Mechanics,
Academia Sinica, Shanghai, P.R. China

J-aggregate formation of Pseudoisocyanine (PIC) can be observed in highly concentrated aqueous solution as well as in molecular monolayers produced on various substrates using the Langmuir-Blodgett technique. While threadlike one-dimensional aggregates are formed in solutions, the aggregates in monolayers consist of two-dimensional crystallites of about 10 μm diameter as revealed by polarization microscopy[1]. Spectroscopic studies at low temperatures appear to have been limited to the range of the characteristic absorption and emission bands in both cases, so far.

We have deposited the PIC derivative 1-methyl-1'-octadecyl-2, 2'-cyanine perchlorate (MOC) mixed with Hexadecane (HD) in ratio 1:2 in single monolayers on the water surface using the usual techniques[2]. Subsequently, the monolayers were transferred to various substrates: (a) sapphire, (b) glass uncoated, (c) glass coated with 1 to 10 layers of arachidic acid. The samples were cooled to 20 K with the aid of a Leybold Kryostat model ROK 10-300. Fluorescence emission and excitation spectra were recorded in a frontal arrangement by a SPEX FLUOROLOG spectrometer equipped with two grating double-monochromators model 1680.

The general behavior of the narrow aggregate emission peak does not markedly depend on the particular substrate. Fig.1 shows it for case (b) (see above). While the peak intensity increases, the halfwidth of the band decreases upon cooling from 300 K to about 100 K, whereafter both attain

308

constant values. The fluorescence quantum yield remains, thus, approximately constant in this wide temperature range. The peak shifts from 593 nm to a constant low temperature position at 585 nm. Our findings are in agreement with recent work with a different dye at higher temperatures[3]. Extension down to 20 K reveals, however, the existence of limiting values.

We found a new intense emission band at longer wavelengths, as shown in Fig.2, the form of which depends on the nature of the substrate. The band starts appearing at about 100 K. On substrates (a), (b) and (c) with up to two arachidic acid layers we observe generally a single broad band of slightly varying shape having a maximum around 670 nm. Beginning with a number of three isolating layers the long wavelength band splits into several components (cf. Fig.2.) with a separation of about 1400 cm^{-1}. The degree of polarization is approximately constant and positive in the whole range (Fig.2, lower part). Irrespective of the nature of the substrate the fluorescence excitation spectra measured at different locations correspond always to the J-aggregate absorption spectrum (Fig.3).

Together with previous evidence obtained with the aid of picosecond spectroscopy[4], our findings from fluorescence excitation spectra and polarization of the emission bands indicate that all bands belong to the same excitonic transition and that the low energy bands appearing at low temperatures represent phonon sidebands. The observed frequency difference appears to correspond to a vibrational mode of the monomer with an energy of 1368 cm^{-1} [5].

ACKNOWLEDGEMENT

Liang Peihui and Zhang Weiqing thank the Alexander von Humboldt Foundation and Prof. F.P. Schafer for providing the opportunity and support to carry out the work at Gottingen.

REFERENCE

[1] D. Mobius, J. Physique, C-10, 441 (1983).
[2] H. Kuhn, D. Mobius, H. Bucher in Physical Methods of Chemistry,
     1, part 3B, A. Weissberger, B. Rossiter, eds, Wiley (1972).
[3] T. Inoue, Thin Solid Films, 132, 21 (1985).
[4] H. -P. Dorn, A. Muller, Appl. Phys., B43, 167 (1987).
[5] E.W. Knapp, P.O.J. Scherer, S.F. Fischer, Chem. Phys. Lett.,
     111, 481 (1984).

Fig. 1 Fluorescence intensity, halfwidth and peak
location of the J aggregate band as a
function of temperature. Single monolayer
of MOC: HD = 1: 2 on glass

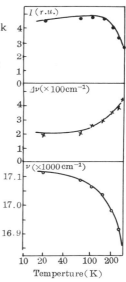

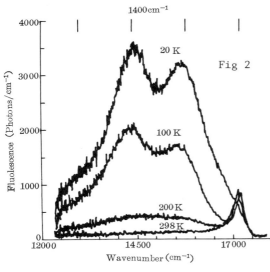

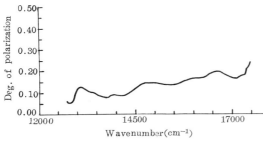

Fig. 2 (upper): Fluorescence
spectra at various temperatures. Glass substrate with
5 monolayers of arachidic
acid and one monolayer of
MOC: HD = 1: 2
(lower): Polarization spectr
um at 20 K

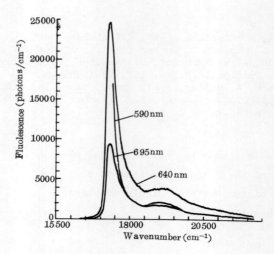

Fig. 3 Fluorescence excitation spectra at 20 K. Curves are
labeled with emission wavelengths. Glass substrate
with 5 monolayers of arachidic acid and one monolayer
of MOC:HD = 1:2

STUDY OF POLYMERIZATION OF LANGMUIR-BLODGETT MONOLAYER
BY SURFACE ENHANCED RAMAN SCATTERING

Chen Gang, Wang Gongming, Liu Liying, Xu Linxiao

Tao Fenggang and Wang Wencheng

Lab. of Laser Physics & Optics, Fudan University

Shanghai, China

The change of the polymerizing related characteristic Raman peak before and after the polymerizing of Langmuir-Blodgett mono-molecular layer has been observed by using surface enhanced Raman scattering technique.

In recent years, more and more attentions have been paid to Langmuir-Blodgett(LB) monolayer because of its potential applications in many fields. Polymerization of LB film is an effective mean to improve its mechanical and thermal stability. It has been reported that IR absorption and attenuated total reflection have been used to study the polymerization of LB film, while the former method needs several tens or even more monolayer to detect the signal and the latter one studies the polymerization via the change of index of refraction of LB film. In this paper, we would report for the first time the studying of polymerization of LB film by surface enhanced Raman scattering.

Synthesized molecules of 11-methacryloylaminiundecanoic acid were used. This molecule can form LB monolayer and has a double carbon bond which can be polymerized under certain condition. Silver island film with mass thickness of about 50 Å was evaporated on glass substrate in high vacuum. After that, the monolayer of 11-methacrylaminiundecanoic acid was transferred onto the silver island film Raman scattering signal of molecules adsorbed on silver island film is known to have enhancement effect, we can observe the Raman modes of 2 layers, 4 layers and 10 layers of LB monolayer, which were identical to that of bulk molecules. Then, those films were exposed to 2537 Å UV radiation of mercury lamp, and measured their Raman signals again. It was found that its characteristic peak corresponding to double carbon bond disappeared and the intensity of another mode which is related to double carbon bond changed also. This result proved that the double carbon bond had broken under UV radiation and had polymerized, which was in consistent with the experimental results of nuclear magnetic resonance and attenuated total reflection.

Since surface enhanced Raman scattering can detect the signal even with monolayer and give the change of carbon double bond, it is a sensitive and direct method to study the polymerization of LB film.

DYNAMICS OF LASER-INDUCED ETCHING OF SI(111) SURFACE OF CHLORINE

Li Yulin, Zheng Qike, Zhang Zhuangjian

Jin Zhongkao and Qin Qizong

Laser Chemistry Laboratory, Fudan University

Shanghai, China

Laser-induced chemical etching of semiconductors has been intensitively investigated[1], owing to its potential application in fabrication of microelectronic devices and the basic scientific interest in the gas-surface interaction. Among the gas-solid systems studied. $Si-Cl_2$[2-6] appears to be particularly attractive. In spite of the previous efforts, our knowledge of the reaction mechanism is very incomplete. In this paper we present the UV and visible laser induced interaction of chlorine molecules with a Si(111) surface. Both mass and translational energy distributions of desorbed reaction products have been measured as a function of laser fluence. The effect of incident molecules' translational energy on the etching reaction is investigated by the use of "seeded" molecular beam technique for the first time. The objective is to achieve a better understanding of the etching mechanism as well as to improve the etching rate.

In the experimental investigation, the laser-induced gas surface interaction is studied with a supersonic molecular beam technique coupled with the time-resolved mass spectrometry. A CW supersonic molecular beam is formed through a three stage differential pumped system. Chlorine molecules are seeded in He, Ar of Ar/He mixture gases. The translational energies of Cl_2 molecules are varied from 10-60 kJ/mol by changing the seeding gas, The Cl_2 beam is incident on a Si(111) surface at an angle of 45 with respect to the surface normal. The laser beams are provided by the third harmonics (355 nm) of a Q-switched Nd:YAG laser, and a dye laser pumped by the SHG of the Nd:YAG laser. The wavelengths of 560 and 355 nm are chosen since gaseous Cl_2 has a strong absorption at 355 nm and very weak absorption at 560 nm, in order to see the role of gaseous Cl atom. The laser beam is focussed and irradiated perpendicularly upon the surface and crossed with the molecular beam on the Si wafer. The laser fluence for the investigation ranges from $10-150 mJ/cm^2$ with a pulse of duration of 10 ns and a repetition rate of 20 Hz. The desorbed reaction products are detected by a quadrupole mass spectrometer which is placed at a right angle to the molecular beam. The effective distance from the surface to the ionizer of the mass spectrometer is 25 cm. The time-of-flight(TOF) spectra of the desorbed reaction products are recorded with a transient recorder interfaced to a computer. Two reaction products, $SiCl^+$ (m/e =63) and $SiCl_2^+$ (m/e =98), were observed at both wavelengths. The TOF spectra of both products can be fitted to a Maxwell-Boltzmann distribution fairly satisfatorily. The relative reaction yield of $SiCl^+$ by integration its TOF spectra has been examined as a function of laser fluence. There appears to be a threshold at laser fluence of 20 and 40 mJ/cm^2 for 355 nm and 560 nm, respectively. Above the threshold the signal of $SiCl^+$ increases linearly with the laser fluence. The temperature of desorbed reaction products T_d has been calculated from the maximum of the TOF signal, t_m. An interesting result is that translational temperature (T_d) is nearly independent on laser fluence at both wavelengths in the range of $20-120$ mJ/cm^2, and the values of T_d are 610±11 K and 820±23 K for 560 nm and 355 nm, respectively.

The signal intensity of $SiCl^+$ under 355 nm irradiation is slightly higher than that under 560 nm laser action less than a factor of two, although the optical absorption cross section of Cl_2 molecules at 355nm is large than that at 560 nm by more than two orders of magnitude. Finally, the effect of the incident Cl_2 molecules' translational energy has been measured. The $SiCl^+$ signal intensities at low translational energy of Cl_2 molecules are very weak; there appears to be a threshold, above which the signal increases rapidly with the normal component of translational energy. This evidence indicates that the initial absorption process for Cl_2 incident with energies lower than the threshold differs from that for molecules with higher energies and the translational energy of incident Cl_2 molecules can be employed to enhance the dissociative absorption of Cl_2 molecules on the Si(111) surface and to increase the etching rate.

Reference

[1] Chuang, T.J.;Surface Sci. Rept,3(1983)1.

[2] Ehrlich,D.J., Osgood,R.M.Jr. and Deutsch,T.F.;Appl.Phys.Lett., 38(1981)1018.

[3] Kullmer, R., and Bauerle, D.;Appl.Phys.,A43(1987)227.

[4] Raller, T., Oostra, D.J., and de Vries, A.E.;J.Appl.Phys., 60(1986)2321.

[5] Okano, H., Horiike, Y. and Sekine, M.;Japan.J.Appl.Phys.,24 (1986)68.

[6] Jackman, R.B., Ebert, H. and Foord, J.S.,Surface Sci.,176(1986) 183.

FOURIER TRANSFORM HETERODYNE SPECTROSCOPY
OF LIQUID INTERFACES

Ka Yee Lee,   Doo Soo Chung,   Sung Rno,   Eric Mazur

Harvard University
Division of Applied Sciences and
Department of Physics, Cambridge, MA 02138, USA

By acousto-optically shifting the local oscillator in a
heterodyne set-up, a spectral resolution of better than
150 mHz can be obtained. Applications of the technique
to the study of interfacial phenomena are discussed.

Recently there has been much renewed interest in the behavior of liquid interfaces far from the critical point.
This is in part because of practical applications in industry, and in part because of recent theoretical interest in
the description of liquids out of equilibrium. [1,2] Laser
light scattering has been used from the early days of the
laser to study properties of liquid interfaces both   far
away and close to the critical point. [3,4] By detecting
the beating signal between the scattered laser light and a
'local oscillator' derived from the same laser, resolving
powers of 10^{15} have been achieved. We present here recent
results obtained with a novel Fourier transform heterodyne
technique. [5.6]

A detailed description of the experimental setup can be
found in previous publications. [5,6] In short, a multimode
4 mW He-Ne laser beam is incident on a liquid-vapor or
liquid-liquid interface, and the inelastic and quasi-elastic scattered light is detected by optical heterodyning.[3,4]
In a conventional heterodyne setup, because the up-shifted
and down-shifted components of the inelastic scattered
light beat at the same frequency with the incident beam,one
cannot distinguish between the two components. In the present setup, to resolve these two components, the local oscillator is frequency-shifted by acousto-optic modulation[5,6] The spectrum is obtained by performing a fast
Fourier transform of the detected beating signal. The tech-

nique has a spectral resolution of better than 150 mHz, and a high signal-to-noise ratio. It therefore allows accurate measurement of quasi-elastically scattered light.

During the past year we have applied this technique to study various interfacial phenomena. First, we have studied the effect of a temperature gradient on surface phonons, also known as capillary waves or ripplons, on liquid-vapor interfaces. [7] Figure 1 shows the fully resolved spectrum of the scattered light from the liquid-vapor interface of water. The unshifted central peak, appearing at the frequency shift of the local oscillator, is due to stray light scattering at the window of the cell, while the two symmetrically shifted side peaks (Brillouin peaks) are due to capillary waves on the surface of the liquid. An asymmetry in the height of the Brillouin peaks was predicted to occur when light is scattered from nonequilibrium bulk liquids; this was verified experimentally a few years ago. [8,9] A recent theoretical paper predicts a similar asymmetry in the spectrum of light scattered from a nonequilibrium interface because of broken time-reversal symmetry. [10] With the above setup, for the first time, we observed and measured this asymmetry in the Brillouin spectrum of light scattered from a liquid interface subject to a temperature gradient. [7] A detailed report will appear in a forthcoming publication.

The present technique can also be applied to determine accurate values for the interfacial tension of liquid-liquid interfaces. Interfacial tension measurements are usually carried out using ellipsometric techniques or the pendant-drop method. [11] The interfacial tension, however, also plays a role in the dispersion of capillary waves. By studying the laser light scattering from the interface under various incident angles, one can measure the dispersion of the capillary waves on the interface, and thus experimentally determine a value for the interfacial tension. [12] Measurements were carried out on cyclohexane-

methanol and water-ether, yielding accurate values for the
interfacial tension for these systems. The advantages of
this technique are that the interface is not disturbed, and
that one can measure much lower interfacial tensions than
previously possible.

Finally, the technique is also well suited to study
quasi-elastic scattering from impurities and surfactants at
liquid interfaces. Although it was predicted more than ten
years ago that impurities at a fluid interface modify the
spectrum of scattered light,[13] to date no optical experiments have been performed on such systems. With conventional
heterodyne techniques, the central peak is always located
at the origin; the $1/f$ noise makes it difficult and un-
reliable to analyze such a central peak. With the present
technique, however, the central peak is shifted away from
the origin, making a quantitative analysis of this peak
possible. Figure 2a and 2b show the quasi-elastic scattering spectra from interfaces with 50 nm and 1 μm particles,
respectively; the spectrum of a clean interface consists
of a single narrow peak.

We have presented here a simple, convenient heterodyne
technique with high resolution. Its ability to provide di-
rectional separation of the Brillouin doublet makes it
suitable for the study of nonequilibrium phenomena. Due to
its high sensitivity, the Fourier transform heterodyne technique also shows promise as a powerful tool to study interfacial tension, and the effects of interfacial impurities
and surfactant layers.

REFERENCES
[1] See, for instance, the review articles in PHYSICS TODAY, 37,(1984)
[2] S.R. de Groot, P. Mazur, Nonequilibrium Thermodynamics (Dover
     Publications, New York, 1984)
[3] R.H. Katyl, U. Ingard, PHYS. REV. LETT., 20, (1986) 248
[4] M.A. Bouchiat, J. Meunier, J. Brrossel, C.R. Acad, Sc. Paris,
     266B (1986) 255
[5] E. Mazur, D.S. Chung, PHYSICA, 147A (1987) 387
[6] E. Mazur, in Laser Spectroscopy VIII, Eds. W. Persson and S.
     Svanberg (Springer Verlag, Berlin, 1987) 390

[7] D.S. Chung, K.Y. Lee, E. Mazur, Journal of Thermophysics, to be published

[8] D. Beysens, Physica, 118A (1983) 250

[9] H. Kiefte, M.J. Clouter, R. Penny, PHYS. REV., B30 (1984) 4017

[10] M. Grant, R.C. Desai, PHYS. REV., A27 (1983) 2577.

[11] A.W. Adamson, PHYSICAL CHEMISTRY OF SURFACES, Fourth Edition (Wiley Interscience, New York, 1982)

[12] R.B. Dorshow, R.L. Swofford, PHYSICS TODAY, 41, No.1, (1988) S49

[13] M.J. Rosen, SURFACTANTS AND INTERFACIAL PHENOMENA (Wiley Inter-Science, New York, 1978)

GENERATION OF HIGH POWER UV FEMTOSECOND PULSES

S. Szatmári*,   F.P. Schäfer

Max-Planck-Institut für Biophysikalische Chemie,
Abteilung Laserphysik, Postfach 2841,
D-3400 Göttingen, Fed. Rep. Germany

*Permanent Address: Research Group on Laser Physics of the
Hungarian Academy of Sciences, JATE University,
Dóm tér 9, H-6720 Szeged, Hungary

We have reported earlier a method for the generation of
ultrashort UV pulses which is based on the use of a hybrid
excimer-dye laser arrangement. In that setup a twin tube
excimer laser (Lambda Physik EMG 150) is used as a pump
source for a subpicosecond dye laser setup and as an amplifier[1,2]. With the use of this technique essentially similar results were obtained for both excimers KrF and XeCl,
if the effect of the additional pulse compression reported
there is not taken into account. The pulse duration directly at the output of the excimer amplifier was measured as
220fs[1] and 370fs[2] for XeCl and KrF, respectively. In
the case of KrF further temporal compression of the amplified pulses resulted in 80fs pulse duration[2]. A similar
pulse compression experiment for XeCl led only to a slight
decrease of the pulse duration from 220fs to 170fs[3]. Recent development of the excimer laser-pumped cascade ps dye
laser setup[4] and the introduction of a new, simple distributed feedback dye laser (DFDL)[5] made it possible to
simplify the construction of our hybrid excimer/dye laser
system, allowing a much stabler operation, while preserving
or improving the original output characteristics.

The experimental arrangement is shown in Fig.1. The EMG
150 Lambda Physik excimer laser is used as a pump laser for
pumping a special subpicosecond dye laser-amplifier arrangement, and as an amplifier for the amplification of the
frequency-doubled output pulses of the dye laser setup. The
oscillator channel of the EMG 150 laser is filled with the
standard XeCl fill, delivering 80mJ, 15ns pump pulses at

308nm. The pump energy is distributed among the various dye cells as indicated in Fig.1. The pump beam coupled out by the first two quartz plates is used for pumping a newly developed, simple dye laser setup[4]. This makes use of two cascade dye lasers, two amplifier stages, and a gated saturable absorber (GSA) in between the amplifiers. The purpose of the dye lasers is to sharpen the leading edge, while the GSA plus the saturated amplifiers cut the trailing part of the pulses. The combined pulse forming effect of the oscillators, saturated amplifiers and the GSA results in an output pulse duration of $\sim$ 8ps at 365nm. These pulses have an energy of typically $\sim$ 4μJ and are then used for pumping the DFDL master oscillator.

The DFDL used here is a tunable, achromatic arrangement first described in [5], utilizing a transmission grating, a microscope objective and a special dye cell with the active medium. In this DFDL the interference fringes, which are necessary for DFDL operation, are created by imaging a coarse transmission grating onto the inner surface of the dye cell by the use of a microscope objective (Fig.2). The main advantage of the arrangement is its easy tunability by translation of the transmission grating and proper choice of the refractive index of the active medium. In this way the whole visible spectrum can be covered.

The output pulse is then amplified in a two stage amplifier of standard design and the amplified and frequency doubled DFDL pulses can be used as seed pulses for two-pass amplification in the excimer amplifier. After amplification, the UV pulses are sent through a double prism pulse compressor consisting of two 60° quartz prisms of 20mm baselength and two dielectric mirrors to use the two prisms in a double pass pulse compressor arrangement. The pulse duration before pulse compression is measured to be $\sim$ 500fs with significant chirp. After compression we have obtained a set of reproducible autocorrelation functions indicating a pulse width

322

around 60 fs.

REFERENCE

[1] S. Szatmári, B. Racz, F.P. Schäfer: Opt. Commun., <u>62</u>, 271 (1987)
[2] S. Szatmári, F.P. Schäfer, E. Müller-Horsche, W. Mückenheim:
    Opt. Commun., <u>63</u>, 305 (1987)
[3] Q. Zhao, F.P. Schäfer, S. Szatmári: Appl. Phys., <u>B46</u>, 139 (1988)
[4] S. Szatmári: Opt. Quant. Electron., (in press)
[5] S. Szatmári, F.P. Schäfer: Appl. Phys., B (in press)
[6] S. Szatmári, F.P. Schäfer: J. Opt. Soc., Am., <u>B 4</u>, 1943 (1987)

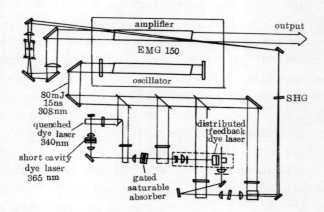

Fig 1   Experimental arrangement

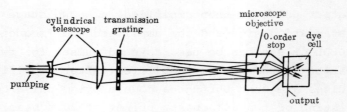

Fig 2   Principle of the new DFDL

Femtosecond Photon Echoes

C.V. Shank, P.C. Becker, H.L. Fragnito, and R.L. Fork

AT&T Bell Laboratories
Holmdel, NJ 07733

The photon echo or time delayed four wave mixing technique has become an important tool for investigating dephasing processes in gases, solids and glasses. The use of coherent optical transients to study such processes for band to band transitions in semiconductors and large molecules in solution has been frustrated by the very rapid dephasing in such systems. Recently advances in pulse compression techniques have permitted the generation of optical pulses as short as 6 femtoseconds [1] making the observation of coherent transients in such systems possible.

Recent hole burning experiments in large molecules in solution [2] suggest that it may be possible to observe photon echoes in these systems. Hole burning experiments have shown that T_2 is on the order of 70 femtoseconds for a single vibronic transition. With a six femtosecond optical pulse the entire vibronic spectrum is excited so the echo will be the sum of all the echoes from the manifold of vibronic states.

Several attempts have been made to observe coherent transient processes in organic dyes including four wave mixing, three pulse echoes, and four wave mixing with incoherent pulses. In the experiments reported here we attempt to observe echoes using a two pulse sequence. The two pulses having wave vectors K_1 and K_2 make a small angle and generate an echo in the momentum matched direction $2K_2$ -K_1. The echo is then separated spatially from the exciting pulses. The energy of the generated echo is measured as a function of relative time delay between the exciting pulses.

The primary utility of a photon echo is to determine the dephasing time, T_2. For a purely inhomogeneously broadened two level transition the echo energy will decay exponentially with the relative time delay, τ , between the exciting pulses.

$$E_{echo} = \exp(-4\,\tau/T_2)$$

For the case of a homogeneously broadened two level transition a polarization free decay will be observed to relax as $\exp(-2\tau/T_2)$. These relations show that an optical pulse much shorter that T_2 must be used to time resolve the echo decay.

The experiments were performed using compressed pulses corrected to third order in a manner described previously. The pulse repetition rate was 10 KHz and the pulse energy was 20 nJ. The energy of the pulse was less than that needed for a π pulse so the echoe's observed in the experiments described here are in the small signal perturbation limit. The pulses were split into two parts using a beam splitter in a modified Michelson configuration to form the two excitation pulses. The two pulses were focused with a 5 cm focal length lens into a flowing stream containing the molecule under study. One pulse was delayed with respect to the other using a stepper motor controlled delay.

The signal detected in the echo direction was directed into a photomultiplier. The detection electronics consisted of a boxcar integrator followed by a phase lock detector. The signal was measured as a function of relative time delay between the pulses.

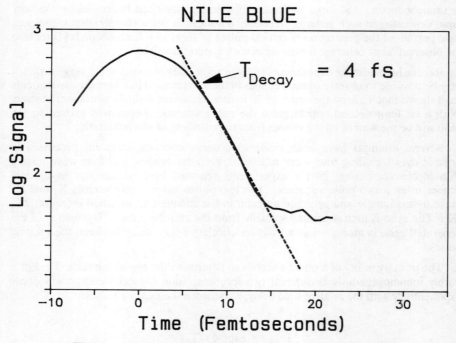

Figure 1. Plot of the log of the echo energy versus time delay.

The measured echo energy for the oxazine dye, Nile Blue is plotted in Figure 1. Note that a very rapid 4 femtosecond time decay is observed. The dynamic range of the experiment is limited by the lack of perfect phase compensation for this very short 6 femtosecond optical pulse. The action of imperfect phase compensation is to produce energy in the trailing edge of the pulse.

How do we reconcile this rapid polarization free decay with the much longer $T_2 \approx 70\text{fs}$ reported in the hole burning experiments? To understand this data we need to consider the photon echo from a system of vibronic levels. Each level in the ground state is coupled to a manifold of vibronic levels having a Franck-Condon overlap with the ground state. For purposes of our discussion let us assume delta function in time excitation pulses and strong inhomogeneous broadening. The echo energy as a function of time delay is given by

$$S(\tau) \approx \left[\sum_{ij} \mu_{oi}^2 \mu_{oj}^2 \cos(w_{ij}\tau)\right]^2 \exp(-4\tau/T_2)$$

where $|i>$ and $|j>$ are excited state vibronic levels and ω_{ij} is the frequency difference between these levels. In such a multilevel system the exponential term can be dominated by the term in the brackets which comes about from the dephasing due to the sum of the echoes coming from the manifold of vibronic levels. Obviously this consideration precludes the determination of T_2 from this measurement.

A number of conclusions can be drawn from this discussion. First it is necessary to take a system point of view in calculating the echo response. Even though hole burning indicates that T_2 is ≈ 70 femtoseconds the echo from a manifold of vibronic levels is dephased much more rapidly. The measured dephasing rate of the manifold of vibronic levels is 4 femtoseconds and is at the limit determined by the spectral linewidth of the S_1 electronic transition.

A more interesting system to study is the polarization dephasing rate of band to band transitions in the direct band semiconductor GaAs. The dephasing rate provides a direct measure of the process of momentum relaxation. At high carrier densities the carrier momentum looses phase coherence primarily through the screened Coulomb interaction between carriers. Both elastic and inelastic carrier-carrier collisions contribute to the momentum dephasing. At low carrier densities electron-phonon interactions are the dominate dephasing process. The density dependence of the polariztion rate provides important information concerning the carrier-carrier interaction.

The experiment was performed in the manner just described. The sample was a thin .1 micron thick sample of GaAs grown by molecular beam epitaxy. Both faces of the sample were antireflection coated. The excitation pulse energy at the sample ranged from .1-.01 nJ per pulse which corresponds to carrier densities ranging from 10^{17} to 10^{18} cm^{-3}. The carrier density was estimated by measuring the number of photons absorbed in the material. The spot size of the focused beam was measured to be 30 microns in diameter.

In Figure 2 we have plotted the log of the echo energy versus the relative time delay between the pulses. At the highest density an exponential decay with a time constant of 3.5 femtoseconds is measured. This is very close to the system response limit. As the density is reduced the echo decay time lengthens and is very clearly resolved.

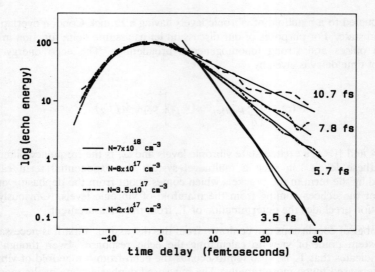

Figure 2. Plot of the log of the photon echo energy as a function of relative time delay between excitation pulses. The time constant of the exponential decay, T_{echo}, is indicated for each carrier density.

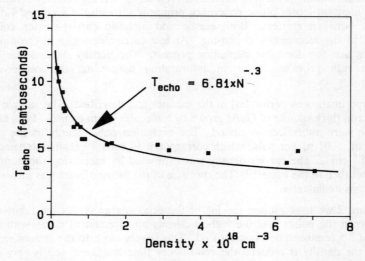

Figure 3. Echo decay time constant as a function of the carrier density for GaAs.

The echo decay constant, T_{echo} is plotted as a function of density in Figure 3. The points are experimental and the solid curve is a power law fit to the data and is given by the expression

$$T_{echo} = 6.81 \times N^{-.3}$$

According to expression (1) the polarization dephasing time is given by the relation $T_2 = 4T_{echo}$. The data reveal that we observe the dephasing time to range from 14 femtoseconds to 44 femtoseconds while the density has been changed from 7×10^{18} to 1.5×10^{17} cm^{-3}.

The dependence of the dephasing rate on density clearly suggests that the dephasing process is dominated by carrier-carrier interactions. The momentum relaxation rate T_m in the limit of large carrier densities is given by:

$$T_m \, N^{-1} S(N)$$

where N is the carrier density and S(N) is a density dependent factor representing the effect of Coulomb screening. In the Thomas-Fermi approximation $S(N) \approx N^{2/3}$ which is close to the experimentally measured $S(N) \approx N^{0.7}$. However, the Thomas-Fermi approximation is not valid in this situation because the excited carrier distribution is nonthermal. Currently a good theory does not exist to describe the process of screening in this highly nonequilibrium population distribution. These experiments provide the first information on nonequilibrium screening processes on such a rapid time scale.

1. R.L. Fork, C.H. Brito Cruz, P.C. Becker, and C.V. Shank Opt. Lett. 12 , 483 (1987).
2. C.H. Brito Cruz, R.L. Fork, W.H. Knox and C.V. Shank, Chem Phys. Lett. 132 , 331 (1986).

TRANSITION RADIATION OF FEMTOSECOND OPTICAL PULSES

Jiang Wenbin, Sun Diechi, Li Fuming

Department of Physics, Fudan University,
Shanghai, P.R. China

With the duration of the optical pulses entering the domain of femtosecond[1,2], their particle behavior becomes more obvious. It has been proved[3,4] that when these ultrashort optical pulses are focused into an electro-optic medium, a nonlinear polarization will be produced by the inverse electro-optic effect (optical rectification) which can be approximately represented by a partical with a dipole moment moving at the group velocity of light which is higher than the phase velocity of propagation in the materials of interest. Thus a Cerenkov-like cone radiation is emitted just like the case of a charged particle moving at the velocity higher than the light velocity in the medium. At the boundary of this radiation, the electric field consists of an extremely fast electrical transient with a correspondingly wide spectral distribution extending well into the far infrared, which can be used as a fast electronic impulse generator or far infrared source.

Besides, when a charged particle crosses the boundary of the vacuum and the medium, a transition radiation will be emitted. The ultrashort optical pulses must have the similar behaviour by the time they propagate through the boundary. We have here taken emphasis on the problem of the transition radiation, discussing the situation when ultrashort optical pulses cross the boundary of the vacuum and the electro-optic material. The calculation of the transition radiation follows the method in Ref[5].

Assume the electric field strength in the free space is

$$\vec{E}=E_0 \delta(x)\delta(y)\exp[-(Z-ct)^2/2c^2\tau^2]\exp[i\omega_0(Z/c-t)]\vec{e}_x, \tag{1}$$

where τ is the duration of the optical pulses and ω_0 the frequency of the laser. In the electro-optic material, the relation between the polarization source P and the complex amplitudes E_j of the field is [6]

$$P_i = (\Sigma_{jk} n_j{}^2 n_k{}^2 \Gamma_{jki} E_j E_k^*)/16\pi \qquad (2)$$

where Γ_{jki} is the electro-optic coefficient and the subscript of E denotes the propagating direction. If the polar axis of the electro-optic material is along $\vec{e}_x$ (Fig.1), the polarization is then completely determined by the Γ_{33} electro-optic coefficient (using contracted notation: $\Gamma_{33} = \Gamma_{333}$), as the incoming optical pulse is also polarized in the direction of $\vec{e}_x$ [3]. From (2) we obtain

$$P_3(t) = n_{333}^2 \Gamma \varepsilon_p \delta(x)\delta(y)\exp[-(Z-vt)^2/v^2\tau^2]/(2\overline{\pi}v\tau) \qquad (3)$$

where n_3 is the optical index of refraction, ε_p the pulse energy and v the propagating velocity of the pulse along $\vec{e}_z$ in the electro-optic material. By Fourier transformation, we have the polarization in the frequency domain from (3) that

$$P_3(\omega) = n_{333}^2 \Gamma \varepsilon_p \delta(x)\delta(y)\exp(-\omega^2\tau^2/4)e^{i\omega z/v}/4\pi v \cdot \qquad (4)$$

According to dipole radiation formula, we can obtain the radiation field at the place P, which is produced by the dipole at M and can be expressed by

$$d\vec{H}_\omega = -(\omega^2/c^2R^2)e^{ikR}[\vec{P}_3(\omega)\times\vec{R}]dxdydz. \qquad (5)$$

Put (4) into (5), and by integrating over the bulk, we obtain

$$\vec{H}_\omega = -(\omega^2 n_{333}^2 \Gamma \varepsilon_p/4\pi vc^2)\exp(-\omega^2\tau^2/4)\int_0^\infty e^{i(\omega z/v+kR)}$$
$$\sin\theta'/Rdz\vec{e}_\phi \qquad (6)$$

The main contribution to the transition radiation is in the area near the boundary of z=0, so there exists

$$R = r - z\cos\theta, \qquad (7)$$

330

and because
$$\sin\theta' = r\sin\theta/R \tag{8}$$
we then obtain
$$H_{\omega\phi} = -(\omega^2 n_3^2 \; \Gamma_{33} \; \varepsilon_p \; r \sin\theta/4\pi v c^2)$$
$$x\exp(-\omega^2\tau^2/4e^{ikr}$$
$$x\int_0^\infty e^{i\omega(1-\beta\cos\theta)z/v}/R^2 dz \tag{9}$$

From (6), (7) and (8), where $\beta = v/c$. Expand (9) into the power series of r. The integration of the term including r^{-1} is given by

$$H_{\omega\phi} = -i(\omega n_{33}^2\Gamma_3 \; \varepsilon_p/4\pi c^2 r)[\sin\theta/(1-\beta\cos\theta)]\exp(-\omega^2\tau^2/4)\cdot$$
$$e^{ikr} \tag{10}$$

where the high frequency term has been neglected.

Hence we obtain the electric field strength of the transition radiation, which is given by
$$E_{\omega\theta} = H_{\omega\phi} \tag{11}$$

Integrating (11) over all the frequency components, we obtain the total electric field strength of the transition radiation, which is given by

$$E_\theta = (n_{333}^2\Gamma_3\varepsilon_p/\sqrt{\pi}c^3\tau^3)[\sin\theta/(1-\beta\cos\theta)][(r-ct)/r]$$
$$\exp[-(r-ct)^2/c^2\tau^2] \tag{12}$$

It is found from (12) that the transition radiation has the similar time behaviour as the Cerenkov radiation [3], i.e., it is also an extremely fast electric transient.

The differential energy spectrum of the transition radiation can be obtained by the formula [1]

$$d^2I/d\omega d\Omega = c[|rE_{\omega\theta}(\omega)|^2 + |rE_{\omega\theta}(-\omega)|^2]/4\pi \tag{13}$$

Put (11) into (13), we obtain

$$d^2I/d\omega d\Omega = (n_{333}^4\Gamma_3^2\varepsilon_p^2\omega^2/32\pi^3c^3)[\sin^2\theta/(1-\beta\cos\theta)^2]\exp(-\omega^2\tau^2/2) \tag{14}$$

It can be seen from (14) that the differential energy spectrum is of e_z symmetry and the forward radiation energy $(0<\theta<\pi/2)$ is greater than the backward radiation energy $(\pi/2<\theta<\pi)$. Integrating (14) over the solid angle with the integral limits of θ from $\pi/2$ to π, we then obtain the total backward energy spectrum of the transition radiation in the free space, which is given by

$$dI/d\omega = (n_3{}^4\Gamma_{33}{}^2\varepsilon^2\omega^2/16\pi^2v^2C)[\beta-2+(2/\beta)\ln(1+\beta)]\exp(-\omega^2\tau^2/2)$$

$$(15)$$

Fig.2 shows the normalized backward energy spectrum with the peak of the spectrum at $\sqrt{2}/\tau$. The shorter is the duration of the optical pulses, the higher the frequency of the peak. For femtosecond optical pulses, the peak of the spectrum is in the region of far infrared, which is the same as that of the Cerenkov radiation of the optical pulses [3].

From (15) we obtain the total energy of the transition radiation

$$I=(n_3{}^4\Gamma_{33}{}^2\varepsilon_p{}^2/16\sqrt{2\pi}^{3/2}v^2C\tau^3)[\beta-2+(2/\beta)]n(1+\beta)] \qquad (16)$$

It can be seen that I is proportional to τ^{-3}. The shorter is the pulse duration, the stronger the transition radiation, which means that the particle behaviour of the optical pulses will get more obvious as the pulse duration gets shorter.

In conclusion we have calculated the transition radiation produced by the femotosecond optical pulses when they cross the boundary of the vacuum and the electro-optic material. The time behaviour as well as the frequency behaviour (relative energy spectrum) is similar to that of Cerenkov radiation calculated by Auston [3].

Thanks to miss J.W. Jiang for her help to prepare this paper.

332

REFERENCE
[1] Fork, R.L., Greene, B.I., Shank, C.V., Appl. Phys. Lett.,
     38(1981) 671.
[2] Sun, D.C., Hu, Y.M., Jiang, W.B., Li, F.M., Acta Optica Sinica,
     6(1986) 865. (In Chinese)
[3] Auston, D.H., Appl. Phys. Lett., 43(1983) 713.
[4] Cheung, K.P. Auston, D.H., Valdmanis, J.A., Kleinman, D.A.,
     Ultrafast Phenomena IV, 409 (Springer-Verlag Berlin Heideberg,
     1984).
[5] Jackson, J.D., Classical Electrodynamics (John Wiley & Sons, 2nd
     ed., 1975).
[6] Bass, M., Franken, P.A., Ward, J.F., Weinreich, G., Phys. Rev.
     Lett., 9(1962) 446.

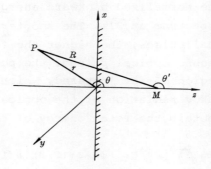

Fig 1 Optical pulse incident into the electro
      optic material is propagating along $\vec{e}_x$
      and its polarization is in the direction
      of $\vec{e}_x$

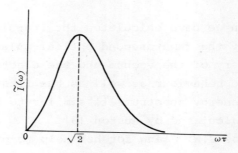

Fig 2 Normalized backward energy spectrum of the
      transition radiation of the femtosecond
      optical pulse $\tilde{I}(\omega)=\alpha dI/d\omega$, where α is the
      normalizing factor with the peak of the
      spectrum at $\omega=\sqrt{2}/\tau$.

OBSERVATION OF DFWM IN A SATURABLE ABSORBER INSIDE THE CPM RING DYE LASER CAVITY

Jiang Wenbin, Sun Deichi, Li Fuming

Department of Physics, Fudan University,
Shanghai, P.R. China

With the emersion of the CPM technique, the duration of the optical pulses has been shortened to femtosecond, and the mechanism of the CPM has attracted lots of attention. It's commonly believed now that the two counter-propagating pulses inside the cavity collide in the saturable absorber, forming an instantaneous phase grating which reflects part of the light from one direction to another, thus to deepen the saturation of the absorber and make the optical pulses much shorter[1,2,3,4,5]. We here use the pulses generated from the CPM ring dye laser incident into the absorber jet. As a result, the conjugate light of the probe is observed, which is the phenomenon of the so-called DFWM. With this method, the phase grating existing in the absorber is indirectly demonstrated.

First, we calculate the intensity of the conjugate light. Assuming A_1 and A_2 are the amplitudes of the two counter-propagating pulses inside of the cavity, and A_3 is the amplitude of the probe, which may respectively be expressed below,

$$A_1 = f(\omega t - k' x) \delta(y) \delta(z) \tag{1a}$$

$$A_2 = f(\omega t + k' x) \delta(y) \delta(z) \tag{1b}$$

$$A_3 = \alpha f(\omega t + k'z + \tau), \quad a < 1 \tag{1c}$$

where $k' = \omega/u$, u is the group velocity of light propagating in the saturable absorber and τ is the delay time between the probe and the pumps. By decoupled wave equation [6]

$$\left(\frac{\partial}{\partial z} + \frac{k}{\omega} \frac{\partial}{\partial t}\right) A_4(z,t) = \frac{i 2\pi \omega \chi}{cn} A_1 A_2 A_3{}^* \tag{2}$$

where χ is the instantaneous third order nonlinear susceptibility, n is the refractive index and A_4 is the amplitude of the conjugate light, ignoring the depletion of the pumps and assuming the conjugate light is very weak, we may obtain

$$A_4(\eta) = \frac{i2\Pi\omega\chi\alpha}{cn} f(\omega\eta - k\,\hat{}x) f(\omega\eta + k\,\hat{}x) f^*(\omega\eta + \tau) \tag{3}$$

where $\eta = t - (k/\omega)z$ is the local time. At the definit delay time, the average intensity of the conjugate light observed is given by

$$I_4 \propto \int_{-\infty}^{\infty} |A_4(\eta)|^2 d\eta = \int_{-\infty}^{\infty} I(\eta) I(\eta + 2x/u) I(\eta + x/u + \tau) d\eta \tag{4}$$

where $I(\eta) = |f(\omega\eta)|^2$. From (4) we note that the intensity is the third order auto-correlative function of the optical pulse. As the third order nonlinear coefficient of the absorber is very small and also the absorber jet is rather thin, the conversion efficiency of DFWM is correspondingly small. Thus the conjugate light will be weak.

The experimental set-up is shown in Fig.2, where the delay time is adjustable. The tight focusing of the pumps and the slight focusing of the probe (to increase the power density) are the special aspects different from that of normal DFWM.

When the probe is in timing with the pumps, i.e. $\tau = 0$, we get the strongest conjugate light which is shown in the photograph (Fig.3). We find that the conversion efficiency of the DFWM is really small, which is in coincidence with the prediction mentioned above. Besides, another factor is worth noticing. It was reported[7] that when DODCI solution was used as nonlinear medium, the conversion efficiency of phase conjugation was 400% for microsecond optical pulses, but it decreased to 50% for picosecond optical pulses, thus the conversion efficiency might be much smaller for femtosecond optical pulses. This conjecture leaves to be proved, of course.

We note from the photograph that there exists the high

order diffraction streak in the congugate light. This phenomenon is concerned with the saturation of the resonant transition among the DODCI energy levels[8]. Obviously the probe and the pumps are all working in the resonant frequency region of the mode locking dye DODCI.

By the way, the thermal relaxation phenomenon is observed during the experiment.

In conclusion, we have observed the DFWM effect which shows indirectly the existence of the phase grating in the saturable absorber inside of the CPM ring dye laser cavity. This method may of course be used to measure any relaxation time of DODCI, which is larger than the duration of the optical pulses but smaller than the interval of the optical pulses, such as the phase relaxation time T_2 of DODCI.

Thanks to Miss J.W. Jiang for helping to prepare this paper.

REFERENCE

[1] Fork, R.L., Greene, B.I., Shank, C.V., Appl. Phys. Lett., 38, 671 (1981).
[2] Garmire, E.M., Yariv, A., IEEE J. Quantum Electron., QE-3, 222 (1967).
[3] Fork, R.L., Shank, C.V., Yen, R., Hirlimann, C.A., IEEE J. Quantum Electron., QE-19, 500 (1983).
[4] Kühlke, D, Rudolph, W., Opt. Quantum Electron., 16, 57 (1984).
[5] Dietel, W., Opt. Comm., 43, 69 (1982).
[6] Janszky, J., Corradi, G., Opt. Comm., 60, 251 (1986).
[7] Tocho, J.O., Sibbett, W., Bradley, D.J., Opt. Comm., 37, 67 (1981).
[8] Reintjes, J.F., "Nonlinear Optical Parametric Processes in Liquids and Gases", Academic Press, Inc. (1984).

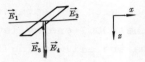

Fig 1 Diagram of DFWM interaction

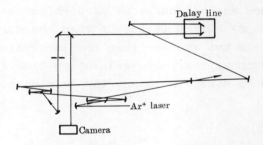

Fig 2 Experimental Set up

Fig 3 Photograph of Conjugate Light

STUDY ON THE INDUCED SPECTRAL SUPERBROADENING OF ULTRAFAST LASER PULSE IN A NONLINEAR MEDIUM

W. H. Qin,   Q. X. Li,   R. Zhu and   Z. X. Yu

Ultrafast Laser Spectroscopy Laboratory
Zhongshan University, Guangzhou, P. R. China

Propagation of high intensity picosecond or subpicosecond laser pulse in a nonlinear meadium can yield an output with a nearly white continuous spectrum [1-3]. This phenomenon, called spectral superbroadening (SSB) or supercontinuum generation, was first observed by Alfano and Shapiro in 1970 [1]. Since then, the superbroadening, as ultrashort pulse with spectral range from ultraviolet to infrared, has been used as a spectroscopic tool for many scientific and technological applications. Recently Alfano and his cooperators observed the induced phase modulation (IPM) [3,5] and cross phase modulation (XPM)[4,5]. The dominant enhancement mechanism for the induced spectral broadening was determined,by Alfano et al,to be caused by an induced phase modulation   process [3,6], not by four-photon parametric   generation. When two pulses of different wavelength are   incident into a   nonlinear meadium simultaneously, couple interactions occur through the nonlinear susceptibility coefficients. These couple interactions can introduce phase modulation and SSB in each pulse due to the presence of the other pulse.

   In this paper, we report our study on the induced spectral superbroadening (ISSB) of ultrashort laser pulse in a 7cm long K   glass. The enhancement of the conversion efficiency and the extension of the spectral range were observed.

Our experimental arrangement is shown in Fig.3. A single 8ps laser pulse at 1060nm generated from a mode-locked Nd glass laser system was used as the pump beam while its second harmonic was used as the probe beam. SF is a spectral filter to select either 1060nm or 530nm, or both wavelength laser pulses. F are color filters and neutral

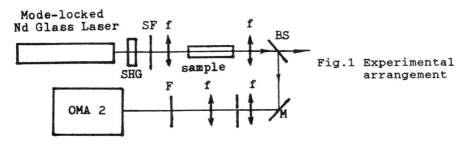

Fig.1 Experimental arrangement

density filters.The laser pulse were weakly focused into a 7cm long K glass. The input energy for the wavelength of 1060nm and 530nm were about 40mj and 2mj respectively. The output beam was sent to an OMA2 for diagnosis. A spatial filter with a 1mm diameter aperture was placed before the OMA2 to distinguish different contributions from either phase modulation or FPPG.

The spectral distribution of ISSB and SSB in the Stokes and anti-Stokes side are displayed in Fig.4. Table 1 give the average enhancement of ISSB over SSB in different wavelength region. Each data was obtained by averaging over 70 points. After removing the spatial filter before OMA2, we obtained the similar results. It shown that the ISSB is produced mainly by the IPM, not by FPPG. There is also an extension of the spectral range, especially in Fig.4 (a). The extension is not so obviously in Fig.4 (b) because of the sensitivity of our OMA system in the ultraviolet.

Table 1 Enhancement of ISSB in different wavelength range

(nm)	410-430	430-450	450-460	460-470
M	4.9	5.9	5.4	4
(nm)	550-560	560-570	570-590	640-680
M	4.2	5.2	5.8	15

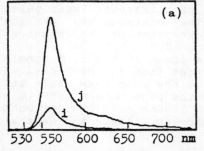 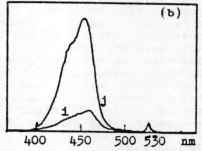

Fig.2 Spectral distribution of ISSB (j) and SSB (i) in the Stokes side (a) and anti-Stokes side (b)

Occasionally, the IPM spectral distribution is not so similar to that of the SPM. As shown in Fig.5(b), the IPM spectral had higher enhancement and some modulation in the long wavelength region. To explain this phenomena, we observed the SPM generated by the pump pulse alone. In most cases, there was no SPM signal except in some shot it produced a small signal as shown in Fig.5 (a). Its

intensity was much smaller than that of ISSB in this
wavelength range. The possiblity for these two phenomenon
is nearly the same. It suggested that the presence of
this small signal would increase the IPM signal a lot.
There are two IPM processes in these shots. When the
intense 1060nm pulse propagates through the meadium, it
causes a refractive index change and, in turn, induced a
phase change. The time variation of this phase will cause
a frequency sweep within the pulse envelope(SPM process).
When anther pulse of different frequency (eg. its second
harmonic) is incident into this disrupted system, its
phase can also be modulated by the time variation of the
nonlinear refractive index originating from the primary
intense pulse.This IPM process enhances the SPM intensity
and extend the SPM broadening range, as shown in Fig.4.
Another IPM process, since our 530nm pulse is not weak
enough and can serve as a pump pulse, as shown in Fig.5
is that the SPM signal near 530nm produced by 1060nm
pulse alone is also enhanced by its second harmonic. It
differs from the previous IPM process in that the
frequency region of SPM is far from the exciting 1060nm
pulse and near the modulating (pump) 530nm pulse. It is
also called a IPM process because the modulated signal is
produced indenpendently by anther incident pulse. There
is no this phenomena in the anti-Stokes side since there
is no SPM signal produced by 1060 nm alone in this
wavelength region.

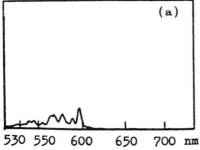
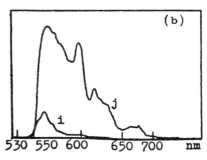

Fig.3   Spectral distribution of SPM produced by 1060nm
        alone (a) and of the ISSB (j) and SSB (i) in the
        Stokes side (b) in the presence of (a)

REFERENCE
1. R.R.Alfano et al: Phys. Rev. Lett. 24 592(1970)
2. R.L.Fork   et al   Opt. Lett. 8, 1 (1983)
3. R.R.Alfano et al: Opt. Lett. 11 626(1986)
4. R.R.Alfano et al   Applied Optics 26 3491(1987)
5. R.R.Alfano and P.P.Ho   IEEE J. of Quant. Electron.
   QE-24 351 (1988)
6. Jamal T. Manassah et al: Phys.Lett. 113A 242(1985)

LASER COOLING AND TRAPPING OF ATOMS

Steven Chu , M.G.Prentiss , A.E.Cable and J.E.Bjorkholm

AT&T Bell Laboratories

Holmdel , New Jersey 07733 , USA

The laser cooling manipulation and trapping of neutral atoms has seen remarkable progress , so it would be impossible to adequately review the work done in recent years . Instead , this paper will summarize the work done on atom cooling and trapping at AT&T Bell Laboratories during the last three years, emphasizing the status of our most recent experimental results .

1. Optical Molasses and Supermolasses

We have previously reported the use of multiple laser beams to create a viscous medium of photons that both cools and confines atoms[1]. The cooling scheme[2] uses the fact that an atom moving in a light field consisting of two counter propagating laser beams tuned below resonance will Doppler shift the beam directed opposite the motion into resonance and shift the co-propagating beam out of resonance . The net result after averaging over many absorptions is that the scattered light creates a damping force directed opposite the motion of the atom. This simple idea , has remarkable consequences. For example , one can show that the damping time for slow moving atoms in this light field is on the order of $\tau \sim 4M/\hbar k^2 \approx 10$ μsec for sodium stoms using the $3S_{1/2} - 3P_{3/2}$ resonance line. The minumum temperature achievable under these conditions has been shown to be $kT \sim \hbar\Gamma/2 \sim 240$ μK in the limit where the single photon recoil energy is small compared to this temperature . Also, the cooling technique provides a means of confining the atoms in space. The motion of the atoms can be modeled as a random walk in a viscous fluid of photons (hence the term optical molasses) analogous to Brownian motion . The frictional force can be calculated to be $\vec{F}_{damping} = -\alpha\vec{v}$ $= 5.7 \times 10^{-18}$ gm/sec for p=1 and $\Delta\nu \sim \Gamma/2$. The Einstein relation then gives a diffusion constant $D = \langle x^2\rangle/2t = kT/\alpha = 5.8 \times 10^{-3}$ cm^2/sec for an infinite viscous fluid . However , the optical molasses extends as far as the overlap of the six laser beams , and an atom that randomly walks to the boundary defined by the overlapping beams will be lost . If we assume an initially uniform concentration of atoms n_0 , the average concentration $\bar{n}$ has been shown to vary as $\bar{n} = n_0 \frac{6}{\pi^2} \sum_{n=1}^{\infty} 1/n^2 \, e^{-Dt[\pi n/R]^2}$ for a spherical boundary of radius R . The presence of the spherical boundary greatly decreases the confinement time of the atoms in the molasses . For R = 0.4 cm , t decreases from 14 sec to 1.6 sec , and for R = 0.1 cm , t goes from 0.9 sec to 0.1 sec . Our initial investigation of optical molasses produced results that were in reasonable agreement with calculations [1] . The measured temperature was in good agreememt with the "quantum limited " temperature $\hbar\Gamma/2$ derived for a one-dimensional two-level system at low laser intensity[4] . Similarly , the confinement time achieved in our initial work was also in agreement with the random walk midel with a spherical boundary . However , soon after the initial work on optical molasses , it became increasingly apparent that the behavior of the atoms was much more complicated than originally thought . Three of the more striking aspects of our more recent discoveries include the following : (i) high intensity molasses $p \gtrsim 1$ (it should not have worked at all!) worked surprisingly well , (ii) the ideal tuning for the longest storage time was on the order of $\Delta \nu \simeq$ 1.5Γ rather than $\Gamma/2$, and (iii) with the proper misalignment of the nominally retroreflected laser beams . The molasses seemed to compress the atoms spatially in to a ball ~2 mn , in diameter , and act like a trap.

The misaligned molasses , dubbed "supermolasses " because it worked so well, was the most intriguing of these effects . The alignment of beams that produces the best result is that of a recetrack for four beams with the other two beams retroreflected . With fine tweaking of the misalignment , atoms within a 2 mm diameter region have been stored for 10 seconds , two orders of magnitude longer than expected from a simple molasses picture . We have not come up with a complete explanation for supermolasses , and the present theoretical treatments provide serious constraints for any plausible model . A careful two-level , one-dimensional treatment of the behavior of atoms in a standing wave[4] shows that the damping force exerted on the atoms for a laser tuned below resonance actually experiences a sign reversal for sufficiently large

p. [5] For $\Delta \nu \sim 2\Gamma$, the upper limit on p should be p $\lesssim$ 0.3 . Furthermore, the momentum diffusion Dp grows linearly with p for p $\gg$ 1 . Experimentally, we find that molasses seems to work for nominally retroreflected beams up to p $\sim$ 1 per beam, and for misaligned beams, intensities as high as p = 5 (the upper limit of our laser) could be used to produce a supermolasses trap.

There is the possibility of trapping the atoms in three dimensional standing lightwaves, as originally suggested by Letokhov, et.al.. [6] Atoms quasi-trapped in a three-dimensional checkerboard of potential wells created by the multiple standing wave interference pattern might diffuse more slowly. If the depth of the potential well U is a few times larger than kT, the diffusion of the atoms out of the mollasses would be greatly reduced. Unfortunately, for pure standing waves in a one-dimensional, 2 level system, U $\lesssim$ kT for frequencies below resonance. The simple standing wave hypothesis also does not explain the polarization tests we have performed on supermolasses. The trap performance was insensitive to the polarization orientations (linear or circular) of the four racetrack beams. On the other hand crossed linear or opposite circular polarization on the retroreflected beams would destroy the trap.

The racetrack configuration suggests that a combination of traveling and standing waves would increase the ratio of U/kT. The damping force for a traveling wave does not undergo a sign reversal and remains fairly effective for p > 1. Furthermore, Dp begins to saturate above p>1 as for a traveling wave, whereas for a standing wave Dp $\propto$ p for large p. This effect was first pointed out by Cook [8]. The observation of channealing when the laser is tuned to red wavelength side of the resonance line[7] and the traveling/standing wave combination gives us hope that a sufficiently complete computer simulation can account for our discovery.

2. Optical Trapping

The optical trapping of atoms was first demonstrated with a "dipole" trap[9]. The basic physics of the trap [10] is analogous to the work done on the magnetic trapping of neutrons [11] and atoms [12]. In the magnetic trap the magnetic moment of the neutron or atom can be in either weak field seeking or strong field seeking states. Since it is possible to design a magnetic field with a local munimum in space , weak field seekers can be trapped. In the case of an electric dipole trap, an electric dipole moment can be induced on the atom by an external electric field. If the time varying electric field is below the natural resonant frequency of the atom , the induced dipole moment is in phase with the driving electric field and the atom will be a strong field seeker. Unlike the case of a static electric or magnetic field[13], it is possible create a local maxima with oscillating electromagnetic fields. Indeed , a single laser beam , sharply focused and tuned below resonance forms the trap successfully demonstrated by our group[9]. Although such a trap is conceptually elegant, the potential well created by the laser is shallow and limited to a small volume. In our first demonstration, 200 mW was focussed to a 20 μm diameter spot (peak power 600 kw/cm^2) to produce a trap 5 x 10^{-3} K deep and 10^{-7} cm^3 in volume.

The scattering force can also be used in laser trapping. Laser intensities on the order of the saturation intensity are all that are required , so large volume traps can be constructed. We report here the first demonstration of a scattering force trap. The work was done in collaboration with E. Raab and D. Pritchard[14], was due to an unpublished suggestion by J.Dalibard. The idea can be introduced as a modification of ordinary mollasses. In molasses, a red detuning of the laser crestes a damping force that resists any motion of the atoms. A net restoring force can be imposed on the atoms by introducing an external magnetic field and using circularly polarized light as shown in Fig.1. For atoms at positive z displacement, the σ^- light is more in resonance with the S = 0→1 transition than the σ^+ light, so there is a net push in the direction -$\hat{z}$. Similarly for atoms at -z, the σ^+ light has a higher absorption probability so there is a net force along +$\hat{z}$. Since the laser is tuned to a frequency below the resonant frequency, molasses-like damping is also present. Simulations based on the actual hyperfine structure of sodium for the F = 2→3 and 1→2 transitions in the D$_2$ line have been made and are reported elsewhere[14].

The three dimensional generalization to the magnetic field can be obtained by using the same quadrupole configuration as the one used by Migdall, et al. [12]. Less than 40 amp-turns in 5 cm diameter coils were needed to produce trapping, corresponding to a magnetic field ≤ 0.2 Gauss in the central 0.8 mm region of the trap. Note that the strength of the magnetic field in this "magnetic molasses" trap is roughly 10^3 times weaker than a pure magnetic field trap of comparable depth. [12], [17]

We have trapped as many as 10^7 atoms for 1/e storage times of 2 minutes at initial densities of 2×10^{11} atoms/cm^3 when the laser is tuned 10-15 MHz below the F = 2-3 and 1-2 transitions. The temperature of the atoms in the trap, as measured by the time-of-flight method, is between 400-900 μK. Tuning the laser to the red of the F = 2$\rightarrow$2 transition, increases the number of atoms by an order of magnitude, but decreases the density to ~2 x 10^{10} atoms/cm^3. The temperature of the atoms for the F = 2$\rightarrow$2 transition is two orders of magnitude higher than for the F = 2$\rightarrow$3 transition.

Fig. 2 shows an example of the decay of the atoms in the trap. It is immediately apparent that the decay of the atoms in the trap cannot be described as a single exponential decay. A good fit can be obtained if the loss of atoms is fit to the solution to the differential equation $dn/dt = -\beta n^2 - 1/\tau * n$. At high densities and early times, the loss of atoms in the trap is diminated by the βn^2 term. This fact explains why it is possible to obtain a larger number of stored atoms in the trap if it is operated at lower density. Work is now underway to determine the βn^2 loss machanism.

Although some aspects of the trap can be understood in a quantitative way[14], there are a considerable number of details that are not understood at this time. For example, the size of the trapping region should be ≥ 1 cm in diameter, but displacement tests suggest that the size is a factor of 5 smaller for the F = 2$\rightarrow$3 transition, but consistent with expectations for F = 2$\rightarrow$2 transition. Also, like supermolasses, the behavior of the trap is very sensitive to small changes in the alignment of the nominally retroreflected beams. Furthermore, the trap appears to be far less sensitive to polarization than expected. Some of

the behavior is probably entangled in the mysterious aspects of supermolasses, and further work is needed. It should also be noted that
damping occured for p > 2 even when the beams were retroreflected

3 Conclusion

Table I summarizes the atom traps demonstrated to date. As one can see,
there has been remarkable progress in the two years since 1984.

Table I. Summary of Atom Traps

	$T_{1/e}$(seconds)	ρ(atoms/cm^3)	N(atoms)
1) Magnetic Trap NBS, 1985 [12]	0.8 sec	10^3	10^4
2) Optical Molasses Bell Labs, 1985, [1]	0.2 sec	10^6	10^5
3) Optical Dipole Trap Bell Labs, 1986 [9]	2 sec	10^{12}	10^3
4) Supermolasses Bell Labs 1986	10 sec	10^7	10^6
5) Magnetic Trap II MIT 1987 [15]	2.5 min	10^6	10^9
6) Magnetic Molasses MIT/Bell Labs, 1987 [14]	2 min	10^{11}	10^7
7) Magnetic Trap,III(hydrogen) MIT 1987 [17]	20min	10^{13}	10^{12}

The lowest temperature obtained is still $kT \sim \hbar\Gamma/2 \sim 240$ μK, [1], but many
cooling schemes have been proposed to reduce the temperature by at
least two orders of magnitude [16]. In addition to better traps and
colder temperatures, the next few years should see an improvements in
neutral atomic beam optics and the development of traps that minimize
the perturbation of the internal degrees of freedom of the atom.
Clearly, the continued development of laser cooling and trapping techniques and the application of these techniques to a wide variety of
physics and chemistry problems has an exciting future. Another exciting
aspect of this field is the unexpected developments. Although the basic
forces that radiation can exert on atoms are well understood, even

seemingly simple experimental realizations of these effects have already produced surprises such as supermolasses. Perhaps a sufficiently clever quantum engineer or mechanic could have anticipated the result, but we have found it easier and more fun to discover the effects in the laboratory.

REFERENCES.

1. S.Chu, L.Hollberg, J.E.Bjorkholm, A.Cable and A. Ashkin, Phys.Rev. Lett., 55, 48 (1985); also in Laser Spectroscopy VII, ed. T.Hansch, Y.R.Shen, (Springer Verlag, Berlin, 1985.) P.14, Methods of Laser Spectroscopy ed.Y.Prior, A.Ben-Reuven, M.Rosenbluh (Plenum, 1986), p.4)

2. T.W.Hansch and A.L.Schawlow, Optics Comm., 13, 68(1975).

3. N.A Fuchs, Mechanics of Aerosols, (Pergamon, Oxford, 1964), pp.193-200

4. J.P.Gordon and A.Ashkin, Phys. Rev. A 21, 1606(1980).

5. A.P.Kazantsev, Zh. Eksp.Teor. Fiz 66, 1599 (1974).

6. V.S.Letokhov, V.G.Minogin and B.D.Pavlik, Optics Comm. 19, 72(1976).

7. A.Aspect, J.Dalibard, A.Herdmann, C.Salomon and C.Cohen-Tannoudji, Phys. Rev. Lett. 57, 1688 (1986); See also the work on channeling reported at this conference.

8. R.J.Cook, Phys. Rev. A 20, 224,(1979).

9. S. Chu, J.E. Bjorkholm, A.Ashkin and A.Cable, Phys.Rev.Lett. 57,314 (1986).

10. A.Ashkin, Phys. Rev. Lett.40, 729 (1978).

11. K.J.Kugler, W.Paul and U.Trimks, Phys. Lett. 72B,422 (1978).

12. A.L.Migdall, J.V.Prodan, W.D.Phillips, J.H.Bergeman and H.J.Metcalf, Phys. Rev. Lett. 54, 2596 (1985).

13. W.Wing, Prog. Quant. Elec.8, 181 (1984).

14. Details are in E.L.Raab, M.Prentiss, A.Cable, S.Chu and D.E.Pritchard, submitted to Phys. Rev. Lett.

15. V.S.Bagnato, G.P. Lafyatis, A.G.Martin, E.L.Raab, R.N.Ahmad-Bitar and D.E.Prichard,Phys.Rev.Lett.58,2194 (1987).

16. S.Stenholm, Rev.of Mod.Phys., 58, 699 (1986).

17. H.F.Hess, G.P.Kochanski, J.M. Doyle, N.Masuhara, D.Kleppner and

T.J.Graytak ,submitted to Phys.Rev. Lett ., (1987).

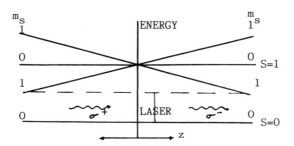

Fig.1. Magnetic molasses for a S=0 to S=1 transition. σ^- light predominates for atoms with z>0 and σ^+ light predominates for atoms with z<0.

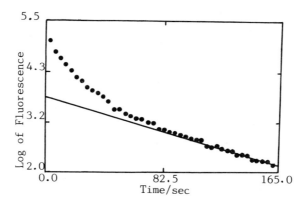

Fig.2. Decay of atoms in the magnetic molasses trap.

FEMTOSECOND ABSORPTION SPECTROSCOPY OF PRIMARY PROCESSES IN BACTERIAL PHOTOSYNTHESIS REACTION CENTERS

S.V. Chekalin, Yu.A. Matveyets, A.P. Yartsev

Institute of Spectroscopy,
USSR Academy of Sciences,
142092 Troitsk, Moscow Region, USSR

The reaction centers (RC) of purple bacteria include 6 molecules of porphyrine nature: 4 bacteriochlorophylls (two of them being the monomers B_L and B_M and two forming the dimer P) and 2 pheophytins H_L and H_M. We have carried out femtosecond absorption spectrometry studies of intact and $NaBH_4$-modified RCs of Rhodobacter Sphaeroides [1] and demonstrated the following:

1) When exciting the RCs with a 300fs pulse at a wavelength of 620nm corresponding to absorption of the pigments P and B (around 10% of all the RCs getting excited), there occurs within 150±100 fs the fast energy transfer process $B^*P_{600} \rightarrow BP^*_{600}$ in those RCs where only the B pigment has undergone excitation.

2) The bleaching of the B_L band observed to take place during the first 3ps after excitation is associated with the formation of the $P^+B_L^-$ radical and not with the electrochromatic shift of the B_L band.

3) The simultaneous bleaching of the B_L and H_L bands during the first 3ps after excitation is due to the transfer of an electron from P^*_{870} to B_L and H_L.

To explain the changes observed in the differential spectra, we have suggested three schemes of the transfer of an electron from P^*_{870} whereby the electron is distributed between B_L and H_L (Fig.1b).

Raising the excitation intensity to around 10^{16} ph/cm^2 causes all the kinetics to change substantially (Fig.2) compared to the case of linear excitation ($\approx 10^{15}$ ph/cm^2). The H band kinetics (ΔA_{750}) features a fast bleaching (on a time scale shorter than the exciting pulse duration), this

bleaching then relaxing to give rise to a kinetics typical
of the transfer of an electron to H_L under linear excitation. The fast bleaching in the P band (ΔA_{870} kinetics) is
observed to diminish slightly even during the exciting
pulse. At the same time, the bleaching maximum in the 870nm
band is observed to shift toward the long-wavelength side.
These changes relax on a time scale of 1-2ps. The B band
(805nm) bleaches as instantaneously as in the case of linear
excitation, but the amount of bleaching and its relaxation
time are much greater. For clarity, Fig.2b shows these "additional" changes separately.

We believe that the fast bleaching of the H band is due
to the transfer of excitation energy to the bacteriopheophytin as a result of the singlet-singlet annihilation of P*
and B* by one of the schemes shown in Fig.1. When only the
pigment P is excited at 620nm in an RC, there occurs the
$PB^*_{600} \rightarrow P^*_{600}$ B energy transfer process which proceeds very
effectively because the energy deficit between the states is
small. The time of this process was measured at 100fs[1].
The transfer of energy from the partially relaxed state
B^*_{800} apparently proceeds not so effectively (according to
[2], the time of this process amounts to 400fs). When the P
and B pigments in an RC are excited simultaneously, the B*
decay channel B*P → BP* is closed. Obviously two versions of
the singlet-singlet annihilation of the excited states P^*_{600}
and B^*_{600} are possible involving partial relaxation of one
of the partners, namely, $P^*_{870}B^*_{600}$ and $P^*_{600}B^*_{800}$, because the
sum of the excitation energies of the partners falls within
the RC absorption band with a maximum at 365nm. The 365nm
band is a combination of the absorption bands of the bacte-
riochlorophyll and bacteriopheophytin. For this reason, when
one of the pigments is excited to a state corresponding to
this band, there can take place a fast transfer of the excitation energy to other pigments, e.g., from B** to H**. This
explains the rapid bleaching of the H band in our experi-

ments. The band shift and diminishing of bleaching in the ΔA_{870} kinetics may be due to the absorption of B^*_{800} in this region. The lengthening (up to 1-2ps) of the bleaching relaxation time of the B band compared to that in the case of linear excitation may be caused by the contribution to the relaxation kinetics from nonannihilating states (e.g., $P^*_{870}B^*_{800}$) or by the absorption of a third quantum by the RC. Obviously all these processes reduce the quantum yield of charge separation.

REFERENCE
[1] Chekalin, S.V., Matveets, Yu.A., Yartsev, A.P., Rev. Phys. Appl., 22, 1761 (1987).
[2] Breton, J., Martin, J. -L., Petrich, J., Migus, A., Antonetti, A., FEBS 209, 37 (1986).

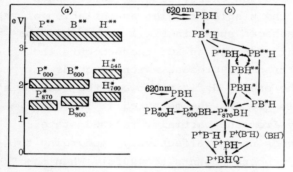

Fig 1 (a) Simplified energy band diagram of RC pigments (b) schemes of processes occurring in RC s at various excitation intensities

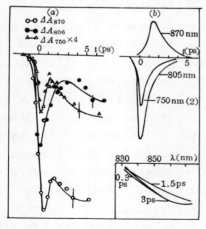

Fig 2 RC absorption kinetics obtained at an excitation intensity of 10^{16} ph/cm^2 (a) real changes (b) additional changes Inset P band shift at various delay times

OBSERVATION OF THE MOTION OF SLOW ATOMS
IN A STANDING WAVE FIELD

Wang Yuzhu, Cai Weiquan, Cheng Yudan, Liu Lian

Laboratory of Quantum Optics,

Shanghai Institute of Optics & Fine Mechanics, Academia Sinica

Shanghai, P.R. China

ABSTRACT

We have observed the motion of slow atoms in a standing wave field. It was found that when the detuning $\Delta \geqslant \gamma$ (the natural linewidth) the radiative force acting on atoms changes its sign in a region of small atom velocity. The velocity bunching effect in an atomic beam has been observed as well.

In recent years the features of radiative pressure have been investigated. Cooling gas atoms has acquired great success. The group of A. P. Kazantsev and the group of C. Cohen-Tannoudji theoretically analysed the moving behaviour of atoms in a standing wave field[1,2]. In strong field when the laser frequency is higher than the resonant frequency of the atom ω_0 (detuning $\Delta = \omega_\ell - \omega_0$ is positive), the atoms are subjected to a damping force and the cooling process occurs. If the laser frequency is lower than the resonant frequency, the atoms will be accelerated and the heating process will occur. The theoretical analyses have been proved by experiments[3]. In the condition of medium strength field and when $\Delta \geqslant \gamma$ (γ is the natural linewidth of the atom) the motion problem becomes more complicated. As far as we know there has not been any experimental or theoretical work on this problem. This paper reports some experimental results in this specific condition.

The motion of atoms is observed by measuring the change of velocity distribution of a collimated sodium atomic beam which undergoes the radiative pressure of the standing wave. If the divergent angle of an atomic beam is small enough the transversal velocity of the atoms in the beam v_t must be small too and the atoms can be seen as slow atoms. The velocity distribution of the atoms is different at different position in the atomic beam. In our experiment the divergent angle of the atomic beam is 1.5×10^3 rad, the most probable velocity is 760m/s. So the atom velocity on the two sides of the beam is $v_t \leqslant \pm 1.14$m/s, and in the central part may be $v_t < \pm 0.2$m/s. The essential elements of our apparatus can be seen in Fig.1. After crossing the atomic beam with

right angle the laser beam is reflected by a flat mirror. In this way a
standing wave is formed. The diameter of the standing wave beam is 4.5
mm. In the detection region the probe laser beam which is perpendicular
to the atomic beam induces resonance fluorescence of the atoms. The
intensity of the fluorescence is proportional to the density of the
atoms. For this reason, the spatial distribution of the fluorescence
represents the profile of the atomic beam. With a camera lens the spatial fluorescence is imaged on the sensitive surface of the detector of
an OMA-II and the spatial profile of the atomic beam is directly shown
on the screen of the OMA monitor. The sensitive surface is composed of
diode array and the final angular resolution of the detection system is
3.0×10^{-5} rad. The exposure time used is 25ms allowing us to collect da-
tum in a very short period of time. After going through an acousto-
optic modulator the laser beam is split into two, one with frequency
ω_ℓ , the other $\omega_\ell \pm \Omega$, Ω is the modulation frequency. We always
use the beam with frequency $\omega_\ell = \omega_0$ -- the atomic resonant frequency
as the probe beam and the beam with $\omega_\ell \pm \Omega$ as the interacting beam.

The involved transition is $(3S_{1/2}, \; F=2, \; M_F=+2 \; -- \; 3P_{3/2}, \; F=3, \; M_f=+3)$
of Na. To avoid optical pumping effect circularly polarized light and
3.7 Gauss magnetic field are used in the interaction region. The experiment starts with block of the acting beam. The first step is finding
the most intense fluorescence when the frequency of the probe beam ω_ρ
is tuned. At this point $\omega_\rho = \omega_0$ and the distribution of the fluorescence represents the initial distribution of the atoms at detection region. Then let the acting beam go into the interaction ragion. The
frequency of it should be $\omega_a = \omega_0 \pm \Omega$. The new distribution of the
fluorescence appears in the detection region and it carries the information of radiative pressure exerted on the atoms.

According to the theoretical analyses [1,2,4] in the condition of
rather strong field and large detuning the force heats the atoms for
a negative detuning and cools them for a positive one. As shown in
Fig.2 our experimental results are similar to that published in reference [3] and agreeable to the theoretical analyses.

In a weak or not too weak standing wave field the force that the
atoms undergo < fr > can be written as[1,4]

$$<fr> = \frac{h\gamma v k^2 v_0^2}{\Delta^3}\{ 1 - \frac{v_0^4}{\Delta^2 [\gamma^2 + (2kv)^2]}\} \qquad (1)$$

where V_0 is the Rabi frequency, v is the velocity of atom, and k is the wave vector. In weak field $V_0 << \Delta$, $<fr>$ is accelerating force if $\Delta > 0$, and damping force is $\Delta < 0$. However, with the increase of the field the second term in Eq.(1) becomes more important and the force becomes damping when $\Delta > 0$ and accelerating when $\Delta < 0$. Fig.3 shows the beam profiles as the field intensity changes. One can see that when $\Delta = -22MHz$ $<fr>$ changes its sign with decreasing saturation parameter, the heating becomes cooling. Here $P=(I/I_s)(\gamma^2/4)/(\Delta^2 + \gamma^2/4)$, where I_s is the saturation intensity ($20mW/cm^2$ for D_2 line of Na) and I is the laser intensity. When P=1.13, there is a small peak at the centre of the profile of the heated atoms. It is a group of very cold atoms[1,3]. In our experiment the turning point I/I_s =2.0 ± 0.2, where the heating changes into cooling, is higher than the result calculated from Eq.(1), I/I_s=1.10. The theory [1] describes the features of $<fr>$ only in the case of $\Delta >> \gamma$, kv. In our condition the detuning is close to γ, kv. Therefore it is not surprising that we observed some effects different from the theory. Reference [1] has predicted that when Rabi frequency $V_0 > (\gamma \times \Delta)^{\frac{1}{2}}$ and $\Delta >> \gamma$, $<fr>$ is a function of v and vanishes at v = 0 and v = $\pm v_c$, where

$$v_c = \frac{\gamma}{2K} [\frac{(V_0/\gamma)^4}{(\Delta/\gamma)^2} - 1]^{\frac{1}{2}} \qquad (2)$$

From Eq.(2) one can find that $<fr>$ has different sign for v > v_c and v < v_c. When $\Delta < 0$, the heating changes into cooling with increased v and at v = $\pm v_c$ the velocity bunching occurs. The prediction is correct for $\Delta >> \gamma$ but not for $\Delta \sim \gamma$. According to Eq.(2) when Δ =-20 MHz, P=0.93, $v_c = \pm 46m/s$, it is impossible to observe the bunching in the velocity region of the atomic beam. On the contrary, we have observed it in the experiments. Fig.4 gives the profile of the beam when $\Delta = -20MHz$, P=0.93. The bunching happens at $v_c \sim 0.2m/s$. That is, $<fr>$ changes its sign at a much smaller velocity, and only those atoms which have very slow velocity are heated. As an overall effect we can still see the cooling when $P \sim 1$ and Δ is negative. All these are not agreeable to the theory. The experimental results shows that when the detuning Δ is close to γ (for example $\Delta = 2\gamma$) the force acting on the atoms depends on the atom velocity and changes its sign at $kv < 0.25\gamma$. This phenomenon is due to the interference between the two laser beams (the forward and the backward) in the stimulated process. Further theoretical and experimental work on this subject are carried forward in our group.

In this work the motion of atoms in a standing wave laser field was experimentally observed. In the condition of strong laser field and large detuning the experimental results are in agreement with the theory. Owing to the complication of the motion of atoms the theoretical analyses were derived under some approximation. Therefore they are applicable only for limited condition. We found that when the detuning Δ is close to γ and $P \sim 1$ the force acting on the atoms changes its sign and this occurs in the region of $kv \ll \gamma$. When Δ is negative and $P \sim 1$ the force still damps the velocity of the atoms. This damping has something to do with the "optical supermolasses"[5] and is the consequence of the retarded dipole force effect. This damping force can be much stronger than the spontaneous emission force $\hbar \gamma k/2$. A theoretical analysis for the case of $\Delta \sim \gamma$ and $P \gtrsim 1$ is needed.

The authors thank Dr. Liao Shiqiang for offering the acousto-optic modulator. This work is supported by science foundation of Academia Sinica. One of the authors, Y.Z. Wang, thanks the International Centre for Theoretical Physics (Trieste) for offering the opportunity of international academic exchange.

Reference
[1] A.P. Kazantsev, V.S. Smirnov, G.I. Surdutovich, D.O. Chudesnikov, V.P. Yakovlev, J. Opt. Soc. Am., B2, 1731 (1985)
[2] J. Dalibard, C.Cohen-Tannoudji, J. Opt. Soc. Am., 2, 1707 (1985)
[3] A. Aspect, J. Dalibard, A. Heidman, C. Salomon, C. Cohen-Tannoudji, Phys. Rev. Lett., 57, 1688 (1986)
[4] A.P. Kazantsev, D.O. Chudesnikov, V.P. Yakovlev, Sov. Phys. JETP, 63, 951 (1986)
[5] S. Chu, M.G. Prentiss, A.E. Cable, J.E. Bjorkholm, Laser Spectroscopy VIII, Springer-Verlag, 58, (1987)

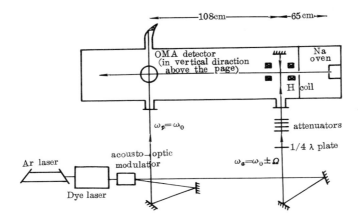

Fig 1 Experimental setup

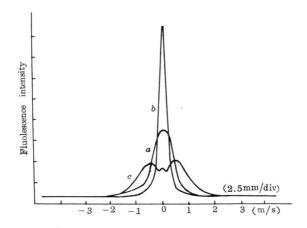

Fig 2 Fluorescence intensity vs position in the atomic beam
The corresponding transverse velocity is given in m/s
Curve a) laser beam off curve b) laser beam on with
a positive detuning (=+50MHz) curve c) laser beam on
with a negative detuning (= 22MHz)

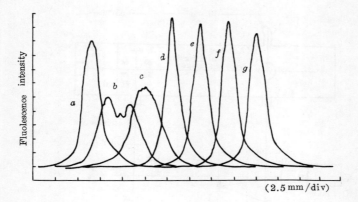

Fig 3 Dependence of the fluorescence distribution on decreasing
P parameter with a negative detuning (+ 22MHz) Curve a)
P=0 b) P=1 13 c) P=0 37 d) P=0 09 e) P=0 015 f) P=
4×10^{3} g) P=1 5×10^{3}

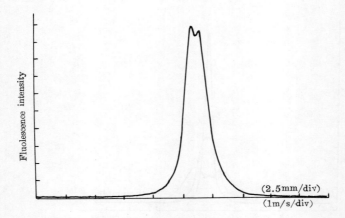

Fig 4 Velosity bunching effect in an atomic beam The correspond
ingtransverse atomic velocity is given in m/s The satura
tion parameter is P=0 93 the detuning is 20MHz

THE INTERRELATION BETWEEN THE OPTICAL PROPERTIES AND THE MBE GROWTH CONTROL OF QUANTUM WELL STRUCTURES

Ping Chen
Surface Physics Laboratory, Fudan University
Shanghai, China

ABSTRACT

The studies on the intensity dynamics of reflection high energy electron diffraction are applied to the MBE growth control of quantum well structures. The correlation between the optical properties and the growth mechanism is established. Several attempts for improving the optical qualities of the MBE grown quantum well structure are discussed.

Several demands are put forwards to the material scientists when the quantum well structures are applied to the device engineering. The quantized energy levels of a quantum well structure are determined by two basic paremeters, the well width and the barrier height. A precise control of these two parameters is then necessary for the construction of a quantum well structure with the desired optical properties. This requires correspondingly during the growth a perfect in-situ control of the growth rate, the growth time and the composition of each layer forming the quantum well structure.

358

It was found that the specular beam intensity of the reflection high energy electron diffraction (RHEED) exhibits a damping oscillation during the MBE growth [1]. The oscillation period is equal to the growth period of one monolayer material(Fig. 1). Thus monitoring the variation of the RHEED specular beam intensity during the growth of quantum well structures makes it possible to measure the

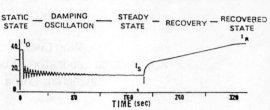

Fig. 1 A typical oscillation spectrum of RHEED specular beam intensity during MBE growth of GaAs (100). The measurement is taken at the out-of-phase condition.

growth rate, the growth time and consequently the composition of the grown layer.

The photo-luminescence (PL) measurement is generally used to characterize the optical properties of quantum well structures. Besides the PL line energy position, the line width and the line intensity are two important characteristic parameters, which reflects the perfection and the efficiency of a quantum well structure. To obtain a narrow and strong PL line, a high purity well structure with abrupt and uniform interfaces or barriers is required.

In order to achieve such a high-quality construction of quantum well structures, on one hand, the extrinsic factors, such as the impurity incoporation, should be considered. On the other hand, the intrinsic factors, i.e. the growth kinetic processes, must be studied. For example, in the growth of the AlGaAs-GaAs quantum well structures the growth kinetic processes are determined through the interplay of the growth parameters such as the Ga or Al flux, the As overpressure, and the substrate temperature. An optimal combination of the growth paremeters results in a smooth growth front which is the starting point to form an abrupt and uniform interface or barrier.

Under certain conditions the intensity of the RHEED specular beam is inversely related to the step density, i.e. the roughness, of the surface (fig. 2). The studies on the intensity dynamics of the RHEED specular beam offers thus an approach to explore the surface kinetic processes during the MBE growth. The optimized combination of growth paremeters can

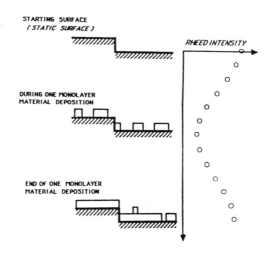

Fig. 2 A schematic illustration of the relation between the RHEED intensity and surface roughness.

be established on the base of a systematic study on the variations of the specular beam intensity as a function of each growth parameter(fig. 3).

360

SQW Al$_x$Ga$_{1-x}$As / GaAs

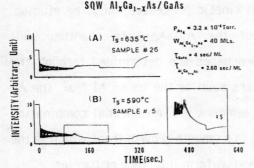

Fig. 3a RHEED oscillation behavior during the growth of quantum well structures at optimal(A) or off-optimal(B) growth condition. (Ref. 2)

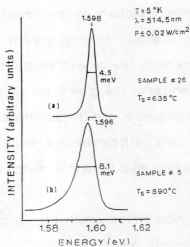

Fig. 3b PL spectra of quantum well structures grown at the condition indentified in Fig. 3a. It is seen that the linewidth of (A) is narrower than that of (B). (Ref. 2)

The concept of the interrupted growth is introduced in attempt to improve the quality of the interface grown, hence the optical quality of the constructed quantum well structure. Fine structures of PL line corresponding to the well width variation of one monolayer step height is observed(Fig. 4).

Fig. 4 PL spectra of quantum well structures grown without interruption (A), with interruption at one interface (B) or at both interfaces(C). Fine structures are observed in (B) and (C). (Ref. 3)

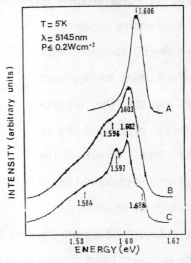

Acknowledgement: The auther is thankful to Professor A. Madhukar for inviting him to perform this study in USC, USA.

References:

[1] J.H. Neave et al., Appl. Phys. A31 1 (1983)

[2] J. Y. Kim, P. Chen, F. Voillot, and A. Madhukar, Appl. Phys. Letts. 50 739 (1987)

[3] F. Voillot, A. Madhukar, J. Y. Kim, P. Chen, N. M. Cho, W. C. Tang, and P. G. Newman, Appl. Phys. Letts. 48 1009 (1986)

Ionic Excimers

R. Sauerbrey, P.J. Wisoff, L. Frey, and S. Kubodera

Department of Electrical and Computer Engineering
Rice University, Houston, Texas 77251-1892

Recently we reported on the excitation of an ionic excimer molecule, $Cs^{2+}F^-$, which is isoelectronic to the rare gas halide XeF molecule, using the soft x-ray flux from a laser-produced plasma (1). Such ionic excimers, $A^{2+}X^-$, have been proposed as possible vacuum ultraviolet (VUV) laser species due to the $A^{2+}X^-(^2\Sigma, B) \to A^+X(^2\Sigma, X) + h\nu$ transition in analogy with the well known rare gas halide transitions (2).

A schematic potential diagram of the $(CsF)^+$ ionic excimer is shown in Fig. 1. The weakly bound ground state corresponds to the atomic states $Cs^+ + F$ representing the first ionization limit of the molecule since an electron has been removed from the fluorine atom. The $Cs^{2+}F^-$ state represents the second ionization limit of CsF and is similar to an inner-shell ionization state of an atom. Just as soft x-rays from laser-produced plasmas have been used for efficient excitation of inner-shell transition of atoms (3,4), molecules such as $Cs^{2+}F^-$ can be excited in the same manner (1). Fluorescence of the $Cs^{2+}F^-(B) \to Cs^+F(X) + h\nu$ transition has also been observed in ion beam excited He/CsF vapor mixtures (5).

The experimental apparatus is shown in Fig. 2. The beam of an injection-controlled KrF laser was focused onto a tantalum target located inside of a heatpipe cell to a power density of up to $10^{11} W cm^{-2}$. A heatpipe cell was operated between 750°C and 900°C to produce CsF and $CsCl$ vapors. The fluorescence from the CsF and $CsCl$ vapor, excited by the laser-produced plasma, was observed through a 0.2 m VUV spectrometer and detected by a photomultiplier. Time-resolved signals were recorded by a digital storage oscilloscope. For time-integrated measurements, a boxcar integrator was used. By setting its 50 ns gate in coincidence with the laser pulse, detection of the recombination light from the tantalum plasma, which appeared some 100 ns after the laser pulse, was avoided.

Fig. 3 shows a typical fluorescence spectrum of CsF after excitation by a laser-produced plasma. A continuum extends from 190 nm to 170 nm and peaks at 185 nm. A second peak is observed at 154 nm and has been identified as the corresponding $Cs^{2+}F^-(D, {}^2\Pi) \to Cs^+F(X, {}^2\Sigma)$ transition (5). No emission on the F_2^* transition at 158 nm was observed. The undulated structure is due to vibrationally-excited $Cs^{2+}F^-$. The temporal shape of the $Cs^{2+}F^-(B-X)$ emission follows the evolution of the exciting laser pulse, indicating a direct excitation mechanism like photoionization and an effective lifetime of the $Cs^{2+}F^-$ state shorter than the laser pulsewidth. Emission from $Cs^{2+}Cl^-$ was observed at 208 nm.

In order to gain information about the kinetic processes involved in laser-plasma excitation of $(CsF)^+$, the behavior of the $Cs^{2+}F^-(B-X)$ fluorescence was studied as a function of several experimental parameters. By variation of the pulse energy of the KrF laser, the effective temperature T_p of the tantalum plasma was changed. The fluorescence intensity over the laser intensity is plotted in Fig. 4. The fluorescence intensity J shows approximately a scaling law of $J \sim I^{3/2}$ for laser intensities I between $10^{10} \, W/cm^2$ and $10^{11} \, W/cm^2$.

In order to understand the behavior of the soft x-ray pumped CsF vapor, a simple kinetic model was compared to the experimental results. In this model it is assumed that only photoionization contributes significantly to the excitation of the CsF vapor. In the case of He/CsF mixtures, the soft x-ray flux of the laser-produced plasma is absorbed due to photoionization of CsF and He. Since the process leading to the population of the $Cs^{2+}F^-$ state corresponds to an inner-shell ionization of a p-electron from the Cs atom, the atomic photoionization cross-section for the $Cs(5p)$ shell can be used as a good approximation (6). The attenuation of the soft x-ray flux from the target was calculated using the total photoionization cross-section of CsF. The excitation of the $Cs^{2+}F^-$ state by electrons was neglected.

For a high Z target material, the spectral characteristic of the pumping soft x-ray flux is assumed to be a blackbody spectrum of a plasma temperature $T_p \approx 15 \, eV$ for a KrF laser intensity of $10^{11} \, W/cm^2$. Photoionization was assumed to be the dominant process in generating electrons. For the high electron densities in the vicinity of the plasma, collisional-radiative recombination effectively limits the electron density. The limiting electron density was calculated by setting the recombination rate equal to the total photoionization rate using a recombination rate of $\alpha = 10^{-25} \, cm^6 \, s^{-1}$ (7).

The observed decrease of fluorescence intensity for increasing CsF-pressure indicates a strong quenching process of the $Cs^{2+}F^-$ state. Using the calculated radiation lifetime for the $Cs^{2+}F^-$ state of $1/A = 1.5 \, ns$ (2), and particle densities of $10^{16} \, cm^{-3}$, the quenching rate constant can be estimated to be $k \approx 10^{-6} - 10^{-7} \, cm^3 \, s^{-1}$. Such high rate constants are typical for reactions of charged particles. De-excitation by electrons thus seems to be most likely as the dominant quenching process.

The results of the model calculations for the $Cs^{2+}F^-$ fluorescence intensity as a function of the pulse energy of the KrF laser is plotted in Fig. 4. This figure shows the calculated plasma temperature for a given laser intensity on the target as calculated from the model of Reference (8). By using plasma parameters which led to a plasma temperature of $10\,eV$ instead of $15\,eV$ for a laser intensity of $10^{11}\,W/cm^2$, good agreement between experimental and theoretical results was achieved. For the following calculations, a black body temperature of $T_e = 10\,eV$ was assumed.

In conclusion, the fluorescence spectrum and kinetics of an inner-shell ionized molecule excited by a laser-produced plasma has been studied for the first time. The soft x-ray flux of a laser-produced plasma proved to be a very suitable excitation technique for the spectroscopic examination of an interesting class of ionic molecules. Application of this method to the $(CsF)^+$ ionic excimer system led to new spectroscopic and kinetic results which still support the feasibility of ionic excimers as new laser candidates in the VUV. This excitation technique should be applicable to other proposed ionic molecular laser systems (9).

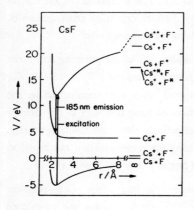

Figure 1: Schematic potential diagram of the cesium fluoride ionic excimer

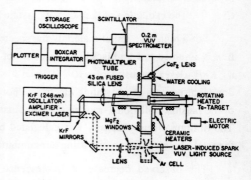

Figure 2: Experimental setup

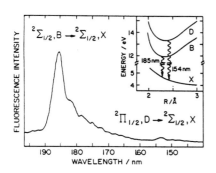

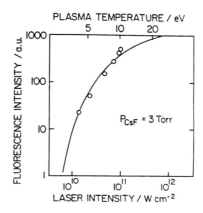

Figure 3: Time-integrated emission spectrum of $Cs^{2+}F^-$

Figure 4: Peak intensities of the $Cs^{2+}F^-$ ionic excimer emission as a function of the KrF laser pulse energy

References

[1] S. Kubodera, L. Frey, P.J.K. Wisoff, and R. Sauerbrey, *Optics Letters*, **13**, p. 446 (1988).

[2] R. Sauerbrey and H. Langhoff, *IEEE J. Quantum Electron.*, **QE-21**, p. 179 (1985).

[3] R.G. Caro, J.C. Wang, J.F. Young, and S.E. Harris, *Phys. Rev.*, **A30**, p. 1407 (1984).

[4] W.T. Silfvast and O.R. Wood II, *J. Opt. Soc. Am.*, **B4**, p. 609 (1987).

[5] F. Steigerwald, F. Emmert, H. Langhoff, W. Hammer, and T. Griegel, *Opt. Comm.*, **56**, p. 240 (1985).

[6] J.J. Yeh and J. Lindau, *Atomic Data and Nuclear Data Tables*, **32**, p. 1 (1985).

[7] J.B. Hasted, *Physics of Atomic Collisions*, New York: American Elsevier Publishing Co. (1972).

[8] P. Mora, *Phys. Fluids*, **25**, p. 1051 (1982).

[9] N.G. Basov, M.G. Voĭtik, V.S. Zuev, and V.P. Kutakhov, *Soviet J. of Quant. Electr.*, **15**, p. 1455 (1985).

OPTICAL SHG STUDY ON POLYMERIZATION OF LANGMUIR–BLODGETT MOLECULAR LAYERS

Yu Gongda
Dept. of Phys., Suchow Univ.,
Suchow, P.R. China

Chen Gang,   Li Le,   Wang Wengchen,
Zhang Zhiming
Dept. of Phys., Fudan Univ.,
Shanghai, P.R. China

Optical SHG (Second Harmonic Generation) has recently been proved to be a promising tool for probing and observing phenomena at surface and interface[1,2]. Here, it is shown that polymerization of organic molecules in LB film can be probed by optical SHG. As the amount of molecules in a monolayer is less, it is difficult to probe the polymerization by using of conventional IR, UV, or NMR method. However, SHG from a surface with inversion symmetry is sensitive to the adsorbates. Besides, during polymerization, transformations of relevant atom-radicals and valence bonds (both inner and inter molecules) would cause the change of the molecular nonlinear polarizability. As molecules in LB film arrange regularly, it would thus result in the changes of the SHG of LB film.

In our experiment, organic molecules of unsaturated long chain monocarboxylic acid (e.g. $CH_2 = \overset{CH_3}{\underset{}{C}} - CONH - (CH_2)_{10} - COOH$) are used as the monomer. They are amphiphilic so as to be able to form LB films onto a silver surface in a Langmuir trough. They contain carbon double bonds and could be polymerized under the exposure of UV radiation by broking the carbon double bonds[3]. The conjugated double bonds in the molecule, and thereby it's π-electron system with higher delocalizability could lead to comparatively large molecular nonlinear polarizability[4]. To examine the intensity and polarization of the SHG respectively, two experimental setup were employed 1) through the excitation of SPW (Surface Plasmon Wave) between the Ag/LB films interface under the Kretschmann's configuration the enhanced reflected SHG intensity was measured vs. the dose of UV exposure during polymerization 2) p-polarized incident laser directly imp-inged upon the free surface at 45 degrees, the s-polarized component of reflected SHG in comparision with the p-compo-nent was measured as well.

SHG experimental results for molecular monolayer and bi-layers during polymerization were given in Fig.1. The varia-tion of SHG from the interface is dramatic even though only one or two layers of molecules are included. For the mono-layer, the SHG intensity of the interface increased during polymerization while that from bilayers/Ag interface also increased at first. It comes from the depletion of the con-jugated double bonds of the molecules, the phase of SHG from which was opposite to that of bare Ag film.That is to say, it contributed from the 1st molecular layer. To check it further, the UV exposure was continued on and then, the SHG intensity of bilayer/Ag interface fell down, while that from the monolayer/Ag interface remained almost constant. Surely, polymerization was saturated in the monolayer and carried on in the 2nd layer of molecules, the phase of SHG from which was nearly the same to that from the bare Ag film. That is to say, the monomer molecular layer adjacent to the silver film surface get polymerized earlier than the other layers.

As to the polarization of SHG, it was shown from the ex-periment that, before polymerization the s-component of re-flected SHG was zero, $I_{p-s}/I_{p-p}=0$, within our experimental accuracy, p-polarized field could only give rise to p-pola-rized SHG[5], but after polymerization, the s-component of SHG became nonzero $I_{p-s}/I_{p-p} \doteq 1/100$. It denoted that, after polymerization, molecules in LB molecular layer crosslinked each other, e.g. Van der Walls bonds between the molecules transformed into chemical bonds, the interaction between the original monomer molecules strengthened, so that the

fundamental field in p-(Z or Y) direction could have influence over the SHG in s-(x) component.

   In summary, we have shown that surface SHG technique is
suitable for probing polymerization of LB molecular layers
even though only one or two molecular layers are included,
as the variation of nonlinear polarizability of the molecular layer comes from the transformation of the corresponding
atom-radicals and valence bonds, especially of conjugated
double bonds and π-electrons. The variation would appear
both in the intensity and the polarization. Moreover, it is
suggested that the monomer molecular layer adjacent to the
silver film surface get polymerized earlier than the other
layers. It is consistent with the previous conclusion of IR
study on polymerization of LB film[6] that polymerizing
possibility and rate are correlated with the molecular pack-
ing and arrangement order.

REFERENCE
[1]   Y.R. Shen et al., Phys. Rev. Lett., 52 (1984) 348.
[2]   Zhan Chen et al., Opt. Comm., 54 (1985) 305.
[3]   Yang X.L. et al., Acta Optica Sinica, 7 (1987) 190.
[4]   T. Rasing et al., Chem. Phys. Lett., 130 (1986) 1.
[5]   I.R. Girling et al., Thin Solid Films, 133 (1985) 101.
[6]   K. Fukuda et al., Thin Solid Films, 99 (1983) 87 & 133 (1935) 39.

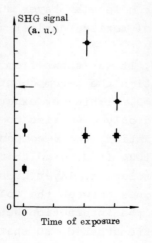

Fig 1 SHG intensity at different
dose of UV exposure
← level of bare silver
■ for interface of Ag/monolayer
◆ for interface Ag/bilayers

WEAK LOCALIZATION OF LIGHT

M. Rosenbluh

Physics Department, Bar-Ilan University, Ramat Gan, Israel

The multiple scattering of light from random systems has recently become a subject of intense activity.[1] The major incentive for experimentation in this new field came from the realization that the concepts of localization and universal fluctuations that are successfully used in the description of electron transport phenomena, can also be applied to the description of multiple light scattering in a random system. This realization has led to numerous investigations of photon transport and scattering in disordered solids and liquids, and to the discovery of a wide range of unexpected phenomena. We present a review of recent results in this exciting new field.

The effects associated with the weak localization of electromagnetic waves occur whenever light is strongly and multiply scattered in a random system. It is convenient to discuss the scattering in terms of photons executing a three-dimensional random walk inside the medium with a transport mean free path, ℓ , corresponding to the average distance necessary in order to scramble the direction of the incident photon's k vector. The essence of weak localization lies in the realization that amongst the large number of possible photon loops through the random system there is a family of loops whose terminal point is on the input plane of the scattering system, and where the exit direction is antiparalell to the direction of incidence. For such loops there is also a probability of traversing the path in the time-reversed direction. This implies that the backscattered trajectories will constructively interfere, and leads to a backscattered peak that is twice as intense as the diffuse background scattering. Classical multiple scattering theory fails to predict such behaviour since it ignores phase correlations of the scattered waves on a scale larger than ℓ . The backscattered peak is an essential feature of weak localization and has recently been demonstrated on a variety of scattering systems[2,3].

In analogy with the universal conductance fluctuations in electron transport, when light is scattered from a solid sample, a characteristic speckle pattern is obtained with large intensity fluctuations. In order to obtain the backscattered peak, an ensemble average over many speckle patterns has to be performed. The backscattered peak is "buried" in the speckle "noise", the analysis of which indeed shows that the speckle statistics is unusual in that it is nonexponential[1].

When similar experiments are performed on a system of dense scatterers suspended in a liquid, the ensemble averaging of the fluctuations is automatic, due to the random Brownian motion of the scatterers. The ensemble averaging in such systems is particularly interesting since the individual scattering events are slightly inelastic due to the random Doppler shifts associated with the moving scatterers. In order to study the dynamics of such systems, we have performed time-resolved measurements of the time evolution of speckle patterns from a suspension of polystyrene spheres suspended in water[4,5]. We find that the quasielastic nature of the scattering greatly accelerates the rate at which correlations between speckle patterns are washed out. A quantitative analysis of data from such experiments has revealed that the time dependence of the intensity-intensity correlation function is totally non-exponential, and that after a very rapid decay to a value of one-half it falls off exceedingly slowly, ultimately approaching zero at the single scattering rate. Furthermore, the initial decay rate is a universal factor of 25 times faster than the decay rate of the single particle scattering correlation function. Theory indicates that this behaviour is due to the strong dependence of the rates at which multiply-scattered photon trajectories of different lengths lose phase as a result of the Brownian motion of the scatterers.

An underlying, and successful, assumption in the analysis of the dynamic behaviour observed in the backscattered direction, has been the photon diffusion approximation; i.e. a gas of photons undergoing diffusion in a random medium with a characteristic transport mean free path. In this approximation the mean trajectory length in the forward direction scales as the square of the sample thickness, s. Surprisingly, our measurements of forward scattered radiation[7] indicate that this approximation breaks down whenever very long scattering trajectories are involved.

In these experiments we measured the frequency spectrum of forward scattered light from various thicknesses of a polystyrene-liquid suspension. From the Fourier transform of the spectrum we obtain the time dependent intensity-intensity correlation function. The frequency spectrum was found to be a Lorentzian, with a half-width that scales linearly with s, contrary to the s^2 prediction based on the photon diffusion approximation. A further serious discrepancy is that the correlation function drops much more slowly than one would expect om the basis of diffusion theory. The clear departure of the forward scattered data from the diffusion approximation can be reconciled through the phenomenological introduction of a new length scale, which reduces the contribution of very long loops to the correlation function. A rigorous theoretical foundation for such a correction to the theory has yet to be formulated. Nonetheless, the justification for the new cutoff length seems to be that the photon diffusion approximation neglects the positive correlations between the time evolution of the phases of different trajectories. This problem worsens as the trajectories get longer due to the exponential growth in the number of possible trajectories with increasing sample length[7].

The realization of the strong localization of light is intriguing, and has been considered theoretically[8]. Strong localization behaviour and the optical Aderson transition is predicted, as the wavelength of the radiation being scattered approaches the transport mean free path of the system. The demonstration of strong localization, however, has been elusive due to the difficulty of producing nondissipative scattering systems with an ultra-short transport mean free path. In two-dimensional systems, however, it is prdicted that strong localization will always occur, provided the sample size is larger than the localization length given by $\exp(l/\lambda)$. The major experimental problem here is to construct a two-dimensional scattering system.

We have made considerable progress torwards this end, and have demonstarted weak localization in a very nearly two-dimensional system, consisting of a bundle of long parallel rods[7]. The light from such a system is scattered into an equatorial plane perpendicular to the rod axis, and therefore the random walk of the photon is confined to this plane. The more perfectly the rods are aligned, the more nearly the system approaches pure two dimensionality and the scattering becomes confined to the equatorial line. The backscattered peak in such a system is a point on this line. Deviations from pure 2-D broaden the equatorial plane and elongate the backscattered point into a vertical line. Quantitative analysis of the light scattering data we have obtained has shown that the dimensionality of our best samples is about 2.05. Prospects for attaining a greater degree of two dimensionality and a much shorter mean free path and thereby observing strong localization effects appear promising.

1) For a comprehensive list of relevant publications see the citations contained in the following references.
2) M. Kaveh, M. Rosenbluh, I. Edrei, I. Freund, Phys. Rev. Lett., 57, 2049(1987).
3) M. Rosenbluh, I. Edrei, M. Kaveh, I. Freund, Phys. Rev. A, 35, 4458(1987).
4) M. Kaveh, M. Rosenbluh, I. Freund, Nature, 326, 778(1987).
5) M. Rosenbluh, M. Hoshen, I.Freund, M. Kaveh, Phys. Rev. Lett., 58, 2754(1987).
6) M. Rosenbluh, M. Hoshen, I. Freund, Phil. Mag. B, 56, 705(1987)
7) I. Freund, M. Kaveh, M. Rosenbluh, to be published.
8) S. John, Phys. Rev. Lett. 53, 2169(1984); P. W. Anderson, Phil. Mag. B 52, 505(1985).
9) I. Freund, M. Rosenbluh, R. Berkovits, M. Kaveh, Phys. Rev. Lett., 61, (1988).

STATISTICAL FRAGMENTATION PATTERNS OF METASTABLE ION:
COMPARISON WITH EXPERIMENT

S.T. Li, H.X. Liu, Z.L. Li and C.K. Wu

Laboratory of Laser Spectroscopy, Anhui Institute of Optics
& Fine Mechanics, Chinese Academy of Science, Hefei, China

It was found experimentally in ultraviolet laser multiphoton ionization -fragmentation (MPI-F) of $Fe(CO)_5$ that high efficient ion-molecule reaction between Fe^+ and $Fe(CO)_5$ produces metastable ion $Fe(CO)_5^+$, then, the ion dissociation into various fragment ions.

$$Fe^+ + Fe(CO)_5 \xrightarrow{h\gamma} Fe(CO)_5^+ \longrightarrow Fe(CO)_n^+, \quad (n=4,3,\cdots,0) \qquad (1)$$

Relative yields of $Fe(CO)_n^+$ $(n=5,4,\cdots,0)$ via mentioned-above reaction process were measured, with a time-of-flight mass spectrometrically detecting system, at 308 nm and 355 nm. Also, the laser intensity dependences of these ion signals were measured at the two laser wavelengths, respectively.

Within the maximal entropy formalism, let P_j $(j=1,2,\cdots,6)$ respresent a distribution of various fragment ions $Fe(CO)_n^+$ $(n=5,4,\cdots,0)$ from the dissociation of the metastable ion $Fe(CO)_5^+$. The P_j is governed by mean energy $\langle E \rangle$ absorbed by the parent ion, under some constraints of the conservations of energy, matter and charge. Subjected to statistical approximation, the fragmentation pattern of maximal entropy was computed by using the usual method of Lagrange multipliers and presented in graphical form, break-down curve (to see Fig.1). By varying the mean energy $\langle E \rangle$ it is found possible to scan the range from low to extensive fragmentation. The branching fractions computed at low, medium and high mean energy $\langle E \rangle$ are given to compare with ones measured experimentally, as shown in Fig.2.

The Lagrange parameter γ_{Fe}, corresponding to Fe element in the fragment ions, is negative at lower energies favoring the larger fragment ions. The influence of γ_{Fe} to the fragmentation pattern decreases rapidly as the mean energy increases, while the statistical fragmentation pattern is primarily affected by γ_{CO}, corresponding to CO ligand in the fragment ions, indicating a proclivity for loss of CO ligand.

It found that the statistical fragmentation pattern at medium mean energy corresponds with the experimental distribution induced by 308 nm laser radiation. However, the high fragmentation computed is very different from that induced by high intensity of 308 nm laser radiation, but similar to that induced by 355 nm laser radiation. As a matter of fact, the mean energy only depends on the excess energy of Fe^+ in the reaction (1) that the total available energy from the initial MPI-F of $Fe(CO)_5$ exceeds the appearance potential of Fe^+ ion. The analysis for $Fe(CO)_5$ MPI-F energetics shows that the excess energy in Fe^+ at 355 nm can reach up to 9.5 eV, but the excess energy in Fe^+ at 308 nm is above 6 8 eV, which means the mean energy in metastable $Fe(CO)_5^+$ at 355 nm is higher than that at 308 nm. This is an acceptable explanation of the comparisons mentioned-above and implies the mean energy is primarily controlled by laser wavelength in $Fe(CO)_5$ MPI-F.

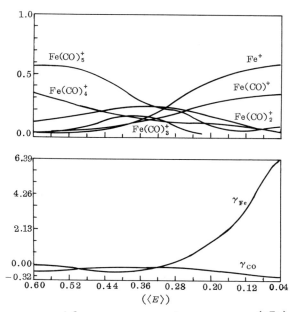

Fig 1 Breakdown curve and Lagrange paramenters versus $(\langle E \rangle)$ computed in the statistical limit for $Fe(CO)_5^+$ metastable ion dissociation

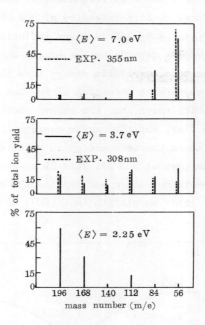

Fig 2 Computed and experimentally measured mass resolution
      fragmentation patterns in which the computation at a
      medium fragmentation was chosen to match a corresponding
      experimental pattern

OXYGENERATION REACTION OF CERIUM WITH XeCl LASER

Zhou Zhengzhuo, Qiu Mingxin

Shanghai Institute of Laser Technology
Yue Yang Road 319, Shanghai 200031, China

Huang Shantang, Bi Gixin, Gu Jialiang

Li Fangling, Shi Jiliang

Shanghai Institute of Organic Chemistry, Academia,
Sinica

The molecule of cerium trichloride solved in pure water was irradiated with photon at 308 nm to become into that of cerium quadrichloride. For this process, the peak absorption wavelength is at 250 nm with the absorption coefficient of 720 $M^{-1}cm^{-1}$, but that at 308 nm is only 12 $M^{-1}cm^{-1}$. Hence, XeCl laser is not enough effective for photo-oxygeneration, but a new agent. Some kind compounds can increase the precipitation efficiency, such as KIO_3, HCl etc.

The excited complex compound, $[CeIO_3^{2+}]^*$, losses its excitation energy quickly due to the collision with the surrounding molecules and the newly obtained ion of Ce^{4+} may also be reduced back to the ion of Ce^{3+} owing to the action with the radical of H*. The radical of IO_3^- can raise the reaction rate of photo-oxygeneration by decreasing the inverse process, because the precipitation of $Ce(IO_3)_4$ takes place.

The existence of radical, IO_3^-, is important for the stability of quadrivalent ion, Ce^{4+}, and for avoiding the inverse reaction, which influences the quantum efficiency. Fig.1 shows the dependence of the relative precipitation rate on the concentration of KIO_3.

Other acid radicals such as CIO_3^-, PO_3^{3-} and BO_3^- instead of IO_3 were tested too. The precipitation was not so effective as that of IO_3 with the same condition.

With the constant concentration of HCl and KIO_3, the precipitation rate of $Ce(IO_3)_4$ was the function of the concentration of $CeCl_3$, as illustrated in Fig.2. Trivalent ion of Ce^{3+} first formed the complex compound with the radical of IO_3 and then the complex compound was excited to the state of 5d which was further oxidized into quadrivalent ion of Ce^{4+}. If there existed enough radicals of IO_3, the precipitation rate increased with the increase of the concentration of $CeCl_3$, as seen in Fig.2.

Fig.3 is the curve on the precipitation rate of $Ce(IO_3)_4$ and the irradiation energy of 308 nm light with a fixed ratio of $CeCl_3:HCl:KIO_3$ of 0.04 M : 2 M : 0,1 M in 2 cc of solution.

With the same dose of the irradiated light intensity at 308 nm and the same concentrations of component parts, the light intensity also has a serious influence on the precipitation rate of $Ce(IO_3)_4$, with the intensity light at 308 nm from 0.1 to 1.4 j/cm^2. There is a maximum where the efficiency is highest. With the increase of light intensity, the light absorption in the solution went up to become heat.

With HNO_3, H_2SO_4 instead of HCl in the experiment, the photo-precipitation was observed, but has not so good results as that in HCl.

In order to observe the element separation with 308 nm light for natural rare-earth mixed compound, 4.5 cc. of natural rare earth chloride with the ratio of 0.01 M : 0.1 M : 2 M for nature rare earth chloride, KIO_3 and HCl, was prepared. Before the light irradiation at 308 nm there were 0.01 and 0.05 for La^{3+} and Ce^{3+} and after that , 0.015 and 0.042 respectively. The separation factor obtained was 5.6.

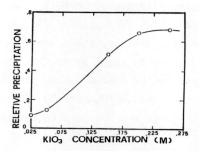

Fig 1 The dependence of the precipitation rate
of cerium on the concentration of KIO_3

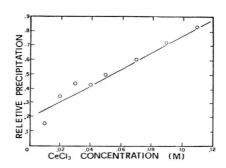

fig 2 The curve of the Ce^{4+} precipitation rate and
the concentration of CeCl$_3$ under the same
other parameters

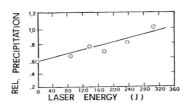

Fig 3 The relation between the Ce^{4+} precipitation rate
and the irradiated light energy at 308 nm with
the same other parameters

MEASUREMENT OF VERDET COEFFICIENT AND MAGNETO-OPTIC SPECTROSCOPY IN TERMS OF BEATS

Zhang Liqun

Beijing Institute of Automation
P.O. Box 3912, Beijing, China

A new method is introduced to measure verdet coefficient V with frequency difference between counter-clockwise laser beam and clockwise. Experiments display that the precision accuracy is up to 10^{-5}. Verdet coefficient V is a function of wavelength λ. From theoretic analysis, get an expression of $V(\lambda)$ and know the nonlinear optic characteristic of Faraday effect. Substance structure and absorption characteristic can be understood by means of magneto-optic dispersion from normal dispersion region to anomalous dispersion region.

As we known Faraday magneto-optic effect can be expressed by

$$\theta == \vec{B} \cdot 1 \cdot V(\lambda) == \frac{\pi}{\lambda} \ (n_+ - n_-)1 \tag{1}$$

where θ is rotation angle of polarization plane, 1, thinkness of sample cell, $\vec{B}$, magnetic induction, n_+ or n_- , index of circular-polarized laser travelling in the direction of $\vec{B}$ or counter to $\vec{B}$. In order to measuring, put the sample into the ring laser cavity. Due to $n_+ \neq n_-$, the optic-cavity-length is difference, i.e. $L_+ \neq L_-$. Hence $V_+ \neq V_-$, we have

$$\Delta v = v_- - v_+ = qc \ \frac{n_+ - n_-}{(\overline{L})^2} = 1 \cdot \tag{2}$$

$\overline{L} = q \ \lambda$, comparing equation (1) and (2), we get

$$\Delta v = \frac{c}{\overline{L}\pi} = \vec{B} \cdot 1 \cdot V(\lambda) = \frac{c}{\pi\overline{L}} \theta \cdot \qquad (3)$$

where $\overline{L} = (L_+ + L_-)/2$, q is longitudinal mode number, c, light velocity.

Since $C/\overline{L} = 10^8$, measurement of Δv is much more sensitive than that of θ. Frequency counts have high precision. There is a combine-beams system out of the cavity, which combines two beams with one direction. The beats are received by an optic-electric device, and displayed in a frequency meter.

Fused quartz glass, quartz, and K_9 glass are used respectively for sample in experiments. The results of experiments agree with theoretic analysis. This method can be used to measure liquid and gas state substance as well.

In the ring cavity, put a dye laser medium (or other adjustable frequency laser medium) and make the frequency adjustable, so we can get laser magneto-optic spectroscopy by means of receiving beats, intension and intension difference as output from the cavity. Then $V(\lambda)$ is expressed analytically by differential equation

$$n_\pm = (1-k) + \sum_{i=1}^{k} \left\{ \left[1 + \frac{4\pi N_i e_i^2}{m_i} \cdot \frac{\omega_{0i}^2 - \omega^2 \mp e_i |\vec{B}| \omega/m_i}{(\omega_{0i}^2 - \omega^2 \mp e_i |\vec{B}| \omega/m_i) + (g\omega/m_i)} \right]^2 + \right.$$

$$+ \left[\frac{4\pi N_i e_i^2}{m_i} \cdot \frac{g_i \omega/m_i}{(\omega_{0i}^2 - \omega^2 \mp e_i |\vec{B}| \omega/m_i)^2 + (g_i \omega/m_i)^2} \right]^2 \right\}^{\frac{1}{4}} \cdot$$

$$\cdot \cos\left[\frac{1}{2} \tan^{-1} \frac{(4\pi N_i e_i^2/m_i)(g_i/m_i)\omega}{(\omega_{0i}^2 - \omega^2 \mp e_i |\vec{B}|m_i)^2 + (g_i \omega/m_i)^2 + 4\pi N_i e_i^2 (\omega_{0i}^2 - \omega^2 \mp e_i |\vec{B}| \omega/m_i)/m_i} \right]$$

where ω_{0i} is resonant frequency, N_i, number of electric particle in per unit volume, m_i mass of electric particle, e_i, an electric particle charge.

Changing cosinusoidal function in expression (5) to sinusoidal function, we can get absorption factor $n_\pm$.

STUDY ON RHODAMINE 6G/XYLENE RED B LASER DYE MIXTURE SYSTEM

Wang Wenyun and Li Li

Changchun Institute of Applied Chemistry, Academia Sinica
Changchun, China

ABSTRACT

The fluorescence spectra, lifetime and lasing characteristics of R6G/XRB dye mixture have been investigated. Experiments show that R6G/XRB syatem has good lasing characteristics.

The output of the nitrogen laser has been shown to be a convenient pump for a wide variety of dye lasers. However, some dyes are difficult or even impossible to pump above threshold with the nitrogen laser, either because their absorption coefficient at 337.1 nm is too low or because the fluorescence intensity for the desired visible flucrescence when pumped by 337.1 nm is too weak. This problem can be overcome by using energy transferring dye mixtures. This paper reports energy transfer lasing between Rhodamine 6G as the donor and Xylene Red B as the acceptor, using nitrogen laser pumping. We have revealed from the EX-CITON DYES CATALOG that no tuning curve is presented for XRB for nitrogen laser pumping, implying the low efficiency of this dye by 337.1 nm excitation. However, our R6G/XRB donor-acceptor pair system can make it possible to give increased output from XRB by pumping with a conventional nitrogen laser.

We investigated the fluorescence spectra of R6G/XRB dye mixture solutions. The results show that the energy transfer from excited R6G to XRB is very efficient. The fluorescene intensity of XRB in mixture solutions is stronger than that of single XRB solutions at the same concentration, thus the lasing threshold of XRB is decreased. Besides, in the mixture solutions the fluorescence peak wavelength has an evident red shift with the increase of XRB concentration. So the dye mixture system can also extend the tuning range.

The fluorescence lifetime of R6G/XRB solutions have been measured directly using time-correlated single-photon counting technique which has excellent sensitivity and good time resolution. From the date we have calculated the energy transfer efficiency E from the excited R6G to XRB. the mathematical expression is $E = 1 - \tau_D/\tau_D^o$ where τ_D^o is the donor lifetime in the absence of acceptor and τ_D is that in the mixture solution. Measurements show that the lifetime of donor decreases with increasing acceptor concentration when the concentration of mixture solutions is about 10^{-3} mol.L^{-1}, apparently obeying the Stern-Volmer relationship for bimolecular quenching. We have calculated the energy transfer rate constant K from this relationship and the critical distance R_o of donor-acceptor pair for which energy transfer from the excited donor to acceptor and emission from the excited donor are equally probable, and have found that the value is within the optimal distance between donor-acceptor pair predicted by Förster theory. So we have come to a conclusion that the main mechanism of energy transfer from R6G to XRB is resonance transfer via dipole-dipole interaction. At last, we measured the laser output characteristics of the R6G/XRB mixtures. Dye solutions were transversly pumped by 337.1 nm radiation from a domestic nitrogen laser (QJD-9) with an output peak power 100 kW at a repetition rate of 10 Hz. The lasing wavelengths of the dye mixture were measured and found that the lasing output could be obtained from 570 to 630 nm when the concentration ratio of R6G/XRB was approximately 2:1.

In conclusion, the R6G/XRB dye system is superior in many respects to XRB in case of nitrogen laser pumping.

ULTRANARROW ABSORPTION RESONANCES OF COLD PARTICLES AND THEIR APPLICATION IN SPECTROSCOPY AND OPTICAL FREQUENCY STANDARDS

S.N.Bagayev, V.P.Chebotayev, A.K.Dmitriyev, A.E.Om, Yu.V.Nekrasov, B.N.Skvortsov

Institute of Thermophysics, Siberian Branch of the USSR Academy of Sciences, 630090 Novosibirsk, USSR

Abstract. The paper reports the obtaining of resonances of width about 10^2 Hz on the $F_2^{(2)}$ line of methane.

One of the mechanisms hampering the attainment of narrow optical resonances by the method of absorption saturation is transit broadening that is associated with the finite time of interaction of a light field with absorbing particles. The transit broadening may be decreased by producing light beams of large cross-section and low divergence, e.g. by using telescopic beam expanders. On the other hand, it has been shown that in a low-pressure gas where a free path length of particles exceeds the field sizes, an effective selection of slow particles occurs, and, hence, resonance shift and broadening decrease due to the second-order Doppler effect.

In the present work the saturated absorption resonances with width about 10^2 Hz have been obtained by using an intracavity telescopic beam expander and optical selection of cold molecules in a gas.

The experiments have been made in methane on the $F_2^{(2)}$ line (F=7 → 6 transition) by using the laser spectrometer based on a He-Ne/CH_4-laser (λ = 3.39 μm) with an intracavity telescopic beam expander. The absorption cell length was 850 cm, the light beam diameter in it 30 cm. The working methane pressure in the cell was 10^{-5}-10^{-6} Torr. The influence of the magnetic Earth field and other external fields on the recorded resonance shape was removed by screening the absorption cell with permalloy. The magnitude of the magnetic field in the absorbing cell did not exceed several

hundredth Oersted. A copper jacket with a pipeline for
liquid nitrogen was put inside the cell. A narrow line of
the telescopic laser was provided in the scheme of spectrometer by its phase locking to a frequency-stable laser with
a narrow line. The resonance in methane was recorded by the
method of synchronous detection of harmonic signal in the
laser power. The first measurements were made at room
temperature of an absorbing gas.

The figure shows the record of the low-frequency compo-
nent of the $(F = 7 \rightarrow 6)$ transition at a pressure of 10^{-5}
Torr. The deviation frequency and amplitude were 230 and 300
Hz, respectively. The recording and averaging times were 10
min and 4 s, respectively. The field in the laser cavity was
10^{-4} W. The signal-to-noise ratio appeared to be determined
by low-frequency fluctuations of the laser intensity due to
angular cavity detuning that exceeded the noises of an InSb
detector by several orders. Elimination of this amplitude
noise by active stabilization of power will allow even in
the nearest future operation at a pressure of about 10^{-6}
Torr and obtaining the resonances with an absolute width of
about 10 Hz. Attainment of such resonances will open up new
possibilities of making precision physical experiments and
allow producing laser frequency standards on cold particles
with reproducibility 10^{-15}-10^{-16}.

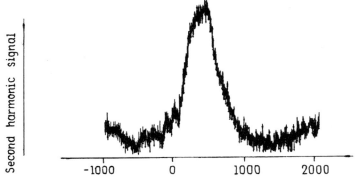

Frequency detuning with respect to stable laser, Hz

THE DYNAMICS OF ION CLOUDS IN PAUL TRAPS: IMPLICATIONS FOR FREQUENCY STANDARD APPLICATIONS

R. Blatt, I. Siemers, M. Schubert, Th. Sauter, and W. Neuhauser

I. Institut für Experimentalphysik, Universität Hamburg, FRG

ABSTRACT

Kinetic energy and spatial distribution of trapped ion clouds have been investigated both theoretically and experimentally. Their dependence on the trap's operating conditions allows evaluation of lineshapes and shifts for frequency standard applications.

The ion storage technique and ultrahigh precision experiments on trapped ions have proved their superior suitability for time and frequency standard applications [1]. Experiments with single cooled ions [2] would seem to be the ultimate tool for ultrahigh resolution spectroscopy, however, their application to frequency standards is usually marred by a low signal—to—noise ratio. On the other hand, experimenting with ion clouds in Paul traps does not allow efficient optical cooling because of spurious heating effects due to the time dependent trapping potential. The accuracy of high resolution measurements in ion clouds is thus limited by 2nd order Doppler shifts [1,3]. This limitation could be overcome if the ions' motion were known. Knowledge of the ions' mean kinetic energy allows the derivation of line shapes and the evaluation of line shifts due to the ions' motion.

Recently, a model has been developed to describe distribution functions of ion clouds in a Paul trap [4]. In the frame of these calculations the ion motion is treated as Brownian motion governed by the trapping potential and the stochastic force due to random collisions with surrounding ions and the background gas. The resulting Fokker—Planck—equation can be solved analytically yielding distribution functions for the spatial and velocity coordinates. Through these, the mean kinetic energy is analytically expressed as a function of the trap parameters.

An experiment has been performed to check these predictions. Clouds of 10^6 Ba$^+$-ions were confined in an rf trap of 2.8 cm cap distance with a drive frequency of 400 kHz, generating well depths of 10—40 eV. In order to derive kinetic energies we observe the Doppler broadened transition $6S_{\frac{1}{2}}$–$6P_{\frac{1}{2}}$ in Ba$^+$. By fitting the theoretically derived line shapes to the observed ones we obtain values for the mean kinetic energy of the ion cloud. In Fig.1 the mean kinetic energy is drawn versus the q_z

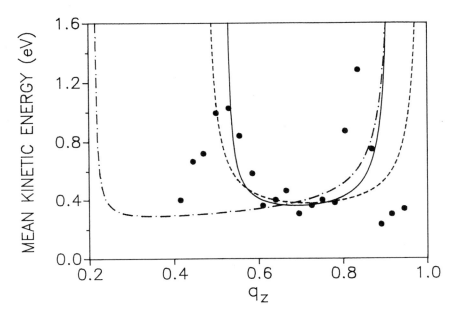

Fig.1 Kinetic energy as function of the trap parameter q_z ($\Omega/2\pi = 400$ kHz).

value, i.e. the trap parameter determined by the ac–voltage [5]. The dots indicate the experimental results, showing a broad minimum for $0.5 < q_z < 0.8$. This is determined by the center of the trap's stability chart. The increasing values for $q_z < 0.5$ and $q_z > 0.8$ indicate that trapping in the time dependent potential becomes less stable for working conditions closer to the boundary of the stability diagram. For $q_z \approx 0.4$ and $q_z \approx 0.95$ stable trap operation is no longer possible. The decreasing values for the kinetic energy close to these q_z values indicate an energy loss due to evaporation cooling [5], which is not contained in the Brownian motion model. The dashed–dotted line in Fig. 1 indicates the theoretically derived dependence if no space charge is considered, the poles marking the boundary of the trap stability diagram. However, the space charge of the ion cloud leads to a shift in the trapping parameters (i.e. the applied ac and dc voltages). Considering only a first order static correction leads to the dashed line, consideration of first order static and dynamic space charge corrections lead to the curve indicated by the solid line in Fig. 1. The mean kinetic energy is also dependent on the trap's driving frequency which was kept constant for the measurements of Fig. 1. Fig. 2 shows values for the mean kinetic energy as a function of Ω^2 (ac and dc voltages kept

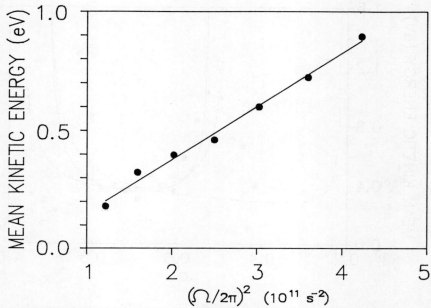

Fig. 2 Kinetic energy as a function of Ω^2 ($a_z = -0.04$, $q_z = 0.5$, Ω: 350-650 kHz).

constant) and the solid line indicates the behavior predicted by the Brownian motion model calculations.

Thus we are enabled to determine kinetic energies of ion clouds from the knowledge of the trap parameters only. According to these measurements we claim an uncertainty of $\approx 5\%$ in the determination of the mean kinetic energy. Second order Doppler shifts in hfs measurements in $^{171}Yb^+$–ion clouds which are currently discussed for frequency standard applications [3] are

$$\Delta\nu/\nu \approx -3 \cdot 10^{-12} \, E_{kin}$$

For $E_{kin} \approx 0.4$ eV and $\Delta E_{kin}/E_{kin} = 5\%$ we obtain an uncertainty due to the second order Doppler effect of $\Delta\nu/\nu \approx 6 \cdot 10^{-14}$ which is comparable to currently used primary standards [1].

1. J.J. Bollinger, J.D. Prestage, W. M. Itano, and D.J. Wineland, Phys. Rev. Lett. **54**, 1000 (1985)
2. W. Neuhauser, M. Hohenstatt, P.E. Toschek, and H. Dehmelt, Phys. Rev. **A 22**, 1137 (1980)
3. R. Blatt, H. Schnatz, and G. Werth, Phys. Rev. Lett. **48**, 1601 (1982)
4. R. Blatt, P. Zoller, G. Holzmüller, and I. Siemers, Z. Physik **D 4**, 121 (1986)
5. I. Siemers, R. Blatt, Th. Sauter, and W. Neuhauser, submitted for publication

FREQUENCY STABILITY MEASUREMENT OF ZEEMAN STABILIZED He-Ne LASER

SEIICHI KAKUMA and KEIICHI TANAKA

Dept. of Applied Physics, Faculty of Engineering,

Hokkaido University, Sapporo 060, Japan

ABSTRACT

Frequency stability of commercial stabilized He-Ne laser as a light source of practical length measuring interferometer is measured by comparison with an iodine stabilized He-Ne laser using beat frequency counting method employing a micro-computer.

1. Introduction

Frequency fluctuation and drift of commercial stabilized 633 nm He-Ne laser were measured. The internal mirror type laser tube is in a transverse magnetic field produced by a pair of parmanent magnet bars and the laser radiats Zeeman split plane polarized triple components, which are orthogonal each other. Zeeman beat frequency produced by a pair of orthogonal components detected and converted into an error signal through a frequency-voltage converter and is used to regulate a current for a heater coil wound around the laser tube. We measured two lasers of A and B with a Zeeman beat of about 563 kHz and 187 kHz, respectively.

2. Method of measurement

The output beams of a measured laser and of an iodine ($^{127}I_2$) stabilized He-Ne laser are mixed, and a photo-detected beat frequency is counted by a frequency counter, which is regulated by a micro-computer.

The gate time is 1s, and the output of the counter is applied to the computer through a GP-IB data bus. Based on the counted data, a short term and long term frequency fluctuation, histograms and power spectram of them are calculated.

3. Experiment

A laser ; I_2 stabilized laser was locked at i-component and an average beat frequency of 69.0 MHz was obtained. A standard deviation of the frequency fluctuation was 0.2 MHz except very rapid increase of the frequency up to 69 MHz for initial one hour. The gate time of the counter was 1s and 15000 data were taken. First half period of 8 hr measurement, fr-

equency fluctuation with period of 10 to 15 min and drift of 0.75 MHz
were observed and the period of the fluctuation gradually increased
to 30 min to 1 hr.
B-laser ; The curve A in Fig.1 shows fluctuation of the beat frequency
obtained by the measurement for 6 hr and a half. The I_2 stabilized laser
was locked at j-component. Curve B is simultaneously measured Zeeman beat
frequency, which was used for stabilizing the B-laser. The histogram C in
Fig.2 shows the distribution of the beat frequency fluctuation of the B-
laser and the I_2 stabilized laser. The histogram D is of the
Zeeman beat frequency. From the data of the histogram C, an average beat
frequency of 20.1 MHz and a standard deviation of 0.08 MHz were obtained.
Based on the data of the histogram D, an average Zeeman beat frequency of
186.974 kHz and a standard deviation of 0.03 kHz were obtained. The frequency stability of B-laser is

$$\pm\ 0.2 \text{MHz} /\ 473 \times 10^6 \text{MHz} = \pm 4.2 \times 10^{-10}$$

4. Long term stability
The same measurement of the A-laser was repeated for three months.
The results showed that the frequency of the A laser monotonically decreased at a ratio of -10 MHz / month. The B-laser also showed similar behavior at a ratio of about $-5 \sim -6$ MHz / month.

5. Wavelength
The wavelength of the A- and B- lasers, which are obtained from the
initially stabilized part of the frequency are

A-laser : 632991295.4 $\pm$ 0.7 fm,
B-laser : 632991400.6 $\pm$ 0.3 fm.

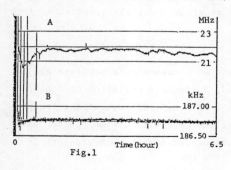

Fig.1

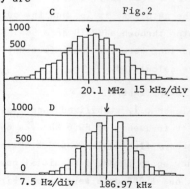

Fig.2

MULTI-WAVELENGTH CW He-Ne LASER AND ITS FREQUENCY STABILIZATION

Zhao Kegong, Ni Yucai

National institute of Metrology, Beijing, China

In this paper, a new progress in the field of He-Ne laser is reported. By using the He-Ne laser tube made in the National Institute of Metrology (NIM), a simultaneous emission of six different visible lines has been obtained.

The lengths of the He-Ne discharge tube in different experiments are 530 and 1000 mm, and the corresponding cavity lengths are 640 and 1200 mm, respectively. The laser cavity consists of two outcoupling mirrors with a curvature of 1 m and a reflectivity of 99.95% over the wavelength range of 600-660 nm. A piezoelectric transducer is used to fine-adjust the cavity length. The laser beam is reflected by a grating of 1200/mm, and easily separated into 6 wavelengths. It was found that there was strong competition among the 6 wavelengths, particularly between 650 and 640 nm. Therefore the power of each line was unstable and could be changed in the range of 10-15%. With the help of a high-resolution prismatic spectrograph and a Kr lamp, the wavelength value in vacuum of all the six lasing lines have been measured to be 611.969, 629.547, 632.989, 633.358, 640.285 and 650.173 nm with an uncertainty of 0.004 nm. Except for the 650 nm line, the results show that all the other wavelengths belong to the transitions of $3s_2-2p_6$, $3s_2-2p_5$, $3s_2-2p_4$, $3s_2-2p_3$ and $3s_2-2p_2$ of Ne, respectively. It has also been proved that the 650 nm laser emission is attributed to the stimulated Strokes-Raman transition between $1s_5$ and $1s_4$ states of Ne pumped by the 633 nm laser radiation.

By using two wide-band cavity mirrors with the reflectivity of 99.95% over the wavelength range of 580-640 nm, the laser can be operated at five simultaneously visible lines of 594, 605, 612, 633 and 640 nm. The 594 and 603 nm lasing lines belong to the transition of $3s_2-2p_8$ and $3s_2-2p_7$, respectively.

By means of frequency-offset locking the frequency of the 633 nm line has been stabilized to an I_2-stabilized 633 nm He-Ne laser at a frequency difference of 21.70 MHz. The output power was not very stable and the frequency stability obtained was only 2×10^{-10} in this case, because of the competition among the 6 lines. With mode-locking operation, the frequency stability can be improved up to 7×10^{-12} and very stable output power can be obtained.

The frequency stability of the other lines are also observed at the same time, while the beat frequency of the 612 nm laser from the multi-wavelength laser with the 612 nm I_2-stabilized laser was measured. The results show that the 612 nm line of the multi-wavelength laser has the same frequency stability with the frequency offset-locked 633 nm line. The multi-wavelength laser will be great significance in metrology, spectroscopy, and particularly in long distance precise measurement.

EFFICIENT ISOTOPE SEPARATION
USING SEMICONDUCTOR LASERS

J.P. Woerdman, A.D. Streater, F. Wittgrefe and E.R. Eliel

Huygens Laboratory, University of Leiden

P.O. Box 9504, 2300 RA Leiden

The Netherlands

ABSTRACT
We have separated the natural isotopes of Rb by means of light-induced drift using AlGaAs diode lasers. Separation efficiency is greatly enhanced by imposing relaxation oscillation sidebands on the diode laser spectrum.

Recently separation of the two naturally occurring Rb isotopes (28% ^{87}Rb and 72% ^{85}Rb) has been achieved[1] using the light-induced drift (LID) effect;[2] enrichment up to 95% ^{87}Rb has been observed. In those experiments a high-power ($\sim$ 150mW) single-frequency dye laser was used. In the case of single-frequency excitation such power is required since most of the Rb atoms are trapped in the non-resonant hyperfine (hf) ground level due to optical hf pumping.[3]. Here we demonstrate that the Rb isotopes can be separated equally well using a low-power (3.3 mW) AlGaAs diode laser which has strong relaxation oscillation sidebands. The presence of these sidebands greatly increases LID by eliminating optical hf pumping.

Experimentally, we use the LID effect to create an isotope-selective "optical shutter" in a cross-shaped capillary cell (Fig. 1). By tuning the optical-shutter diode laser to various frequencies within the RbD$_2$ line profile, the rate of Rb atoms entering the probe capillary is varied via the frequency dependence of the LID effect in an isotope-selective way. The isotopic composition in the probe capillary is determined from absorption spectra which are recorded by scanning the frequency of a weak single-frequency diode laser.

We used an AlGaAs diode laser with relaxation oscillation sidebands for the optical shutter; such sidebands may occur spontaneously or be induced by optical feedback[4] or microwave modulation. An example of such sideband spectrum is shown in Fig. 2. In this case the sideband spacing (3.2 GHz) nearly coincides with the groundstate hf splitting of ^{85}Rb (3.05 GHz); note also that there is appreciable power in the diode laser spectrum near the second-order sidebands, close to the ground-state hf splitting of ^{87}Rb (6.84 GHz). Figure 3 shows the percentage of ^{87}Rb, observed in the probe capillary, as a function of the frequency of the central mode of the shutter laser (its spectrum has been shown in Fig. 2). The power of the shutter laser inside the capillary is 3.3±0.2 mW (0.19 W/cm^2). A maximum percentage of ^{87}Rb of 78% is observed. The shape of the "isotope separation spectrum" in Fig. 3 agrees roughly with the first derivative of the absorption spectrum of natural Rb in the small-field limit;[1,3] this indicates that indeed optical hf pumping has been (largely) eliminated due to backpumping by the sidebands. Calculations based on a rate-equation model[5] show that in the case of *single*-frequency excitation (i.e., no sidebands) at an intensity of 0.19 W/cm^2, 99.3% of the atoms would accumulate in the non-resonant hf ground level. In this case the LID effect would be greatly reduced because only 0.7% of the atoms interact resonantly with the laser.

Due to their much lower cost, better reliability and much higher wall plug efficiency diode lasers are much more attractive than dye lasers. A direct application of the present work might be separation of radioactive Rb isotopes which are used for medical diagnostics.[6] Since development in the field of diode lasers is rapid, especially towards the visible region of the spectrum, in the future many other elements could in principle be separated using diode lasers. When considering other elements note that appreciable hf pumping occurs only for hf splittings larger than the Doppler width, i.e., ≥ 1 GHz, i.e., the same range as relaxation oscillations, so that the sideband technique could be useful again.

This work is part of the research program of the Foundation of Fundamental Research on Matter (FOM) and was made possible by financial support from the Netherlands Organization for Scientific Research (NWO).

REFERENCES

1. Streater, A.D., Mooibroek, J., and Woerdman, J.P. Opt. Commun. <u>64</u>, 137 (1987).

2. Werij, H.G.C., Haverkort, J.E.M., and Woerdman, J.P., Phys. Rev. <u>A33</u>, 3270 (1986), and references therein.

3. Hamel, W.A., Streater, A.D., and Woerdman, J.P., Opt. Commun. <u>63</u>,32 (1987).

4. Miles, R.O., Dandridge, A., Tveten, A.B., Taylor, H.F., and Giallorenzi, T.G., Appl. Phys. Lett. <u>37</u>, 990 (1980).

5. Streater, A.D., and Woerdman, J.P., submitted to J. Phys. B.

6. Mulders, J.J.L., and Steenhuysen, L.W.G., Opt. Commun. <u>55</u>, 105 (1985), and references therein.

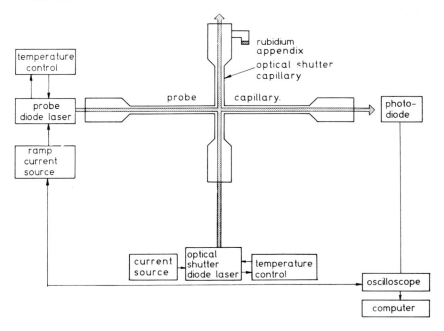

Fig. 1. Experimental arrangement. The cell has been filled with 8 Torr Ar. The capillaries have a diameter of 1.5 mm.

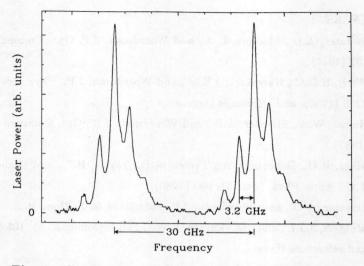

Fig. 2. Spontaneous relaxation oscillation sidebands of AlGaAs laser (Philips CQL 10A).

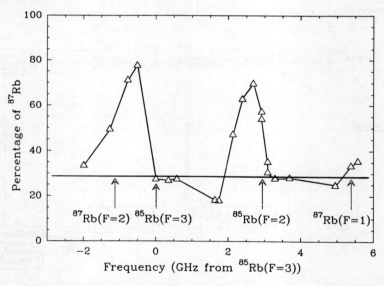

Fig. 3. Percentage of ^{87}Rb measured in the probe capillary as a function of the frequency of the central method of the optical shutter diode laser, at a power of 3.3 ± 0.2 mW.

MULTI-BEAM CIRCULARLY POLARIZED HOLOGRAPHY

J. POLITCH and J. BEN URI

Technion - I.I.T. - Haifa, Israel

On recording a hologram interaction between reference and object beams will occur only when the electric fields of the two waves are parallel to each other, as none can occur between components which are orthogonal. No record will be obtained in the holographic material of the polarization of the recording waves.

However polarization effects can be important and this problem has been discussed in many papers during the last two decades.

Kogelnik [1] has analyzed the electric field inside the hologram using a grating discussing the two cases of polarization in the plane of the grating and perpendicular to it and any intermediate problem can be solved by resolving into components. Applying two reference beams [2] we can produce such a grating and actually Som et al [3] has experimented with this method. However we think that the object beam should not be perpendicular to the holographic plate and much better results may be obtained by assuming a suitable Bragg angle and adjusting the angles of the reference beams so that the results may prove to be much better. Experiments to this effect will be carried out and the results shown.

Lohman [4] and Kurtz [5] have also applied two reference beams perpendicularly polarized and their methods have been discussed quite in detail in some papers [6,7].

The present authors have suggested to use circularly and elliptically polarized beams [8,9] and this system seems to have many advantages. Using circularly polarized beams for the reference beam and also for the object beam will improve the reproduction efficiency by appr. 45 to 50% which is quite desirable.

We are not limited to one reference beam and to one object beam and using one or two for each will have many advantages. Let us assume that we shall use two beams each for the object and reference. The results will be demonstrated during presentation of the paper. The beams can be circularly or elliptically polarized in the clockwise or anti-clockwise directions; they can be arranged so that some will be clockwise and other counter clockwise, or even linearly polarized.

Such holographic system can be of great advantage for many engineering problems, e.g. testing photoelastic materials (birefringent property) [10]. Examples where some of the beams illuminate from the front side while the other from the back side. The two reference beams can also vary the angle between them for up 20 to 40° with reference to the normal of the recording plane.

As all the beams are circularly polarized we shall obtain information on isoclinics, isochromatics and polarization preservation. If e.g. one of the object beams is counter-clockwise while the other clockwise - only one isoclinic line will be obtained. When the reference beams are linearly polarized - the state of polarization will be preserved in the hologram as well.

It will be shown that the advantages of such systems are in the information content, the signal to noise ratios of the reconstructed images and also in efficiencies of the reconstructed images.

Details will be shown during presentation of the paper.

References:

1. H. Kogelnik - Bell S.T.J. 48 1969 p 2909

2. J. Politch & J. Ben Uri - Optik 38 1973 p 368

3. S.C. Som & R.A. Lessard - Appl. Phys. Lett. 17 1970 p 381

4. A.W. Lohman - Appl. Optics 4 1965 p 1667

5. C.N. Kurtz - Appl. Phys. Lett. 14 1969 p 59

6. K. Gasvik - Optik 30 1973 p 47

7. G. Windischbauer et.al. - Optik 37 1973 p 385

8. J. Politch, J. Shamir & J. Ben Uri - Appl. Phys. Lett. 18 1970 p 496

9. J. Politch, J. Shamir & J. Ben Uri - Opt. & Las. Techn. 3 1971 p 226

10. J. Politch & A.A. Betser - Proc. ICO-11 Conf. Madrid 1978 p 639

RING LASER OPTICITY METER

Zhang Liqun

Beijing Institute of Automation,

Beijing, P.R. China

Optically active matter can be detected by frequency difference of R- & L- circularly polarized light when we utilize ring laser to form a new method of measuring optical activity and opticity rotation spectroscopy. In measuring the rotation angle of a polarized light, the new method has the advantages of high sensitivity, high precision. In fact, we have $C/L \doteq 10^9$ in formula (6) below. The ring laser opticity meter is so sensitive that it can be also used to measure magnetic opticity, electric opticity, and to verify the very important view in unifying weak and electromagnetic forces into a single "electroweak" force.

It is known that

$$\theta = \alpha \cdot d \qquad (1)$$

where θ denotes rotation angle, α, specific rotation, and, d, the thickness of optically active sample.

We have

$$\alpha = \frac{\P}{\lambda} (n_L - n_R) \qquad (2)$$

λ being wavelength, n_L or n_R, index of L- or R- circularly polarized laser. For pure liquid, we have

$$\alpha = [\alpha]_\lambda^T \cdot \rho \qquad (3)$$

where ρ being density. For solution, we have

$$\alpha = [\alpha]_\lambda^T \cdot C$$

C being the concentration of solution density. Let $n(\lambda) = n_L - n_R$ which is a complicated function of wavelength with plain, signal Cotton effect and complex Cotton effect patterns.

During measurement, optically active sample is put in a ring laser cavity, which consists of four reflects. Circularly polarized laser lights resonate in the cavity. We have

$$\nu_L = \frac{qc}{L_L} = \frac{qc}{L+(n_L-1)d} \tag{3}$$

$$\nu_R = \frac{qc}{L_R} = \frac{qc}{L+(n_R-1)d} \tag{4}$$

c being light velocity; L, geometric cavity length; g, longitudinal mode number; ν_L and ν_R frequencies of L- and R- circularly polarized laser; L_L and L_R, optical cavity length of L- and R- circularly polarized laser.

We have $L_L \neq L_R$ and $\nu_L \neq \nu_R$ becuase of $n_L = n_R$. The lasers of frequency ν_L and ν_R transmit from output mirror, pass through a polarizing prism and is received by a photoelectric detector. The frequency difference $\Delta \nu = \nu_R - \nu_L$ is given by

$$\Delta \nu = qc \frac{L_L - L_R}{(\overline{L})^2} = \frac{\Delta n(\lambda)d}{\overline{L}} \nu \tag{5}$$

where $\overline{L} = (L_L+L_R)/2$, $\nu = \frac{qc}{\overline{L}} = \frac{c}{\lambda}$ combining (5), (1) and (2) we have

$$\Delta \nu = \frac{c}{\pi L} \cdot \theta \tag{6}$$

We can measure α, $[\alpha]_\lambda^T$ and C of opticity sample by signal $\Delta \nu$, optical activity spectroscopy can be measured by ring dye laser or other wavelength tunable laser.

In our experiment, a quartz singal crystal is used. We detect $\Delta \nu$ between left and right circularly polarized laser lights in RL. We detect $\Delta \nu$ between iodine stabilized He-Ne laser and our ring laser. We find $\Delta \nu = 317$MHz ($\lambda = 0.6328$ μm).

IMPROVED RADEMACHER FUNCTIONS AND RADEMACHER TRANSFORM

Joseph BEN URI

Technion - Israel Institute of Technology, Haifa - Israel

Digital image processing has grown considerably during the last decade including many applications in digital processing in general. The standard procedure is to use the well known Fourier transform. There is no particular reason to use the sin-cos Fourier transform on digital computers. On the contrary recently developed digital transform methods could be more useful. It turns out that methods based on some called Rademacher, Walsh, Hadamard, Haar transforms are simpler and faster in operation. Also quite a reduced memory system is usually required.

The simplest of these digital functions and one of the first suggested is the Rademacher function serie which is actually a digital deformation of the sin function. The basic Rademacher[1] functions are based on such similarity. But there is no reason why actually the cos function should not be also adopted for digital processing methods and the present paper is actually a suggestion in this direction. Such a system of Rademacher sin curves and the now suggested Rademacher cos curves, which simulate the sin and cos family of the Fourier series can be a base for a new transformation method as will be shown. Actually these basic car and sar functions (see fig.1.) are the building blocks for the more complicated Walsh, Hadamard, Haar etc. functions and may prove of very great help in digital transformation methods.

The car and sar functions follow a limited shift theorem and can therefore be treated more or less in a similar way as the cos and sin functions. This means that we can simulate a kind of Euler's equation and write

$$\rho^{ipt} = car(p,t) + i.sar(p,t) \quad \text{and} \quad \rho^{-ipt} = car(p,t) - i.sar(p,t)$$

where p is the order number of the Rademacher functions - similar to frequency in the trygonometric ones - and t stands for the running ordinate. The difference is that the order number is not actually a frequency in the sense of trygonometric functions, but a power of 2. The ρ function can take only binary values of +1 and -1.

From these equations we find similar to the Euler's one that

$$car(p,t) = 1/2[\rho^{ipt} + \rho^{-ipt}] \quad \text{and}$$
$$sar(p,t) = 1/2[\rho^{ipt} - \rho^{-ipt}]$$

and therefore we can now write for the "extended" Rademacher transform,

$$[rad](\rho) = \int\limits_{-\infty}^{\infty} \left[\sum_{n=-\infty}^{\infty} a_n \rho^{i(np_o\alpha)} \right] \cdot \rho^{-i(p\alpha)} d\alpha$$

assuming a Rademacher series to be

$$f(\alpha) = \sum_{n=-\infty}^{\infty} a_n \rho^{i(np_o\alpha)}$$

The standard Rademacher function is actually an incomplete function but the present ρ function is a complete one - also an advantage.

The treatment of the DRT (Discrete Rademacher Transform) is now simple and more or less similar to the DFT.

The "extended" Rademacher functions can be made to building blocks of the more complicated digital functions as Walsh, Haar, Hadamard, Slant functions, and other, but this must be left to a more extended paper to be published soon.

```
+++++++++++++++++----------------    sar(1,t)
++++++++--------+++++++++---------    sar(2,t)
++++----++++----++++----++++----    sar(3,t)
++--++--++--++--++--++--++--++--    sar(4,t)
```
fig.1.
```
++++++++----------------++++++++    car(1,t)
++++--------+++++++++--------++++    car(2,t)
++----++++----++++----++++----++    car(3,t)
+-+-+-+-+-+-+-+-+-+-+-+-+-+-+-+-    car(4,t)
```

The Editors regret that the manuscripts of the following invited papers had not been received by the time the proceedings were compiled:

1. Development of New Solid State Laser Materials
   by Peter F. Moulton
   (Schwartz Electro-Optics Inc Concord USA)

2. Laser Transitions in $LiYF_4$
   by Hans P. Jenssen
   (Crystal Physics & Optical Electronics Laboratory, MIT, Cambridge, USA)

3. Solitons in Optical Fibers and the Solition Laser
   by L. F. Mollenauer
   (AT & T Bell Lab., Holmdel, USA)

4. Femtosecond Relaxation Processes in Semiconductors and Large Molecules
   by C. L. Tang
   (Cornell University, Ithaca, USA)

5. Spectroscopy of Chromium-Doped Tunable Laser Materials
   by Richard C. Powell
   (Dept. of Phys., Oklahoma State University, Stillwater, USA)

6. TeraHz Spectroscopy with Ultrashort Electrical Pulses
   by D. Grischkowsky
   (IBM Watson Research Center, Yorktown Heights, USA)

7. Optical Nonlinearities and Photocarrier Dynamics in Small Semiconductor and Metal Particles
   by D. Richard, Ch. Flytzanis
   (Laboratoire D'Optique Quantique, Ecole Polytechnique, Palaiseau, France)

8. Advances in Beta Barium Borate and Other Boron-Oxygen NLO Crystals
   by Chen Chuangtian
   (Fujian Institute of Research on the Structures of Matter, Academia Sinica, Fuzhou, PRC)

9. Two Photon Masers
   by S. Haroche
   (Laboratoire de physique de l' E. N. S., Paris, France)

10. Diode Laser Pumped Solid State Lasers
    by Byer, R
    (Standford University, USA)